International Models of Occupational Health and Safety:

Strategies, Systems, and Success Stories

Kaan Koçali
Istanbul Gelisim University, Turkey

IGI Global
Scientific Publishing
Publishing Tomorrow's Research Today

Vice President of Editorial	Melissa Wagner
Director of Acquisitions	Mikaela Felty
Director of Book Development	Jocelynn Hessler
Production Manager	Mike Brehm
Cover Design	Jose Rosado

Published in the United States of America by
IGI Global Scientific Publishing
701 East Chocolate Avenue
Hershey, PA, 17033, USA
Tel: 717-533-8845 | Fax: 717-533-7115
Website: https://www.igi-global.com E-mail: cust@igi-global.com

Library of Congress Cataloging-in-Publication Data

LCCN: 2025039350 (CIP Data Pending)
ISBN13: 979-8-3373-6370-7.
Isbn13Softcover: 979-8-3373-6371-4
EISBN13: 979-8-3373-6372-1

British Cataloguing in Publication Data
A Cataloguing in Publication record for this book is available from the British Library.

Table of Contents

Detailed Table of Contents

This chapter explores the approach of the International Labour Organization (ILO) to occupational health and safety (OHS) from a historical perspective. It begins by examining the ILO's founding vision in 1919, the role of OHS within its overall mission, the development of international conventions and standards over time, and the significance of Conventions No. 155 and 187. Subsequently, the impacts of the ILO's 2022 recognition of a safe and healthy working environment as the fifth fundamental principle and right at work will be examined, along with debates concerning its binding nature for countries that have not ratified the relevant conventions Following this, the focus shifts to the ILO's new OHS strategy introduced in 2023, highlighting its key components and preventive approach. Finally, the intersection of OHS with the 2030 Sustainable Development Goals (SDGs) is addressed, with particular attention to issues such as health, gender equality, decent work, and poverty reduction.

This chapter examines international models of Occupational Health and Safety (OHS) with an emphasis on its critical role in enhancing socio-economic welfare through industry safety, ethical practices, employee trust, productivity, and national economic growth. It covers sectors including high-risk industries like construction, where construction safety remains a challenge, as well as expanding fields such as IT, healthcare, and education. The chapter synthesizes research on occupational stress and its impact on workplace safety, integrating findings on green building hazards with ergonomic strategies. It highlights the importance of safety management systems and worker engagement in reducing accidents. The use of predictive analytics, including advanced machine learning, is discussed as a powerful tool for hazard detection and risk mitigation, supporting adaptive OHS systems for sustainable workplace health improvement

Occupational Health and Safety (OHS) systems across the world have evolved in diverse ways, shaped by historical, socio-economic, regulatory, and cultural contexts. Understanding these models is essential for policymakers, industries, and global organizations seeking to benchmark practices, improve workplace safety, and foster sustainable work environments. This chapter presents a comparative overview of major international OHS models—European Union, United States, Japan, Australia–New Zealand, Canada, Nordic countries, and Singapore. For each model, the chapter outlines key strategies, regulatory systems, structural features, and notable success stories. The analysis concludes by highlighting global best practices and strategic insights that can guide countries and organizations in enhancing workforce safety, resilience, and well-being.

This chapter contributes to the book by providing a structured, comparative analysis of successful international occupational health and safety models, moving beyond regulatory descriptions to focus on why and how certain systems effectively reduce occupational accidents and diseases. Unlike chapters that concentrate on single-country cases or specific regulations, this contribution synthesizes diverse regional experiences—developed and developing alike—to extract transferable success factors and policy-relevant lessons. The chapter aligns closely with the book's core themes of strategies, systems, and success stories, offering both analytical depth and practical implications. Its emphasis on prevention, learning from success, and adaptability ensures relevance for academics, policymakers, and practitioners seeking evidence-based pathways to improve OHS performance across different institutional and socio-economic contexts.

Shubhashree Sahoo, Sphoorthy Engineering College, India
P. Selvakumar, Department of Science and Humanities, Nehru Institute of Technology, Coimbatore, India
R. Shobarani, Dr. M.G.R. Educational and Research Institute, India
T. C. Manjunath, Rajarajeswari College of Engineering, Bengaluru, India
Subramaniam Seshan Iyer, Datta Meghe Institute of Management Studies, India
Rishabh Patel, Datta Meghe Institute of Management Studies, India

Risk assessment is a cornerstone of public health, clinical medicine, and pharmaceutical development, serving as a structured approach for identifying, evaluating, and managing the potential hazards associated with drugs, chemicals, or environmental exposures. In the context of drug safety and efficacy, risk assessment enables stakeholders—regulators, healthcare providers, manufacturers, and researchers—to make informed decisions by quantifying potential harms and weighing them against anticipated therapeutic benefits. The science of risk assessment has evolved significantly, driven by advancements in toxicology, pharmacology, epidemiology, and computational modeling. This comprehensive exploration outlines the fundamental principles, methodologies, and practical applications of risk assessment, emphasizing its central role in safeguarding public health and promoting responsible innovation in medicine. Risk assessment refers to the systematic process of evaluating the likelihood and severity of adverse outcomes resulting from exposure to a particular agent under defined conditions.

Rabia Balcı, Occupational Health and Safety Program, Amasya University, Turkey

Leading indicators in occupational health and safety (OHS) are proactive measures that provide early warning of potential incidents, complementing traditional lagging indicators such as injury rates. This chapter explains the concept of leading indicators and their importance in fostering a prevention-focused safety culture. It discusses how leading indicators can be developed and integrated into an organization's OHS management system, particularly in alignment with the ISO 45001:2018 standard for OHS management. Through a review of recent literature and case examples,

in healthcare settings. Encompasses the collection, structured, and ranging electronic health (EHRs), laboratory test results, and vital signs to clinician notes, wearable device data, and even social determinants of health. Patients are categorized based on their risk levels, allowing clinicians to prioritize interventions.

Organizational success relies on workplace safety, yet complex modern environments often make traditional training insufficient. AI offers a transformative solution by combining real-time monitoring, immersive simulations, adaptive learning, and predictive analytics to enhance safety training. It reduces human error, strengthens resilience, improves risk prediction, and ensures regulatory compliance. Applications such as automated inspections, behavioural monitoring, VR simulations, and predictive maintenance lower accidents while fostering a proactive safety culture. Gamified learning, wearables, and IoT devices further personalize training and boost engagement. Despite challenges like bias, privacy, and high costs, proper strategies and human oversight enable ethical, scalable adoption. With AI, organizations gain a cost-effective, future-ready system that protects workers, improves compliance, and drives sustainable excellence.

Digitization in healthcare is bringing new ways to improve safeguarding patients and results through the utilization of advancements like AI, EHRs, CDSS, and IoMT. However, digital evolution requires more than just technology; its impact improves when it coincides with interaction, a collaborative atmosphere, and favorable work

Preface

Occupational health and safety (OHS) has evolved from being a narrowly defined set of workplace regulations to a broad, multidimensional discipline encompassing public health, human resources, risk management, and sustainable development. In the context of globalization, industrial transformation, and technological acceleration, OHS has become a central element of modern organizational strategy and national development agendas. The COVID-19 pandemic, climate-related disasters, and the growing prevalence of remote and digital work environments have further demonstrated that the well-being of workers is inseparable from the resilience of economies and societies. This book, International Models of Occupational Health and Safety: Strategies, Systems, and Success Stories, provides a comprehensive and comparative perspective on how OHS systems are designed, implemented, and improved across diverse geographical, cultural, and industrial settings.

THE GLOBAL CONTEXT OF OCCUPATIONAL HEALTH AND SAFETY

In recent decades, global initiatives—such as the International Labour Organization's (ILO) conventions, the World Health Organization's Healthy Workplace Framework, and the United Nations Sustainable Development Goals (SDGs)—have collectively redefined the principles and priorities of occupational health and safety. These frameworks emphasize not only the prevention of occupational accidents and diseases but also the promotion of a holistic safety culture that integrates physical, psychological, and social well-being. Yet, the implementation of these principles varies widely among countries and sectors, reflecting differences in economic development, institutional capacity, governance models, and cultural attitudes toward risk and responsibility.

This diversity underscores the need for a global comparative approach. By examining successful OHS models and strategies from multiple continents, this book aims to identify both common principles and contextual variations that can inform future policy and practice. The chapters collectively demonstrate that there is no single "best" system; rather, there are adaptable frameworks shaped by local realities, stakeholder engagement, and continuous learning.

RELEVANCE AND POSITION IN THE CONTEMPORARY WORLD

Today's world of work is characterized by complexity and interdependence. Global supply chains, transnational corporations, and remote digital work have blurred the boundaries of traditional workplace environments. Consequently, occupational risks are no longer confined to factories or construction sites—they extend into offices, homes, and digital platforms. At the same time, emerging technologies such as artificial intelligence, robotics, and wearable sensors are transforming the way safety is monitored and managed. This transformation presents both opportunities and challenges: while automation and data analytics can enhance hazard detection and decision-making, they also raise ethical, psychological, and skill-related concerns.

Within this dynamic context, the study of international OHS models becomes crucial. Understanding how countries integrate legal frameworks, institutional coordination, training systems, and cultural values into safety management can provide valuable lessons for global adaptation and resilience. This book positions itself as a bridge between academic research, policy formulation, and practical application in the field of occupational health and safety.

TARGET AUDIENCE

This book is intended for a wide range of audiences. It will be particularly valuable to:

- Academicians and researchers in the fields of occupational health, industrial engineering, human resource management, and public policy who seek to explore comparative and interdisciplinary perspectives.
- Practitioners and safety professionals aiming to benchmark and enhance their organizational safety systems based on international best practices.
- Policy makers and regulators involved in developing or revising national occupational safety frameworks.
- Students and early-career professionals who wish to gain a holistic understanding of how OHS systems function at both organizational and systemic levels across different countries.

- The book balances theoretical foundations with practical insights, making it both a scholarly reference and a field guide for real-world application.

OVERVIEW OF THE CHAPTERS

This book offers a comprehensive and integrative examination of occupational health and safety (OHS) from a global, comparative, and innovation-driven perspective. Designed to serve scholars, policymakers, practitioners, and graduate students, the volume brings together normative frameworks, international governance systems, technical risk management tools, emerging global challenges, and digital transformation dynamics shaping the future of worker safety. The chapters are structured in a progressive manner, beginning with foundational principles and advancing toward cutting-edge technological and psychosocial considerations.

The volume opens with a normative and rights-based perspective in "The Safe and Healthy Working Environment" as a Fundamental Principle and Right: An Assessment in the Context of the ILO and UN 2030 SDGs." This chapter positions occupational health and safety within the broader framework of international labor standards and sustainable development. By examining the evolution of the recognition of a safe and healthy working environment as a fundamental principle and right at work, the chapter establishes the ethical and legal foundation upon which contemporary OHS systems are constructed. It further links worker safety to the United Nations Sustainable Development Goals, particularly those related to decent work, inequality reduction, and sustainable economic growth.

Building upon this conceptual foundation, International Models and Success Stories in Occupational Health and Safety: Global Strategies and Systems for Worker Safety provides a global overview of institutional models and governance mechanisms that have successfully enhanced occupational safety outcomes. Through comparative analysis, the chapter identifies structural components—such as regulatory enforcement capacity, tripartite collaboration, safety culture integration, and data transparency—that characterize high-performing systems.

A detailed comparative policy analysis is presented in Occupational Health and Safety Strategies of European Union, United States, Japan, Australia–New Zealand, Canada, Nordic Countries, and Singapore. By examining diverse regulatory traditions and enforcement models, this chapter provides insights into how national context, legal culture, and economic structure influence safety performance. Complementing this macro-level analysis, Learning from Success: International Best Practices in Reducing Occupational Accidents and Diseases synthesizes evidence-based prevention strategies that have demonstrably reduced accident rates and occupational illnesses across sectors.

The book then shifts to core management systems and operational tools. Risk Assessment and Hazard Prevention Techniques Across Industries offers a structured overview of hazard identification methodologies, risk evaluation models, and prevention strategies applicable to manufacturing, construction, logistics, healthcare, and service industries. The subsequent chapter, Leading Indicators in Occupational Health and Safety and Integration with ISO 45001, focuses on performance measurement and proactive safety management. It discusses the strategic use of leading indicators, predictive metrics, and continuous improvement frameworks aligned with ISO 45001 standards.

Emerging global risk landscapes are addressed in two dedicated chapters. Navigating the Intersection of Climate Change and Occupational Health and Safety analyzes climate-related occupational risks, including heat stress, extreme weather exposure, environmental hazards, and green transition challenges. Learning from COVID-19: Building a Robust Pandemic Preparedness Management System draws lessons from the global pandemic and proposes structured preparedness models for biological risk governance, crisis coordination, and institutional resilience.

The digital transformation of work and its regulatory implications are explored in Occupational Health and Safety in the Age of Industry 4.0: Legal and Regulatory Responses to Digital Risk Environments. This chapter examines cyber-physical systems, automation, platform-based employment, and algorithmic management, highlighting emerging legal and compliance challenges in digitalized work environments.

Artificial intelligence and advanced analytics form a dedicated thematic cluster within the volume. The Role of AI and Machine Learning in Predictive Safety Analytics discusses predictive modeling, real-time monitoring systems, and data-driven risk detection mechanisms that enable proactive intervention. Transforming Employee Safety Training Through Artificial Intelligence Innovations explores adaptive learning platforms, simulation technologies, and AI-enhanced behavioral safety training models. Extending digital transformation into sector-specific application, Digital Transformation and Healthcare Safety: Role of Technology, Communication, and Work Environment in Patient Outcomes analyzes the integration of digital tools within healthcare systems and their impact on patient and worker safety outcomes.

The final section of the book emphasizes sectoral applications and human-centered dimensions of safety governance. Creating a Culture of Safety: International OHS Models in Hospitality examines safety culture frameworks within service-intensive industries, highlighting leadership, employee engagement, and behavioral safety practices. Psychosocial Risk Management among First Responders: A Comparative Analysis of Burnout Prevention Strategies in Canada and the United States focuses on mental health resilience, burnout prevention, and psychosocial risk mitigation in high-stress occupations. Concluding the thematic exploration, Nonverbal Incivility in the Workplace: Rethinking Organizational Politics investigates subtle interpersonal dynamics, micro-behaviors, and organizational culture factors that may indirectly influence occupational risk and employee well-being.

Collectively, the chapters in this volume present a multidimensional and forward-looking perspective on occupational health and safety. By integrating rights-based foundations, comparative governance models, technical risk management systems, emerging global challenges, digital transformation, artificial intelligence applications, and psychosocial considerations, the book provides a comprehensive framework for understanding and advancing OHS systems in a rapidly evolving world of work. The volume contributes to academic scholarship by bridging policy analysis, management science, technology studies, and organizational behavior, while also offering practical insights for institutional leaders and safety professionals seeking resilient and sustainable safety strategies.

SIGNIFICANCE AND CONTRIBUTION OF THE BOOK

This book contributes to the global discourse on occupational health and safety by providing a comparative, evidence-based analysis of models and practices. Its interdisciplinary approach bridges the gap between policy, research, and implementation, offering a roadmap for building resilient and sustainable safety systems. The diverse contributions emphasize the shared responsibility of employers, employees, governments, and society in promoting a culture of prevention and care. Furthermore, by documenting success stories, the book not only highlights achievements but also provides replicable insights for continuous improvement. It invites readers to reflect on how safety can be reimagined as an investment in human capital, organizational productivity, and social well-being.

CONCLUSION

In an era marked by rapid change, uncertainty, and global interconnectivity, occupational health and safety remains a cornerstone of sustainable development. The models and experiences presented in this volume remind us that safety is not a static goal but a dynamic process rooted in learning, adaptation, and cooperation. As editor, I hope this book will inspire researchers, practitioners, and policymakers to pursue innovative strategies that safeguard not only workers but also the societies they help build. The lessons shared across nations demonstrate that advancing occupational health and safety is ultimately an investment in humanity's collective progress.

Acknowledgment

I would like to express my deepest gratitude to all the contributors who have devoted their expertise, time, and effort to this book. Their valuable chapters and scholarly insights have made this publication a comprehensive and meaningful contribution to the global understanding of occupational health and safety. Special thanks are extended to the reviewers and colleagues who provided constructive feedback throughout the editorial process, ensuring the academic rigor and quality of this work. I am also grateful to the editorial team at IGI Global for their professional guidance and continuous support during each stage of publication. Finally, I would like to acknowledge my institution, Istanbul Gelişim University, for fostering an environment that encourages academic collaboration and innovation.

Chapter 1
"The Safe and Healthy Working Environment" as a Fundamental Principle and Right:
An Assessment in the Context of the ILO and UN 2030 SDGs

Ayşenur Öktem Özgür
http://orcid.org/0000-0002-9011-9704
Karamanoglu Mehmetbey University, Turkey

ABSTRACT

This chapter explores the approach of the International Labour Organization (ILO) to occupational health and safety (OHS) from a historical perspective. It begins by examining the ILO's founding vision in 1919, the role of OHS within its overall mission, the development of international conventions and standards over time, and the significance of Conventions No. 155 and 187. Subsequently, the impacts of the ILO's 2022 recognition of a safe and healthy working environment as the fifth fundamental principle and right at work will be examined, along with debates concerning its binding nature for countries that have not ratified the relevant conventions Following this, the focus shifts to the ILO's new OHS strategy introduced in 2023, highlighting its key components and preventive approach. Finally, the intersection of OHS with the 2030 Sustainable Development Goals (SDGs) is addressed, with particular attention to issues such as health, gender equality, decent work, and poverty reduction.

DOI: 10.4018/979-8-3373-6370-7.ch001

INTRODUCTION

In addition to biological and physiological variables, human health is influenced by social, environmental, behavioral, and political elements that are influenced by the environments in which people live, work, and maintain social relationships. To effectively address health issues, this multifaceted interplay calls for comprehensive, multi-sectoral, and institutionally robust policy measures. Occupational safety and health (OSH) policies, in particular, aim to improve the overall health and well-being of both workers and the wider community (WHO and CSDH, 2008; ILO, 2023a).

The pace of change in working life is accelerating, driven by demographic shifts, economic transformations, technological advances, and global crises (World Bank, 2019; UNDP, 2021). In this rapidly evolving landscape, safeguarding and promoting decent work for all has become a key priority for everyone involved in OSH.

Since its establishment in 1919, the International Labour Organization (ILO) has prioritized protecting workers from workplace accidents and occupational illnesses. To support this goal, it has developed international standards and conventions (ILO, 2021a). In 2022, the ILO formally recognized a safe and healthy working environment (SHWE) as a fundamental human right. This milestone has broadened the scope of OSH, moving it beyond purely technical considerations to encompass human rights, social justice, and sustainable development. As a result, the ILO's global OSH strategy adopted in 2023 seeks to reduce workplace accidents and illnesses, improve the quality of work life, promote business sustainability, and lessen the socio-economic burden, especially in low-income regions (ILO, 2023b).

This is an important milestone in the construction of a sustainable, inclusive, and fair working life. Moreover, this approach directly or indirectly overlaps with the United Nations' 2030 Sustainable Development Goals (SDGs) and contributes to the strengthening of policies to improve the quality of working life on a global scale (SDG & BOHS, 2022). In this framework, OSH stands out as a fundamental policy area that both increases the welfare of workers and contributes to the creation of a sustainable, inclusive, and equitable working life.

From this perspective, the chapter initially examines the original vision of the ILO established in 1919, the role of OSH within the organization's broader mission, the international conventions and standards that have evolved over the years, and the significance of Conventions No. 155 and 187. It then evaluates the consequences of recognizing a SHWE as the fifth fundamental principle and right by the ILO in 2022, and analyzes the binding implications of this decision for countries that have yet to ratify the pertinent conventions. Next, the newly introduced global OSH strategy for 2023 is explored, with an emphasis on its key elements and preventive measures. Lastly, the connection between OSH and the 2030 SDGs is examined, particularly focusing on its relevance to no poverty (SDG 1), health and well-being

(SDG 3), gender equality (SDG 5), decent work and economic growth opportunities (SDG 8), and the reduction of inequalities (SDG 10).

1. ILO's FOUNDING VISION AND APPROACH TO OSH

Fatalities and injuries in the workplace result not only in financial losses but also impose a significant human toll that impacts the lives of hundreds of millions. In this context, it is not surprising that OSH has been one of the main priorities of the ILO since its establishment in 1919. The safeguarding of workers' health and safety is a key constitutional objective of the ILO. The Preamble of the "*ILO Constitution (1919)*" declares that 'the protection of workers against diseases, ailments and injuries connected with their work' is among the issues that need urgent attention. Conversely, the "*Declaration on the Aims and Objectives of the ILO (Philadelphia Declaration, 1944)*" acknowledges it as a 'solemn obligation' for the Organization to create initiatives that ensure sufficient protection of workers' life and health across all occupations (ILO, Constitution; ILO, 2019a, p. 13). In fact, protecting the health and safety of workers is an important element of the ILO's constitutional objectives.

Following the Second World War, political and economic structures were transformed on a global scale and international organisations such as the United Nations (UN) and the World Health Organization (WHO) were established. Although work-related health and safety risks persisted in all countries, the new international order significantly influenced the ILO's approaches to improving working conditions and OSH. During this period, the impact of work on workers' health gained more importance in developed countries and rapid progress was made in disciplines such as occupational hygiene, occupational medicine, and safety engineering. In addition, with the spread of social welfare policies, publicly funded research institutions were established and scientific studies in the field of OSH increased (ILO, 2019a, p. 16). In 1950, ILO and WHO initiated co-operation by establishing the first "*Joint Committee on Occupational Health*". The intersection of the mandates of the two institutions over time paved the way for the ILO to move towards a holistic safety and health programme based on preventive measures, rather than approaching OSH only from a medical perspective (ILO, September, 12, 2025).

Another important feature of the ILO's post-war work was the development of "*Codes of Practice and Guidelines on OSH*" in response to the increasing number of developing countries joining the ILO. While these codes are not legally binding, they provide guidance to ensure acceptable standards of practice, in addition to national laws and regulations. These guidelines, which differ based on the type of hazard and industry, are developed with technical specificity and encompass various fields such as construction, forestry, mining, and the iron and steel industry, as well as risks like dust, noise, vibration, ionizing radiation, and airborne viruses. Codes of practice serve as practical guidelines for public institutions, employers, workers, and relevant safety organizations (ILO, 2004, p. 5; Demirbilek, 2005, p. 24).

2. EVOLUTION OF GLOBAL OSH CONVENTIONS AND STANDARDS

The reorganization of the ILO following World War II saw a swift rise in membership, leading to developed nations becoming a minority by 1970. In response to this change, the ILO has started technical support programs to improve national policy-making and the capabilities of labor inspections. In 1975, the "International Labour Conference (ILC)" established the basis for the management systems approach by deciding to create OSH policies at both the national and corporate levels. This approach was concretised by the *"Occupational Safety and Health Convention"* No. 155, adopted in 1981, and provided a comprehensive policy framework (ILO, September, 3, 2025; ILO, 2005a, p. 4). The Convention calls for *"the development, implementation, and periodic review of national OSH policies to prevent work-related accidents and health risks, while striving to effectively reduce hazards in the workplace"*. Subsequent OSH conventions also require the establishment of national policies and focus mainly on protective and preventive measures (ILO, 2005a).

In 1998, the ILO adopted the *"Declaration of Fundamental Principles and Rights at Work"*, which applies to all workers in all member states, regardless of their level of economic development or ratification status of the fundamental conventions. The Declaration guarantees four fundamental rights and confirms the obligations of ILO membership. These rights covered *"freedom of association and the effective right to collective bargaining"*; *"the elimination of all forms of forced and compulsory labour"*; *"the effective abolition of child labour"*; and *"the elimination of discrimination in respect of employment and occupation"* (Kellerson, 1998, p. 223). In 1999, the concept of "decent work" gained prominence in a report presented to the ILC by Somavia, prompting the ILO to launch the "Decent Work Agenda" to reinforce the social aspects of globalization and provide a human-centered response to the challenges facing the world of work (Sengenberger, 2001, p. 39). ILO defines decent work as *"productive work for women and men in conditions of freedom, fairness, security, and human dignity"*. Decent work refers to a comprehensive approach that includes productivity, fair income, job security, social protection, the right to personal development and participation, equal opportunities, and equal treatment (ILO, 1999; Somavia, 2000, p. 3). These two initiatives represent important milestones in the recognition of the right to a safe and healthy workplace as a fundamental principle and right.

In 2003, the ILO adopted the "Global Strategy on OSH", recognizing that effective OSH systems require prioritizing the issue at both national and international levels and establishing ongoing improvement mechanisms in collaboration with all stakeholders. The Strategy emphasizes the central role of international labour standards in OSH and identifies awareness raising, technical assistance, development of national programs, protection of vulnerable groups, and addressing the socio-economic benefits of OSH as key areas for action. Building a preventive OSH culture at the national level and putting systematic approaches into practice at the enterprise level are the main goals (ILO, 2004, p. 3). In the same year, the "World Day for Safety and Health at Work" was celebrated for the first time (ILO, September, 12, 2025) and the theme of creating and sustaining a safety and health culture was emphasized; by 2005, "prevention of occupational accidents and diseases" became the main focus (ILO, 2005b, p. ii).

The ILO's 2006 "Promotional Framework for OSH" (Report 4 (2B) document contains important provisions on OSH. This document contains the text that forms the basis of the ILO's *"Promotional Framework for OSH Convention (ILO Convention No. 187)"*, adopted in 2005 and entered into force in 2006. The document emphasizes that member states should establish national OSH policies and programs that improve over time, put them into action, and check their progress regularly. Furthermore, it is important to establish national profiles reflecting the current state of OSH and to support these with training, awareness, and inspection activities. In this process, the ILO aims to strengthen international cooperation by encouraging the sharing of information and experience (ILO, 2006).

The *"Declaration on Social Justice for a Fair Globalisation"* adopted in 2008 emphasized the importance of the Decent Work Agenda; global challenges such as financial crises, unemployment, informal employment and inadequate social protection increased the necessity of the agenda (Rantanen et al., 2020, p. 4). Subsequently, the 'Global Jobs Pact' adopted in 2009 recognized the importance of respect for fundamental rights and principles for human dignity and development, and emphasized that governments and social partners should develop policies in line with decent work principles in cooperation (ILC, 2009; ILC, 2019, p. 4). This process laid the groundwork for the ILO's efforts to promote a SHWE to turn it into concrete and comprehensive policies.

The ILO developed the "just transition" approach in 2015, taking into account the impacts of climate change and sustainable transformation processes on the world of work. This approach aims to manage the socio-economic impacts that may arise during the transition to low-carbon and environmentally friendly economies through social dialogue and participatory policies, and to ensure job security, social protection, quality employment, inclusion, and OSH (ILO, 2015, pp. 4, 15-16).

To improve development cooperation in accordance with the 2030 SDGs, the ILO started the "Safety and Health for All" Program in 2015. Through four strategic components — "increasing knowledge", "developing national policy and legal frameworks", "strengthening institutional capacity", and "promoting safe working environments" — the program seeks to improve OSH globally (ILO, 2023c, p. 2). Through the Vision Zero Fund, a crucial part of the initiative, the program guarantees the prevention of occupational accidents and diseases in low- and middle-income countries, with a particular focus on women, youth, migrant workers, and workers in high-risk sectors (ILO, 2023d, p. 25). This program, which was implemented in 15 countries between 2016 and 2020, reached over 60 million workers (ILO, 2020a, p. 15).

Lastly, a major step toward recognizing "a SHWE as a fundamental principle and right" was taken with the adoption of the *"ILO Centenary Declaration on the Future of Work"* in June 2019. The Declaration emphasized the importance of comprehensive OSH measures during the COVID-19 pandemic and affirmed that SHWE are a core component of decent work. The ILO aimed to increase workplace resilience and protect worker health by providing employer guidelines, training materials, and practical tools during the pandemic. It also focused on issues such as the participation of individuals with disabilities and the applicability of international standards (ILO, 2020a, p. 38).

3. RECOGNIZING A SHWE AS A FUNDAMENTAL RIGHT

The right to protection against workplace accidents is guaranteed by the *"1948 Universal Declaration of Human Rights"* and the *"1966 Covenant on Economic, Social and Cultural Rights"*. This has been a central issue since the ILO was established (ILO, 2021a, p. 132). With the "1998 Declaration", ILO member states promised to respect and implement fundamental rights, no matter their level of economic development (Kellerson, 1998, p. 223; ILO, October, 10, 2025). While the declaration originally included four fundamental principles, the fifth right, "a SHWE", was added to the scope of fundamental rights at the 110th ILC Session in 2022 (ILO, 2022a, pp. 2, 3). Thus, OSH gained a stronger legal and political status. This decision, from the "ILO's 2019 Centenary Declaration" process, officially recognized SHWE as a key part of decent work (ILO, 2020b, p. 38; Hilgert, 2024, p. 96).

The ILO (2022a) says that the "right to a safe and healthy working environment" is a basic principle and right in the workplace (p. 2). This shows the ILO's commitment to upholding basic human values. With the recognition of OSH as a new right, the "Conventions No. 155 and 187" in this field gained the status of fundamental agreements. Conventions No. 155 and No. 187 complement each other, together embodying the essential principles, obligations, and rights needed to secure a safe and healthy workplace. They include *"broad provisions that apply to all sectors and all workers, regardless of the specific hazards they may face"* (ILO, 2023a, p. 4). This right is binding for all members, even those that have not ratified the conventions. It must be implemented according to the obligations that come with ILO membership. The ILO's mission is to encourage its members to apply these principles. It aims to support the development of national OSH policies and systems. The ILO also seeks to promote a preventive culture through effective OSH management systems (ILO, 2023a, p. 1).

The "principle of prevention" is central to OSH. This principle is explicitly reflected in Conventions No. 155 and No. 187. Convention No. 155 identifies *"the primary goal of a national OSH policy as preventing occupational accidents and work-related health hazards, aiming to minimize the inherent risks in the workplace as much as possible*". The principle of prevention is the main focus of OSH. This principle is clearly shown in Convention No. 187, which calls for the ongoing improvement of OSH. This helps prevent occupational diseases, accidents, and work-related deaths. Accordingly, "member States are expected to implement the following three key components, in consultation with the most representative organizations of employers and workers": *"a national OSH policy, a national OSH system*, and *a national OSH programme"*. The Convention also emphasizes the creation and reinforcement of a national culture of prevention in OSH, where the "principle of prevention" is prioritized and all stakeholders—state /government, employers, and workers—actively engage within a framework of clearly defined rights and responsibilities (ILO, 2022a, p. 7; ILO, 2023a, p. 5).

Today's multifaceted uncertainties—demographic aging, digitalization, climate change, economic insecurity, and social violence—threaten the well-being of individuals and the stability of socio-ecological systems, leading to declines in human development indicators (UNDP, 2022, pp. 8-13). In this context, the recognition of the right to a SHWE as a fundamental right represents a significant transformation in the ILO's normative structure and constitutes a significant threshold in the context of global human development (Hilgert, 2024, p. 97).

4. THE ILO's OSH STRATEGY AFTER THE 2022 RESOLUTION

Building on the 2022 resolution, the ILO Governing Body approved the "Global OSH Strategy for 2024–2030" in November 2023. It promotes a preventive OSH culture and aims for continuous improvement. The strategy focuses on reducing occupational deaths, injuries, and illnesses. It also aims to enhance the quality of working life, boost business productivity and sustainability, and alleviate socio-economic burdens. The strategy emphasizes its dedication to social justice by safeguarding workers' health and lives and by fostering fair and dignified working conditions (ILO, 2023e, p. 3).

The strategy is based on four core principles: "(1) adherence to international labour standards, (2) social dialogue and stakeholder engagement, (3) a human-oriented and gender-transformative approach, and (4) ensuring health and safety at every stage of life". The Strategy emphasizes the interaction and complementarity between OSH and other fundamental labour rights - in particular non-discrimination and freedom of association (ILO, 2023e, p. 6). Tripartite stakeholder engagement and social dialogue mechanisms function as key tools in creating a SHWE (Rombouts & Zekić, 2020, p. 340). Furthermore, the strategy, which focuses on prevention throughout the life cycle with a human-oriented and inclusive approach, is envisaged to be integrated and implemented in training, job design, and skills development processes (ILO, 2023e, p. 12).

The Strategy encourages ILO members to take action in three complementary areas:

- First, strengthening all components of national OSH systems to enhance "effective governance", "ensure reliable data" and "evidence-based research", and "improve service capacity".
- Second, prioritizing OSH on the national and global policy agenda and mainstreaming OSH policies by ensuring policy integration consistent with Conventions No. 155 and 187. This approach supports the development of a preventative OSH culture by integrating OSH with the environment, health, education, social and economic development, other fundamental rights, violence and harassment in the workplace, and job security.
- Third aims to improve workplace OSH management systems within the framework of the Guiding Principles (ILO-OSH 2001) through gender-sensitive guidelines (ILO, 2023e, p. 9, 13).

The effectiveness of all OSH policies depends on considering changing working conditions. An aging population, the increase in migrant and refugee workers, vulnerable groups working in atypical employment forms, and the widespread prevalence of informal employment are contributing to increased work-related diseases and risks. The number of workers in the informal economy globally exceeds 2 billion, comprising more than half of the world's workforce (ILO, 2024a, p. 30). This group often works in high-risk occupations and is vulnerable to a wide range of hazards (Takala, 2019, p. 3). Therefore, gradually expanding the scope of OSH to include prevention, protection, health promotion, care, rehabilitation, compensation and return to work services to all workers is a priority, especially for agriculture, the informal sector and the self-employed (ILO, 2003, pp. 10-12; Rantanen et al., 2017, p. 8; Takala, 2019, p. 19). In this context, the Strategy highlights new OSH priorities, reflecting the challenges and opportunities of a changing world of work. It also continues to focus on issues in high-risk sectors such as agriculture, construction, and mining, and considers current issues such as climate change, just transition, and the integration of new technologies and work processes. Furthermore, it promotes initiatives that respond to demographic change and the diversifying needs of the workforce, while recognizing the elimination of inequalities in the workplace as a fundamental condition for achieving social justice (ILO, 2023e, p. 3). The comprehensive approach of the 2023 strategy is crucial for providing a practical policy roadmap or implementation framework for the adoption of SHWE at the national level.

5. THE RELATIONSHIP BETWEEN OSH AND THE 2030 SDGs

"Transforming Our World: The 2030 Agenda for Sustainable Development", adopted by the UN General Assembly on September 25, 2015, includes 17 SDGs and 169 goals aimed at *"eradicating poverty, protecting the planet, and ensuring peace and prosperity for all"*. This agenda, a continuation of the "Millennium Development Goals", offers a holistic framework that balances economic, social, and environmental dimensions. It prioritizes eradicating poverty, reducing inequalities, achieving gender equality, protecting human rights, building inclusive and just societies, and economic growth based on decent employment opportunities, and strengthening environmental protection (UN General Assembly, 2015: UN, October, 2, 2025; Gafney, 2014).

OSH is seen as a fundamental element of prosperity and development in the context of both the ILO's Decent Work Agenda and the UN's 2030 SDGs (ILO, 2023f). With the 2022 decision, OSH has been addressed beyond a technical issue, within the framework of human rights and social justice, and has been directly linked to many SDGs to create a sustainable and inclusive work life (SDG & BOHS, 2022).

"The Global Coalition for Safety and Health at Work (GOSH)" report, published in July 2025, recommended adding the goal of "significant reductions in work-related diseases and deaths by 2050" to the SDGs. The report also calls for governments to advocate for OSH at the UN, for the WHO to prioritize OSH, and for the reactivation of the ILO-WHO Joint Committee on OSH. The fact that *"approximately 3 million work-related deaths and 395 million occupational accidents"* occur each year further enhances the importance of OSH. Undoubtedly, effective OSH practices are directly linked to poverty reduction, access to healthcare, decent work, and universal health coverage, and are a fundamental condition for a sustainable, equitable, and resilient economic structure (GOSH, 2025, pp. 2–3).

A number of goals and initiatives pertaining to worker health and safety are part of the SDGs strategy. The right to a SHWE is especially related to SDGs 1, SDG 3, and SDG 8. Moreover, gender-specific risks such as increased women's participation in the workforce, working conditions during pregnancy, sexual harassment, and sectoral discrimination need to be specifically addressed within the scope of SDG 5 (IndustriALL, 2023; Sperper, 2021; Kelani, 2024). In this situation, the concept of a safe and healthy work environment should include social and psychological aspects in addition to physical safety. Additionally, given the high concentration of low-income, informal, and migrant workers in high-risk occupations, SDG 10 highlights the significance of acknowledging OSH as a universal right (ILO, 2021a; ILO, 2023e, p. 1). In light of this, this chapter investigates the connection between SDGs 1, 3, 5, 8, and 10 and OSH.

5.1. SDG 1 (No Poverty) and OSH

Occupational diseases deepen poverty both directly and indirectly by restricting individuals' participation in economic life, leading to loss of productivity, reduced tax revenues, and increased costs for health and social services. Consequently, this situation reduces family incomes, pushing many households below the poverty line (GOSH, 2025, p. 4). Studies on chronic and temporary poverty reveal that illness or injury of the individual providing household income is a key determinant in the perpetuation and spread of poverty. Moreover, declines in labor productivity are observed along with effects such as income loss, increased healthcare expenditures, and asset liquidation. In this context, effective OSH management plays an important role in protecting both individual and social welfare, maintaining productivity, and ensuring financial security (Sen, 2003; Hulme & Shepherd, 2003; Lund & Marriott, 2011; Atake, 2018; Zhao et al., 2025; GOSH, 2025).

While SDG 1, "no poverty", contains many goals, two indicators are directly linked to OSH: "*SDG indicator 1.1.1 - Working poverty rate (percentage of employed living below US$2.15 PPP)*" and "*SDG indicator 1.3.1 - Proportion of population covered by social protection floors/systems*" (ILOSTAT). Today, extreme working poverty remains a persistent structural problem globally, particularly in low-income countries. While moderate poverty rates have decreased by over 5% since 2015, extreme working poverty still affects approximately 240 million workers, 7% of the global workforce. This situation negatively impacts not only individual well-being but also economic development and social equality goals (ILO, 2025a, p. xiii). Due to the prevalence of informal employment, more than half of the global workforce lacks social security and OSH measures (Rantanen et al., 2017, p. 11). The majority of these losses occur among vulnerable workers in the poorest countries. Moreover, according to the latest ILO data, by 2023, only 52.4% of the world's population received at least one form of social protection, while the remaining 47.6% had no coverage at all. Only 37.4% of the global workforce has effective access to protection programs such as occupational accident insurance (ILOSTAT, 2025a).

In summary, SDG 1 calls on member states to establish comprehensive national social security systems for everyone, setting a clear goal of achieving broad coverage for the poor and vulnerable. Beyond its focus on ending poverty, social protection also supports SDG 2 on ending hunger, SDGs 3, 5, 8, and 10, as well as SDG 16, which promotes peace, justice, and strong institutions (ILO, 2024b, p. 2).

5.2. SDG 3 (Good Health and Well-Being) and OSH

In the current work environment, employment is no longer just about fulfilling the economic needs of individuals; it has evolved into one of the key factors influencing health-related quality of life (ILO, 2021b). However, the increasing competition driven by globalization has brought the pursuit of lower costs to the forefront, leading to the spread of adverse working conditions that threaten workers' health (Mariappanadar, 2024, p. 1). The fact that occupational health problems are one of the leading causes of preventable deaths in developed countries (ILO, 2021b, p. 2) demonstrates that OSH is a societal responsibility, as well as an individual one. In this context, SDG 3, Health and Well-Being, aims to ensure that everyone, including workers, lives a healthy life and promotes their well-being (Mariappanadar, 2024, p. 1). The goal encompasses components such as "reducing morbidity and mortality in vulnerable groups," "preventing communicable and non-communicable diseases," "reducing health risks," "ensuring access to universal health services," and "strengthening health systems." Although SDG 3 does not directly focus on social determinants of health, such as working conditions or income levels, the impact of these factors on health is widely recognized in the literatüre (WHO Commission on Social Determinants of Health & World Health Organization, 2008, p. 1662; Solar & Irwin, 2010, p. 16; Berkman et al., 2014, p. 153).

Working adults spend a significant part of their day working and doing related activities. The physical and social conditions experienced during this time directly affect an individual's physical, mental, and social health. Individuals who work under high stress for long periods cannot devote enough time to their family, social life, and personal interests. This negatively impacts life satisfaction and psychological resilience. Restructuring working life with a human-centered and health-oriented approach is essential for supporting individual well-being and ensuring the sustainability of organizational productivity (Adams, 2019, p. 583; GOSH, 2025, p. 4; Dumitriu et al., 2025, p. 2).

Organizations that have incorporated SDG 3 into their development strategies have achieved notable progress in reducing occupational accidents and diseases by adopting comprehensive and preventive approaches to worker health protection. OSH policies developed in this way focus on eliminating risks and promoting both the physical and mental well-being of workers. This approach fosters a human-centered work culture and, over time, enhances employee well-being while supporting the creation of sustainable and healthy workplaces (Mariappanadar, 2024, p. 1; Milea et al., 2025, p. 9).

5.3. SDG 5 (Gender Equality) and OSH

Gender inequality is a structural problem that directly impacts women's health and safety at work and in other areas of life (IndustriALL, October 16, 2025). Women face more "psychosocial risks such as stress, burnout, violence, discrimination, and harassment" compared to men. This, along with their domestic care responsibilities, leads to a double burden (ILO, 2013, p. 10). Observed gender gaps in "labor force participation, unemployment, and employment opportunities" show the structural barriers women encounter in the labor market. These include fragmented and gender-segregated labour markets, the unequal distribution of unpaid care responsibilities, and gender-based violence and harassment. In this context, gender-based inequalities indicate missed economic and social chances for sustainable development (ILO, 2017, p. 2; ILO, 2019b, p. 12, 13; ILO, 2025a, p. 8).

Research shows that women are at high risk of physical and psychological violence both in and outside the workplace (Sigursteinsdóttir et al., 2020; Eurofound, 2023; Nelson et al., 2024). Only 30% of women worldwide are employed in managerial positions (ILOSTAT, 2025b). The concentration of women in jobs that involve direct contact with the public or are done individually, along with their roles in low-paid and low-status positions, raises the risk of sexual violence and harassment. Men, in contrast, are more likely to face physical attacks (ILO, 2009, p. 1; Eurofound, 2023; ILO, 2013, p. 4; ILO, October 17, 2025). Psychological violence, bullying, and mobbing are being recognized more and more as OSH issues around the world. Many countries are creating laws to help prevent these risks. The *"ILO Convention on Violence and Harassment at Work (No. 190)"* encourages employers to carry out risk assessments that consider gender stereotypes and various types of discrimination. It also highlights the importance of having female employees actively involved in these processes. This method is essential for creating gender-equality OSH policies (ILO, 2019c, p. 4). OSH policies that ignore gender differences create unequal effects on women and men and reproduce existing inequalities at the institutional level (ILO, 2014; TUC, 2017, pp. 3, 4). Therefore, OSH policies must go beyond just preventing risks and take an inclusive approach that supports gender equality.

Current practices focus on visible risks in male-dominated sectors, but they do not sufficiently address the psychosocial risks, musculoskeletal disorders, and specific health needs faced by women (TUC, 2017, p. 6; IndustriALL, October 16, 2025). Work equipment and personal protective equipment are often designed for male body sizes (ILO, 2013, p. 35). Most international OSH legislation, instead of addressing risks from a gender perspective, obscures gender differences and operates under the assumption of "neutral" application. This method does not achieve true equality because it overlooks the unique needs of both male and female workers (ILO, 2013, p. 3).

The ILO (2013) guideline *"Gender Mainstreaming in OSH"* and the 2023 OSH strategy aim to connect the gender perspective with OSH policies by guiding policymakers and practitioners. The guide aims to ensure equal priority in OSH processes by taking into account the unique gender-based differences between women and men. It offers recommendations for effectively incorporating gender into areas such as information gathering, standard setting, legislation development, policy formulation, and preventative strategies. This approach demonstrates the ILO's commitment to integrating gender equality into OSH policy development and its contribution to achieving SDG 5 (ILO, 2013, p. 1; ILO, 2023e, pp. 13, 15). This integrated relationship between decent work and sustainability also forms the basis of the ILO's "2019 Centenary Declaration". In its report, the ILO highlighted the need for more investment in decent and sustainable work as one of its key goals, in line with the 2030 Agenda.

5.4. SDG 8 (Decent Work and Economic Growth) and OSH

The strong connection between decent work and sustainability forms the basis of the ILO's "Future of Work Agenda". The ILO's "2019 Centenary Declaration" emphasizes that decent work will be increasingly linked to sustainable development, and this connection will be an essential reference in future policies (ILC, 2019, pp. 3, 4). Decent work and fundamental social rights are at the heart of sustainable development, or "meeting the needs of the present without compromising the needs of future generations" (Rombouts and Zekić, 2020, p. 330). SDG 8 aims *"to promote sustainable economic growth, full and productive employment, and decent work for all"*.

The rapid transformation of work, the workforce, and the work environment makes the role of the OSH field in supporting decent work crucial. Decent work refers to *"a form of employment that respects the fundamental rights of workers and provides safe and healthy working conditions"* (Schulte et al., 2022, p. 2). In this context, SDG 8.8, a sub-goal of SDG 8, aims to *"ensure that all workers — particularly migrants, women, and those in precarious employment— are provided with safe and equitable working conditions*". This goal is closely linked to OSH practices (UN General Assembly, 2015, p. 20). "*SDG indicator 8.8.1 refers to the goal of reducing fatal and non-fatal occupational injuries*". However, ensuring the achievement of SDG 8.8.1 and maintaining fair working conditions critically depend on the freedom of association and the effective right to collective bargaining. Therefore, "*SDG indicator 8.8.2 - Level of national compliance with labor rights relates to the right to freedom of association and collective bargaining*". Indeed, the 2023 OSH Strategy places special emphasis on this right.

According to the ILO's "World Employment and Social Outlook: Trends to 2025" report, decent work and productive employment are of fundamental importance for achieving the 2030 SDGs. However, progress has stalled over the past decade, and economic growth remains weak and uncertain, particularly in low-income countries. Globally, key labor market indicators —such as employment, labor force participation, unemployment, and informality— have shown little improvement. Approximately 240 million people continue to work for extremely low wages, and 57.7% of the global workforce remains employed in the informal sector. Migrant workers are particularly concentrated in low-paid and insecure jobs, making them more vulnerable to OSH risks. Moreover, a large share of the global workforce cannot adequately benefit from social security mechanisms (ILO, 2024c, p. 1; ILO, 2025a, p. xvi; ILOSTAT, 2025c). On the other hand, "the global average for SDG indicator 8.8.2", which measures "national compliance with fundamental labor rights (freedom of association and collective bargaining), was 4.9 in 2023" (ILOSTAT, 2025d). This indicator shows that freedom of association and collective bargaining continue to face significant challenges worldwide.

Work-related accidents and occupational diseases remain a major global concern. According to the latest data from the ILO, around 2.93 million workers worldwide die each year due to work-related causes, with the vast majority of these fatalities (89%) attributed to occupational diseases (ILO, 2023b, p. 2). The most common causes of death are cardiovascular diseases, cancers, and respiratory disorders. In addition, climate change and digitalization are increasingly posing threats to OSH. Climate change is creating increasing OSH risks, particularly for outdoor workers and vulnerable groups. Air pollution affects more than 1.2 billion workers each year and causes over 860,000 deaths, while the livelihoods of those employed in sectors such as agriculture, fisheries, and forestry are threatened by ecosystem degradation (ILO, 2018, p. 2; ILO, 2024d, pp. 5, 9). In this context, Table 1 presents example regulations and practices related to the impacts of climate change.

Table 1. Practices on the Effects of Climate Change on OHS

Several countries have prioritized heat stress in their national OSH policies for the 2022–2028 period to protect workers from the increasing risks associated with rising temperatures. – Chile has focused on preventing occupational hazards linked to extreme heat, while Japan has developed specific indicators aimed at reducing heatstroke cases. – Belgium highlights the implementation of technical measures and the use of personal protective equipment during periods of excessive heat. – France has made the monitoring of fatalities and serious injuries caused by heat stress a key policy priority. – Spain's 2023–2027 OSH Strategy concentrates on improving working conditions and strengthening the protection of vulnerable workers, particularly under extreme heat conditions (ILO, 2024d, p. 22).
OSH regulations in certain countries focus on a variety of measures to protect workers. These include; – heat adaptation programs (e.g., Singapore, South Africa), – hydration and fluid supplementation initiatives (e.g., Austria, Chile, India), – ventilation system improvements (e.g., India, Argentina, Morocco), – scheduled rest breaks in shaded or air-conditioned areas (e.g., Saudi Arabia), – mandatory OSH training (e.g., Oman, South Africa), and – the provision of personal protective equipment (PPE) (e.g., Bahrain, Fiji, Italy). Additionally, some countries have adopted more comprehensive regulations that explicitly define workplace measures, risk assessment procedures, and appropriate prevention and mitigation strategies (ILO, 2024d, pp. 22, 25).
A notable example of inclusivity under the Decent Work Project took place in 2023, when the ILO, in collaboration with the Ministries of Labour in Tunisia, Morocco, and Egypt, trained labour inspectors on climate change, OSH, and gender. The sessions explored the impacts of climate change on women's health, discussed occupational safety measures, and emphasized the importance of gender-responsive policies. After completing the program, the trained inspectors were tasked with leading local awareness-raising activities to promote these issues within their communities. This project has been implemented to address climate change and gender inequality in the workplace, and to support the 2030 SDGs (ILO, 2025b, p. 3)
Safe Work Australia recommends implementing the hierarchy of controls to protect workers from air pollution. This includes measures such as, – elimination (relocating work to areas with clean air, or enabling work from home or alternative sites), – substitution (e.g., conducting work indoors whenever possible), – engineering and administrative controls (e.g., using air purifiers, increasing rest breaks, or reducing physical workload), and – the use of PPE. The overarching aim is to systematically and effectively mitigate health risks associated with air pollution in the workplace (Safe Work Australia, 2025).

Conversely, the digitalization of workplaces has shifted traditional work toward remote and hybrid models and has expanded the prevalence of platform-based employment. The COVID-19 pandemic notably accelerated this transformation (Elsamani & Kajikawa, 2024). Although digital work creates employment opportunities for groups such as people with disabilities, older adults, and those with caregiving responsibilities, it also introduces a range of health and safety risks, including poor ergonomics, extended working hours, social isolation, psychosocial stress, and exposure to digital harassment (Yoo et al., 2024; Yarrow & Davies, 2025; EU-OSHA, 2025). Platform and freelance workers are particularly vulnerable due to regulatory gaps, insufficient safety measures, and job insecurity. This situation poses challenges for employers in maintaining OSH standards. While existing OSH regulations provide general protection, legal frameworks that specifically address the unique risks associated with digitalization remain limited. Many countries are adopting national strategies on digitalization and artificial intelligence (AI). Although these strategies rarely address OSH issues directly, they increasingly promote a worker-centered approach to digital transformation (ILO, 2025c, pp. 30, 36). Table 2 presents examples of practices related to the impacts of digitalization on OSH.

Table 2. Practices on the Effects of Digitalization on OHS

Some countries are preventing worker burnout through "right to disconnect" laws, which enable employees to refrain from digital communications outside of working hours. – Luxembourg implemented this legislation in 2023, while in Spain, the 2024 Tekniker collective agreement secured this right (Brunnerová et al., 2024; Gathongo & Ndimurwimo, 2025).
Certain guidelines emphasize leveraging digital technologies in OSH to enhance risk management. – In Argentina, the *Libro Blanco* seeks to strengthen workplace risk prevention through the use of digitalization, AI, and data analytics. – Canada's WorkSafeBC offers tools that allow employers to track injury and compensation data and to make informed safety planning based on performance analysis. – In France, the potential of AI for risk prevention is being examined, and under the *Digital Labour Code*, safety, health, and ethical standards are defined for robot-assisted work environments (ILO, 2025c, pp. 47, 48).
Social partners are implementing various initiatives in the areas of digitalization and OSH. – The ASEAN Employers' Confederation emphasizes ethical principles and governance in the use of AI. – BusinessEurope focuses on training and skills development to address the psychological impacts of digitalization and remote work. – CEEMET examines the effects of technologies such as collaborative robots (cobots), smart PPE, and ergonomic solutions on OSH (ILO, 2025c, p. 41).
Regional trade unions are strengthening workers' rights by providing training and guidance on digital rights, data security, AI governance, and algorithmic transparency. – Public Services International conducts training programs on digital rights and OSH across Africa and the Middle East (ILO, 2025c, p. 41).

Regardless of how comprehensive legal regulations and policy frameworks in OSH may be, various practical challenges arise during implementation. In low- and middle-income countries, the most common issues include environmental hazards (extreme weather, radiation, toxic substances), shortages of financial and material resources, political instability, lack of PPE and local guidance, limited technical expertise, insufficient data, and low safety awareness and inadequate training among both leaders and workers. Regions affected by conflict or economic development constraints are particularly prone to significant challenges in the effective implementation of OSH practices (Abu Aisheh et al., 2021; Baghdadi, 2024; Abdi & Hareru, 2024; Kunodzia et al., 2024; Arntz et al., 2024). These practical challenges highlight the gap between legal frameworks and on-the-ground implementation, underscoring the need for policymakers and employers to develop additional measures. Effective inspection mechanisms, regional capacity-building, and equitable access to technology will play a crucial role in overcoming these barriers. Therefore, addressing the impacts of these issues comprehensively is essential to achieving SDG 8.8.

5.5. SDG 10 (Reduced Inequalities) and OSH

In keeping with the principle of "leaving no one behind" in the sustainable development process, it is essential to collect reliable, timely, and detailed disaggregated data to monitor the conditions of traditionally excluded or vulnerable social groups (WHO, 2016, p. 23). This requirement also applies to efforts aimed at ensuring safe and healthy working conditions for all workers. The fundamental goal of SDG 10 is to reduce all forms of inequality based on factors such as income, opportunity, social status, age, gender, disability, origin, or migration status. The disadvantaged position of migrant workers in OSH, as well as the hidden inequalities they experience, is particularly highlighted (ILO, October, 25, 2025; The Global Goals, October, 25, 2025; UN General Assembly, 2015, p. 21).

Looking at the future trajectory of the SDGs, significant challenges remain in achieving them. While each SDG has its own specific objectives, the overarching implementation challenges affect all of them. SDG 10, which focuses on reducing inequalities within and between countries, also faces substantial obstacles. Five key factors have been identified as hindering progress on SDG 10 (Bhandari, 2024, pp. 2, 5):

1. "*Economic inequality:* Differences in income and wealth are among the primary obstacles to achieving equality.
2. *Social exclusion:* Marginalized groups, including minorities, women, and people with disabilities, face barriers in accessing resources and opportunities, which further exacerbate economic inequalities.
3. *Political barriers:* A lack of political will and ineffective governance systems hinder efforts to combat inequality and implement inclusive policies.
4. *Unequal access to education and healthcare:* Limited access to quality education and healthcare perpetuates social and economic disparities.
5. *Global economic interdependence:* Globalization, through increased trade, finance, and technology flows, can deepen inequalities between countries".

Within the context of the right to a SHWE, economic inequality can restrict access to OSH services. Workers in low-paid or precarious jobs, in particular, often struggle to obtain adequate protection and safe working conditions. Social exclusion increases the risk of discrimination and exposure to hazardous working conditions, particularly for women, migrants, and other vulnerable groups. Data from the ILO indicate that these populations are at higher risk of occupational accidents and diseases. Political and governance deficiencies further impede the implementation of effective OSH policies, making it difficult to protect disadvantaged workers (Biermann et al., 2022, p. 797; Bhandari, 2024, p. 3). Furthermore, unequal access to vocational training and healthcare services increases the risk of occupational accidents and diseases and perpetuates socio-economic inequalities (Çoban et al., 2018, p. 127). The risks faced by workers are increasing in certain countries and sectors due to globalization. Low-income and migrant workers, in particular, often do not benefit adequately from protective measures (Singla, 2023, p. 373). In this context, equal access to a SHWE is closely linked to SDG 10's goal of reducing inequalities. The ILO's Decent Work agenda supports this objective by promoting safe and non-discriminatory working conditions for all workers.

CONCLUSION

The ILO's 2022 decision, by recognizing Conventions No. 155 and No. 187 as fundamental, reaffirms the commitment to protecting workers' health and lives, promoting decent work, and advancing social justice, while aligning these efforts with the 2030 Agenda for Sustainable Development. However, in market-based economies, economic growth and workers' access to safe and healthy working conditions are often at odds. While capitalist systems encourage innovation, they can also produce inequalities and precarious work environments, particularly in terms of OSH. Indeed, informal workers are especially disadvantaged, lacking legal protections and social security, which makes them highly vulnerable to workplace accidents and occupational diseases. Policies focused primarily on growth tend to prioritize capital, leaving many workers marginalized.

Even the most comprehensive OSH laws and policies encounter practical obstacles, particularly in low- and middle-income countries. Environmental risks, limited financial and material resources, inadequate training, and low safety awareness often impede the full implementation of OSH measures. While technological innovations provide potential solutions, disparities in infrastructure and access limit their effectiveness. This highlights the necessity for policymakers and employers to adopt supplementary measures, such as enhanced monitoring, capacity development, and equitable technology access.

To accelerate progress toward providing safe and healthy working conditions in line with the SDGs addressed in this chapter (SDG 1, 3, 5, 8, 10), these structural challenges must be addressed. OSH policies should not rely solely on market-based labor approaches; they must also consider multiple intersecting crises such as demographic change, digitalization, global climate change, and gender inequality. Closing legislative gaps and ensuring equal safety standards for all workers is particularly critical for temporary workers, platform workers, domestic workers, migrant workers, and those employed in the informal economy. Furthermore, strengthening the capacity of labor organizations to represent workers in non-standard forms of employment and adapting social protection systems to cover these workers are essential steps for reducing OSH risks and preventing work-related illnesses.

Consequently, as highlighted in the 2023 OHS Strategy, all stakeholders — including policymakers, managers, entrepreneurs, employer organizations, unions, and employees— must take responsibility for advancing decent work. This involves considering the ongoing transformations in social, economic, and environmental systems, while embracing social justice as a fundamental principle. To ensure that societies evolve into more sustainable, equitable, just, and environmentally friendly structures, social justice, supported by social dialogue, should guide policy decisions and practices in accelerating progress toward the SDGs.

REFERENCES

Abdi, M. A., & Hareru, W. K. (2024). Assessment of factors affecting the implementation of occupational health and safety measures in the construction industry in Somaliland. *Advances in Civil Engineering*, *8324294*(1), 8324294. Advance online publication. DOI: 10.1155/2024/8324294

Abu Aisheh, Y. I., Tayeh, B. A., Alaloul, W. S., & Jouda, A. F. (2021). Barriers of occupational safety implementation in infrastructure projects: Gaza Strip case. *International Journal of Environmental Research and Public Health*, *18*(7), 3553. DOI: 10.3390/ijerph18073553 PMID: 33805581

Adams, J. M. (2019). The value of worker well-being. *Public Health Reports*, *134*(6), 583–586. DOI: 10.1177/0033354919878434 PMID: 31600480

Arntz, M., Findeisen, S., Maurer, S. E., & Schlenker, O. (2024). *Are We Yet Sick of New Technologies?* The Unequal Health Effects of Digitalization.

Atake, E. H. (2018). Health shocks in Sub-Saharan Africa: Are the poor and uninsured households more vulnerable? *Health Economics Review*, *8*(1), 26. DOI: 10.1186/s13561-018-0210-x PMID: 30317395

Baghdadi, A. (2024). Navigating occupational safety and health challenges in sustainable infrastructure projects: A comprehensive review. *Frontiers in Built Environment*, *10*, 1414366. DOI: 10.3389/fbuil.2024.1414366

Berkman, L. F., Kawachi, I., & Glymour, M. M. (Eds.). (2014). *Social epidemiology* (2nd ed.). Oxford University Press.

Bhandari, M. P. (2024). What is next for the sustainable development goals: What are the challenges concerning SDG 10—Reduced inequalities? *Sustainable Earth Reviews*, *7*(1), 23. DOI: 10.1186/s42055-024-00093-8

Biermann, F., Hickmann, T., Sénit, C. A., Beisheim, M., Bernstein, S., Chasek, P., Grob, L., Kim, R. E., Kotzé, L. J., Nilsson, M., Ordóñez Llanos, A., Okereke, C., Pradhan, P., Raven, R., Sun, Y., Vijge, M. J., van Vuuren, D., & Wicke, B. (2022). Scientific evidence on the political impact of the Sustainable Development Goals. *Nature Sustainability*, *5*(9), 795–800. DOI: 10.1038/s41893-022-00909-5

Brunnerová, S., Ceccon, D., Holubová, B., Kahancová, M., Lukáčová, K., & Medas. (2024). Collective bargaining practices on AI and algorithmic management in European services sectors. FriedrichEbertStiftung Competence Centre on the Future of Work, Brussels.

Çoban, O., Çoban, A., & Mirza, E. (2018). The impact of socioeconomic factors on access to health services in Turkey: A logit analysis. *Journal of Selçuk University Social Sciences Vocational School, 21*(2), 117–131. DOI: 10.29249/selcuksbmyd.422243

Demirbilek, T. (2005). *İş güvenliği kültürü*. Dokuz Eylül Publishing.

Dumitriu, S., Bocean, C. G., Vărzaru, A. A., Al-Floarei, A. T., Sperdea, N. M., Popescu, F. L., & Băloi, I.-C. (2025). The role of the workplace environment in shaping employees' well-being. *Sustainability (Basel), 17*(6), 2613. DOI: 10.3390/su17062613

Elsamani, Y., & Kajikawa, Y. (2024). How teleworking adoption is changing the labor market and workforce dynamics? *PLoS One, 19*(3), e0299051. Advance online publication. DOI: 10.1371/journal.pone.0299051 PMID: 38502670

EU-OSHA. (2025). *Smart digital systems for all: Supporting inclusion and diversity in workplace safety and health*. European Agency for Safety and Health at Work. Retrieved November, 10, 2025, from https://osha.europa.eu/en/oshnews/smart-digital-systems-all-supporting-inclusion-and-diversity-workplace-safety-and-health

European Foundation for the Improvement of Living and Working Conditions (Eurofound). (2023). *Violence in the workplace: Women and frontline workers face higher risks*. Retrieved October 17, 2025, from https://www.eurofound.europa.eu/en/commentary-and-analysis/all-content/violence-workplace-women-and-frontline-workers-face-higher-risks

Gafney, O. (2014). Sustainable development goals: Improving human and planetary wellbeing. *Global Change, 82*, 20–23. Retrieved October 2, 2025, from http://www.igbp.net/news/features/features/sustainabledevelopmentgoalsimprovinghumanandplanetarywellbeing.5.62dc35801456272b46d1f7.html

Gathongo, J. J. K., & Ndimurwimo, L. (2025). A Critical Analysis of the Proposed Law on the Employees' Right to Disconnect in Kenya. [Published-on.]. *Potchefstroom Electronic Law Journal*, •••, 28.

Global Occupational Safety and Health (GOSH). (2025). *Occupational health and the SDGs: A policy brief from the GOSH Coalition*. The Society of Occupational Medicine.

Hilgert, J. (2024). Safety and health at work as fundamental rights: A comparative-historical study of the ILO's strategy of realistic vigilance. *International Labour Review, 163*(1), 95–115. DOI: 10.1111/ilr.12401

Hulme, D., & Shepherd, D. (2003). Conceptualizing chronic poverty. *World Development, 31*(3), 403–423. DOI: 10.1016/S0305-750X(02)00222-X

ILO. (2023c). *Transformative Change and SDG 8: The Critical Role of Collective Capabilities and Societal Learning*. International Labour Office.

ILOSTAT. (2025a). *SDG indicator 1.3.1 – Proportion of population covered by social protection floors/systems (%).* Retrieved October 16, 2025, from https://rshiny.ilo.org/dataexplorer40/?lang=en&id=X01_A

ILOSTAT. (2025b). *SDG indicator 5.5.2 - Proportion of women in managerial positions (%) – Annual.* Retrieved October 17, 2025, from https://rshiny.ilo.org/dataexplorer77/?lang=en&segment=indicator&id=SDG_0831_SEX_ECO_RT_A

ILOSTAT. (2025c). *SDG indicator 8.3.1 – Proportion of informal employment in total employment by sex and sector (%) – Annual.* Retrieved October 18, 2025, from https://rshiny.ilo.org/dataexplorer50/?lang=en&segment=indicator&id=SDG_0111_SEX_AGE_RT_A

ILOSTAT. (2025d). *SDG indicator 8.8.2 - Level of national compliance with labour rights (freedom of association and collective bargaining) – Annual,* Retrieved October 19, 2025, https://rshiny.ilo.org/dataexplorer35/?lang=en&segment=indicator&id=SDG_0882_NOC_RT_A

Industri, A. L. L. (2023, April 28). Safety for all – the need for gender-responsive OHS. Retrieved October 16, 2025, from https://www.industriall-union.org/safety-for-all-the-need-for-gender-responsive-ohs

International Labour Conference (ILC). (2019). *ILO centenary declaration for the future of work adopted by the conference at its one hundred and eighth session.* Geneva.

International Labour Organization (ILO). *C155 – Occupational Safety and Health Convention, 1981 (No. 155).* Retrieved September 3, 2025, from https://normlex.ilo.org/dyn/nrmlx_en/f?p=NORMLEXPUB%3A12100%3A0%3A%3ANO%3A12100%3AP12100_INSTRUMENT_ID%3A312300%3ANO

International Labour Organization (ILO). (1999). *Decent work.* International Labour Office, International Labour Conference, 87th Session, Report of the Director General. Geneva: International Labour Office.

International Labour Organization (ILO). (2003). *Safety in numbers: Pointers for a global safety culture at work.* International Labour Office.

International Labour Organization (ILO). (2004). *Global strategy on occupational safety and health.* International Labour Organization.

International Labour Organization (ILO). (2005a). *Promoting safety and health at work: ILO report for World Day for Safety and Health at Work*. International Labour Organization.

International Labour Organization (ILO). (2005b). *Promotional framework for occupational safety and health*. International Labour Office.

International Labour Organization (ILO). (2009). Recovering from the crisis: A global jobs pact. *International Labour Conference*.

International Labour Organization (ILO). (2009). *Violence at work – A major workplace problem*. Retrieved October 17, 2025, from https://www.ilo.org/resource/violence-work-major-workplace-problem

International Labour Organization (ILO). (2013). *10 keys for gender sensitive OSH practice: Guidelines for gender mainstreaming in occupational safety and health*. International Labour Organization.

International Labour Organization (ILO). (2014). *The gender dimension: Integrating the gender perspective in OSH policies*. Retrieved October 19, 2025, from https://www.ilo.org/resource/article/gender-dimension-integrating-gender-perspective-osh-policies

International Labour Organization (ILO). (2015). *Guidelines for a just transition towards environmentally sustainable economies and societies for all*. International Labour Office.

International Labour Organization (ILO). (2017). *World employment and social outlook: Trends for women 2017*. International Labour Organization.

International Labour Organization (ILO). (2018). *World employment social outlook: Greening with jobs*. International Labour Organization.

International Labour Organization (ILO). (2019, May 21). *100 years of making work safer and healthier*. Retrieved September 12, 2025, from https://www.ilo.org/resource/article/100-years-making-work-safer-and-healthier

International Labour Organization (ILO). (2019a). *Safety and health at the heart of the future of work: Building on 100 years of experience*. International Labour Office.

International Labour Organization (ILO). (2019b). *A quantum leap for gender equality: For a better future of work for all*. Retrieved from https://www.ilo.org/publications/major-publications/quantum-leap-gender-equality-better-future-work-all

International Labour Organization (ILO). (2019c). *Eliminating violence and harassment in the world of work: ILO Convention No. 190, Recommendation No. 206, and the accompanying Resolution.* Geneva: International Labour Organization.

International Labour Organization (ILO). (2020a). *Safety + health for all: An ILO flagship programme: Key facts and figures (2016–2020).* International Labour Office.

International Labour Organization (ILO). (2020b). *In the face of a pandemic: Ensuring safety and health at work.* International Labour Office.

International Labour Organization (ILO). (2021a). *World social protection report 2020–22: Social protection at the crossroads – in pursuit of a better future.* International Labour Office.

International Labour Organization (ILO). (2021b). *WHO/ILO joint estimates of the work-related burden of disease and injury, 2000–2016: Global monitoring report.* World Health Organization.

International Labour Organization (ILO). (2022). *Violence and harassment at work has affected more than one in five people.* Retrieved October 17, 2025, from https://www.ilo.org/resource/news/violence-and-harassment-work-has-affected-more-one-five-people

International Labour Organization (ILO). (2022, June 10). International Labour Conference adds safety and health to fundamental principles and rights at work. Retrieved October 10, 2025, from https://www.ilo.org/resource/news/ilc/110/international-labour-conference-adds-safety-and-health-fundamental

International Labour Organization (ILO). (2022a). *A safe and healthy working environment is a fundamental principle and right at work.* International Labour Organization.

International Labour Organization (ILO). (2023a). *The fundamental conventions on occupational safety and health: An overview of the Occupational Safety and Health Convention, 1981 (No. 155) and the Promotional Framework for Occupational Safety and Health Convention, 2006 (No. 187), 2023.* International Labour Organization.

International Labour Organization (ILO). (2023b). *A call for safer and healthier working environments.* International Labour Office.

International Labour Organization (ILO). (2023c). *Safety + health for all: Accelerating the realization of fundamental principles and rights at work.* International Labour Office.

International Labour Organization (ILO). (2023d). Safe and healthy working environments for all – ILO introductory report: Realizing the fundamental right to a safe and healthy working environment worldwide, *23rd World Congress on Safety and Health at Work,* 27–30 November 2023, Sydney, Australia. Geneva: International Labour Office.

International Labour Organization (ILO). (2023e). *The ILO global strategy on occupational safety and health and its plan of action.* International Labour Organization.

International Labour Organization (ILO). (2024a). *World employment and social outlook.* International Labour Office.

International Labour Organization (ILO). (2024b). *Bridging social protection and occupational health to advance sustainable development goals.* International Labour Office.

International Labour Organization (ILO). (2024c). *ILO global estimates on international migrant workers: International migrants in the labour force.* International Labour Organization.

International Labour Organization (ILO). (2024d). *Ensuring safety and health at work in a changing climate.* International Labour Organization.

International Labour Organization (ILO). (2025, April 28). *The right to occupational safety and health: Still unrealized.* Retrieved October 25, 2025, from https://ilostat.ilo.org/the-right-to-occupational-safety-and-health-still-unrealized/

International Labour Organization (ILO). (2025a). *World employment and social outlook: Trends 2025.* International Labour Office.

International Labour Organization (ILO). (2025b). *Decent Work for Women in Egypt, Tunisia and Morocco: Phase III.* International Labour Organization Cairo Office.

International Labour Organization (ILO). (2025c). *Revolutionizing health and safety: The role of AI and digitalization at work.* International Labour Office.

International Labour Organization (ILO). *ILO Constitution.* Retrieved September 11, 2025, from https://normlex.ilo.org/dyn/nrmlx_en/f?p=1000:62:0:NO:62:P62_LIST_ENTRIE_ID:2453907:NO

Kelani, Z. A. (2024). Invisible women: Data bias in a world designed for men. *Journal of International Women's Studies, 26*(5), 1–6. https://vc.bridgew.edu/jiws/vol26/iss5/18

Kellerson, H. (1998). The ILO declaration of 1998 on fundamental principles and rights: A challenge for the future. *International Labour Review, 137*(2), 223–227. https://researchrepository.ilo.org/esploro/outputs/journalArticle/The-ILO-Declaration-of-1998-on/995274862902676

Kunodzia, R., Bikitsha, L., & Haldenwang, R. (2024). Perceived factors affecting the implementation of occupational health and safety management systems in the South African construction industry. *Safety (Basel, Switzerland), 10*(1), 5. DOI: 10.3390/safety10010005

Lund, F., & Marriott, A. (2011). *Occupational health and safety and the poorest.* University of KwaZulu-Natal, School of Development Studies.

Mariappanadar, S. (2024). Improving quality of work for positive health: Interaction of Sustainable Development Goal (SDG) 8 and SDG 3 from the sustainable HRM perspective. *Sustainability (Basel), 16*(13), 5356. DOI: 10.3390/su16135356

Milea, A., Moraru, R. I., & Cioca, L. I. (2025). Occupational risk management through the lens of the Sustainable Development Goals (SDGs): An integrated approach to promoting sustainability in the workplace. *Sustainability, 17*(5), 1864. DOI: 10.3390/su17051864

Nelson, S., Ayaz, B., Baumann, A. L., & Dozois, G. (2024). A gender-based review of workplace violence amongst the global health workforce—A scoping review of the literature. *PLOS Global Public Health, 4*(7), e0003336. DOI: 10.1371/journal.pgph.0003336 PMID: 38954697

Rantanen, J., Muchiri, F., & Lehtinen, S. (2020). Decent work, ILO's response to the globalization of working life: Basic concepts and global implementation with special reference to occupational health. *International Journal of Environmental Research and Public Health, 17*(10), 3351. DOI: 10.3390/ijerph17103351 PMID: 32408597

Rombouts, S. J., & Zekić, N. (2020). Decent and sustainable work for the future? *UCLA Journal of International Law and Foreign Affairs, 24*(2), 317–358. https://www.jstor.org/stable/48609838

Safe Work Australia. (2025). *Managing the risks from air pollution: Advice for PCBUs.* Retrieved November, 9, 2025, from https://www.safeworkaustralia.gov.au/safety-topic/hazards/working-heat/managing-risks-air-pollution-advice-pcbus

Schulte, P. A., Iavicoli, I., Fontana, L., Leka, S., Dollard, M. F., Salmen-Navarro, A., Salles, F. J., Olympio, K. P. K., Lucchini, R., Fingerhut, M., Violante, F. S., Seneviratne, M., Oakman, J., Lo, O., Alfredo, C. H., Bandini, M., Silva-Junior, J. S., Martinez, M. C., Cotrim, T., & Fischer, F. M. (2022). Occupational safety and health staging framework for decent work. *International Journal of Environmental Research and Public Health, 19*(17), 10842. DOI: 10.3390/ijerph191710842 PMID: 36078562

Sen, B. (2003). Drivers of escape and descent: Changing household fortunes in rural Bangladesh. *World Development, 31*(3), 513–534. DOI: 10.1016/S0305-750X(02)00217-6

Sengenberger, W. (2001). Decent work: The International Labour Organization agenda. *Dialogue and Cooperation, 2*(2001), 39–55. Available at: https://library.fes.de/pdf-files/iez/global/02077.pdf

Sigursteinsdóttir, H., Rafnsdóttir, G. L., & Einarsdóttir, T. (2020). Threats and physical violence in female-dominated workplaces in times of an economic crisis. *International Journal of Workplace Health Management, 13*(4), 377–392. DOI: 10.1108/IJWHM-01-2019-0009

Singla, N., Sood, K., Grima, S., & Yadav, U. S. (2023). Target 8.8: Protect labor rights and promote a safe working environment. *The impact of climate change and sustainability standards on the insurance market*, 373–392. https://doi.org/DOI: 10.1002/9781394167944.ch24

Solar, O., & Irwin, A. (2010). A conceptual framework for action on the social determinants of health (Social Determinants of Health Discussion Paper 2). World Health Organization.

Somavia, J. (2000). *Perspectives on decent work statements by the ILO Director-General*. International Labour Office.

Takala, J. (2019). Burden of injury due to occupational exposures. In *Handbook of Disability* (pp. 1–22). Work and Health. DOI: 10.1007/978-3-319-75381-2_5-1

The Global Goals. Reduce inequality. Retrieved October 25, 2025, from https://globalgoals.org/goals/10-reduced-inequalities/

TUC. (2017). Gender in occupational safety and health. Trades Union Congress. Retrieved October 17, 2025, from https://www.tuc.org.uk/research-analysis/reports/gender-occupational-safety-and-health

United Nations. (UN) General Assembly. (2015). *Transforming our world: The 2030 agenda for sustainable development (Resolution A/RES/70/1).* https://sdgs.un.org/2030agenda

United Nations Development Programme (UNDP). (2021). *The changing nature of work.* UNDP Accelerator Labs.

United Nations Development Programme (UNDP). (2022). *Human development report 2021/2022: Uncertain times, unsettled lives: Shaping our future in a transforming world.* New York, NY.

United Nations (UN). (2015). *Sustainable development goals: Transforming our world: The SDGs 2030 agenda for sustainable development.* United Nations. United Nations (UN). *The 17 goals.* Retrieved October 2, 2025, from https://sdgs.un.org/goals#implementation

WHO Commission on Social Determinants of Health & World Health Organization. (2008). *Closing the gap in a generation: Health equity through action on the social determinants of health: Commission on social determinants of health final report.* World Health Organization.

World Bank. (2019). *World development report 2019: The changing nature of work.* World Bank.

World Health Organization (WHO). (2016). *World health statistics 2016: Monitoring health for the SDGs, Sustainable Development Goals.* World Health Organization.

World Health Organization (WHO) & Commission on Social Determinants of Health (CSDH). (2008). *Closing the gap in a generation: Health equity through action on the social determinants of health: Final report of the Commission on Social Determinants of Health.* World Health Organization.

Yarrow, E., & Davies, J. (Eds.). (2025). *Worker Well-being, Human Factors and the Gig Economy: Interdisciplinary Perspectives.* CRC Press. DOI: 10.1201/9781003462897

Yoo, H., Yang, M., Song, J. H., Yoon, J. H., Lee, W., Jang, J., Yoon, M., & Kang, M. Y. (2024). Investigation of working conditions and health status in platform workers in the Republic of Korea. *Safety and Health at Work, 15*(1), 17–23. DOI: 10.1016/j.shaw.2024.01.002 PMID: 38496284

Zhao, X., Wang, X., Zhou, R., & Zhang, C. (2025). The impact of health shocks on vulnerability to poverty: Evidence from rural households in China. *Frontiers in Public Health, 13*, 1657269. DOI: 10.3389/fpubh.2025.1657269 PMID: 40951400

Chapter 2
International Models and Success Stories in Occupational Health and Safety:
Global Strategies and Systems for Worker Safety

Krishna Kumar Ravichandran
http://orcid.org/0000-0002-7948-0291
KPR Institute of Engineering and Technology, India

ABSTRACT

This chapter examines international models of Occupational Health and Safety (OHS) with an emphasis on its critical role in enhancing socio-economic welfare through industry safety, ethical practices, employee trust, productivity, and national economic growth. It covers sectors including high-risk industries like construction, where construction safety remains a challenge, as well as expanding fields such as IT, healthcare, and education. The chapter synthesizes research on occupational stress and its impact on workplace safety, integrating findings on green building hazards with ergonomic strategies. It highlights the importance of safety management systems and worker engagement in reducing accidents. The use of predictive analytics, including advanced machine learning, is discussed as a powerful tool for hazard detection and risk mitigation, supporting adaptive OHS systems for sustainable workplace health improvement

DOI: 10.4018/979-8-3373-6370-7.ch002

1. INTRODUCTION

Occupational Health and Safety (OHS) has emerged as one of the most critical pillars of modern industrial development, economic sustainability, and workforce welfare across the world (International Labour Organization [ILO], 2023. As nations continue to evolve technologically, demographically, and economically, the protection of workers from injuries, illnesses, and psychosocial risks has transformed from a regulatory obligation into a strategic imperative. Today, OHS frameworks not only safeguard physical well-being but also enhance productivity, strengthen organizational culture, support innovation, and contribute to long-term national competitiveness.

The concept of OHS has changed dramatically over the past century. Earlier models were predominantly reactive, primarily focusing on accident reporting, compensation, and basic hazard control measures. Over time, growing awareness, industrial expansion, and the psychological and social complexities of work environments (Leka & Jain, 2019) pushed OHS toward more advanced preventive and systemic models.(Quinlan, 2015) Modern OHS systems emphasize proactive risk identification, integrated health promotion, and continuous improvement supported by scientific research, standards, and evidence-based practices.

This chapter explores the evolution of international OHS models, compares different national approaches, and presents successful case examples that illustrate how organizations worldwide have achieved remarkable outcomes through robust safety strategies. It draws attention to human factors, technological innovations, and socio-cultural influences that shape workplace safety in diverse environments. It also highlights the global movement toward harmonized standards, digital transformation, and policy-driven reforms that aim to create safer and healthier workplaces for all.

1.1. The Global Emergence of OHS as a Strategic Priority

The global emphasis on OHS is rooted in both economic and ethical dimensions. Organizations increasingly recognize that workplace injuries and illnesses impose significant direct and indirect costs, including medical expenses, productivity loss, legal liabilities, reduced morale, and reputational damage. At the macro level, governments realize that national productivity and human capital development are closely tied to workplace safety.

A significant milestone in the global recognition of OHS came with the formation of international bodies such as the International Labour Organization (ILO), which established conventions and recommendations that guide national legislative frameworks. The World Health Organization (WHO) further expanded the scope by linking workplace health to broader public health concerns.(World Health Organization

[WHO], 2021) Over the decades, these institutions influenced countries to transition from compliance-oriented regulations toward preventive risk management systems.

1.2. Historical Development of OHS Across Regions

While the basic principles of workplace safety have universal relevance, the evolution of OHS varies widely across continents due to differences in industrial maturity, cultural norms, governance systems, and economic capacity.

1.2.1 Europe: The Foundation of Systematic OHS Governance

Europe has long been regarded as the global benchmark in workplace safety governance. The European Union (EU) employs a harmonized legislative framework emphasizing worker participation,(European Agency for Safety and Health at Work [EU-OSHA], 2020) systematic risk assessment, and preventive control strategies. The Framework Directive 89/391/EEC introduced the concept of general prevention principles, laying the foundation for industry-specific regulations across member states.

European nations such as Germany, Sweden, and the Netherlands emphasize a holistic approach to OHS, integrating physical safety, mental health, and social well-being. Their success is attributed to strong unions, comprehensive regulatory systems, widespread adoption of safety management frameworks, and the cultural acceptance of worker empowerment.

1.2.2 North America: Innovation and Standardization

The United States and Canada pioneered several modern approaches to OHS, driven by strong regulatory institutions such as OSHA (Occupational Safety and Health Administration) and NIOSH (National Institute for Occupational Safety and Health). These agencies introduced the concepts of hazard communication, industrial hygiene, and behavioral safety, supported by large-scale research and data exploration.

North America's contribution to global OHS includes advancements in ergonomics, psychosocial risk research, safety culture theory, and technological innovations. Many internationally adopted safety methodologies—such as Safety Management Systems (SMS), job safety analyses, total worker health frameworks, and leading indicator approaches—originated in this region.

1.2.3 Asia-Pacific: Rapid Industrial Growth and Expanding OHS Systems

The Asia-Pacific region presents a diverse OHS landscape. Countries like Japan, Singapore, South Korea, Australia, and New Zealand have solid, mature systems emphasizing technological integration, rigorous training, and continuous improvement. These nations are known for applying lean management principles to safety, focusing on hazard elimination, ergonomic design, and discipline in operational practices.

Emerging economies in Asia, including India, Vietnam, and Indonesia, face challenges due to large informal sectors, limited resources, and varying enforcement capacities. Yet, these countries are increasingly adopting digital and low-cost innovations to improve OHS outcomes.

1.2.4 Middle East and Africa: Evolving Frameworks Through Modernization

Several Middle Eastern countries have strengthened their safety infrastructures in parallel with expanding industrial sectors such as oil, gas, and construction. Investments in mega-projects have led to increased adoption of international standards, advanced technology, and worker welfare programs. (Fukuzawa et al. 2020)

African countries exhibit wide variation, with some regions adopting strong OHS frameworks and others facing significant resource constraints. International partnerships, capacity-building programs, and multinational corporation involvement play major roles in improving workplace conditions.

1.3. The Modern OHS Paradigm: A Systemic, Multidimensional Approach

OHS in the 21st century is no longer confined to the prevention of physical injuries. Modern workplaces encompass psychosocial challenges, technological complexities, environmental hazards, and workforce diversity that require advanced, multidimensional models. Several key principles define the contemporary global OHS paradigm:

1.3.1 Prevention Over Reaction

Instead of responding to incidents, organizations now focus on identifying risks in advance. This shift is anchored in proactive risk assessment, predictive analytics, and early warning systems.

1.3.2 Systems Thinking

Safety is influenced by interactions between people, processes, environments, and organizational structures. (International Organization for Standardization [ISO], 2018) System-based models highlight the importance of leadership, culture, communication pathways, engineering controls, and continuous improvement.

1.3.3 Worker Participation and Empowerment

Modern OHS models recognize workers as partners in safety. Their involvement in hazard reporting, decision-making, and intervention design enhances ownership and effectiveness.

1.3.4 Integration of Health and Well-being

Physical safety and mental well-being are interdependent. Programs that address fatigue, stress, work–life balance, and psychosocial risks have become essential.

1.3.5 Technology as a Strategic Enabler

Digital transformation—AI, IoT, wearables, automation, and real-time analytics—plays a central role in modern safety strategies.

1.4. Importance of International Models and Success Stories

Studying international models allows organizations and policymakers to understand proven frameworks, identify adaptable concepts, and avoid repeating past mistakes. Success stories, whether from multinational corporations or national safety bodies, provide evidence-based insights into what works in real-world environments.

They demonstrate how leadership commitment, safety culture, and innovation converge to create strong, resilient OHS systems. They also show how different countries adapt safety solutions to their cultural, economic, and technological contexts.

2. HUMAN AND ORGANIZATIONAL FACTORS IN GLOBAL OHS

Human and organizational factors lie at the heart of occupational health and safety performance across all industries. Regardless of how advanced technologies or regulatory systems become, the success of any OHS framework ultimately depends on the behaviors, perceptions, culture, and interactions of individuals within the workplace. This section examines the global understanding of human factors, explores organizational influences that shape safety outcomes, and highlights how different nations and industries integrate these elements into their OHS strategies.

2.1. Human Factors as Determinants of Safety Outcomes

Human factors refer to the psychological, behavioral, cognitive, and physical characteristics that influence how individuals interact with work systems. These factors contribute significantly to both safe and unsafe outcomes.

2.1.1 Psychosocial Risks and Mental Health

Workplaces pushed OHS toward increasingly recognize psychosocial risks as central determinants of worker well-being. These risks include:

- high job demands
- lack of control
- role conflict
- poor communication
- organizational injustice
- bullying and harassment
- fatigue and burnout

Global studies consistently show that chronic psychosocial stress contributes to increased accidents, (Sun et al., 2022) reduced productivity, emotional disorders, and long-term health problems.

In the construction industry, for example, workers often face unpredictable schedules, tight deadlines, and physically demanding tasks. In healthcare, emotional exhaustion and compassion fatigue are prevalent due to continuous exposure to suffering, rapid decision-making, and shift rotation. Office environments too are not immune—high workloads, digital overload, and sedentary behavior create invisible but significant risks.

2.1.2 Cognitive Factors and Decision-Making

Worker attention, memory, risk perception, and situational awareness determine how safely tasks are executed. Poor cognitive capacity caused by fatigue, stress, or information overload can increase the likelihood of errors. (Liang et al., 2022)

Industries such as aviation, nuclear power, and oil and gas, where decision accuracy is critical, have invested heavily in human factors engineering. They use structured programs such as:

- crew resource management
- cognitive ergonomics
- human error analysis frameworks
- training simulations

These approaches emphasize teamwork, communication, and mental readiness.

2.1.3 Physical and Ergonomic Factors

Repetitive movements, awkward postures, heavy lifting, and vibration exposure can lead to musculoskeletal disorders (MSDs)—one of the most common global occupational health concerns. (Ahmadi et al., 2022) Ergonomic design has therefore become a central component of OHS in manufacturing, logistics, and office environments. Many nations adopt ergonomic standards based on ISO guidelines to address workstation design, tool usability, and physical load demands.

2.2. Organizational Influences on Safety Performance

While human factors determine individual-level risk, organizational systems shape broader patterns of safety behavior. A strong organizational foundation is essential to sustain preventive practices and reinforce worker engagement.

2.2.1 Safety Culture and Leadership Commitment

Safety culture reflects the shared values, beliefs, and behaviors that define how an organization perceives and manages safety. Organizations with strong safety cultures typically exhibit:

- visible leadership involvement
- open communication channels
- proactive hazard reporting

- trust and accountability
- consistent reinforcement of safety behaviors

Leadership commitment is repeatedly identified as a driving force in reducing accidents. (Sampson et al., 2014) When leaders prioritize safety, allocate resources, and participate in safety activities, it signals to workers that safety is integral—not optional.

Across high-performing industries worldwide, leaders commonly engage in:

- safety walk-throughs
- incident learning sessions
- cross-functional safety committees
- continuous improvement programs

This alignment between leadership and workforce creates a psychologically safe environment where concerns can be raised without fear.

2.2.2 Training, Competency, and Worker Engagement

Employee competency is essential for executing tasks safely. High-quality training programs ensure that workers understand risks, procedures, and the correct use of equipment. Modern OHS frameworks integrate:

- hands-on simulations
- scenario-based learning
- digital learning platforms
- refresher training schedules
- skill verification assessments

Worker engagement further enhances competency. Organizations that empower workers through involvement in decision-making, safety talks, audits, and toolbox meetings have higher compliance and incident prevention rates.

2.2.3 Communication Systems and Information Flow

Clear communication is critical in high-risk environments. Breakdowns in communication—between supervisors, team members, or departments—frequently contribute to accidents. Effective communication systems include:

- standardized operating procedures

- reporting channels for near-misses
- multilingual safety signage in international environments
- digital communication tools such as mobile OHS apps
- shift handover protocols

In global organizations with multicultural workforces, language diversity adds complexity, making clear, consistent communication essential.

2.3. The Interaction of Human and Organizational Factors

A core principle of modern safety management is the understanding that human and organizational factors do not operate in isolation. Instead, they interact continuously, shaping safety behavior and system performance.

2.3.1 Behavioral Safety Models

Behavior-based safety (BBS) programs are widely used to encourage safe actions through observation, coaching, and feedback. However, the global evolution of BBS underscores that it must be integrated with organizational systems; focusing solely on worker behavior without addressing systemic influences is ineffective.

2.3.2 Safety Climate as a Diagnostic Tool

Safety climate—the perceived priority of safety at a given point in time—serves as a useful indicator of an organization's safety health. It reflects worker perceptions of leadership, communication, rules, and incident handling. A strong safety climate correlates with lower accident rates across industries and countries. (Winge et al., 2019)

2.3.3 Integration with Engineering and Technology

Human and organizational factors increasingly intersect with engineering and technological systems. For example:

- Wearable sensors monitor fatigue and alert workers when thresholds are exceeded.
- Automation reduces repetitive strain while requiring new skillsets.
- AI-based analytics provide insights into behavioral patterns and organizational processes.

This integration reinforces that OHS is a socio-technical system where human capabilities and technological advancements complement each other.

2.4. Global Insights on Human and Organizational Factors

Different countries approach human and organizational factors based on cultural norms, economic capacity, and institutional maturity.

- **Japan** emphasizes discipline and continuous improvement, with workers actively participating in Kaizen-based safety activities.
- **Germany** prioritizes engineering controls and strong social partnership between employers and unions.
- **Singapore** uses a tripartite model involving government, industry, and workers to promote a shared responsibility culture.
- **New Zealand and Australia** apply risk-based regulatory models focusing on consultation and worker empowerment.
- **Nordic countries** integrate psychosocial well-being into mainstream OHS policies.

These global variations offer valuable lessons in developing safety systems that reflect both universal principles and local realities.

2.5. The Strategic Importance of Human and Organizational Factors. (Wachter & Yorio, 2014)

Human and organizational factors represent the foundation upon which all other safety strategies are built. Even the most advanced technologies, equipment, or regulatory systems are ineffective without engaged workers, competent leadership, and supportive organizational structures.

Strong emphasis on these factors ensures:

- consistent risk mitigation
- stronger safety culture
- higher productivity
- improved morale and retention
- reduced legal and compliance risks

Organizations that successfully integrate human and organizational dimensions into their OHS systems demonstrate higher resilience, adaptability, and long-term sustainability.

3. INDUSTRY-SPECIFIC CHALLENGES AND INNOVATIONS IN OCCUPATIONAL HEALTH AND SAFETY

Industries differ widely in their risk profiles, work environments, organizational cultures, and regulatory obligations. Understanding these differences is crucial for designing effective OHS strategies that respond to the distinctive nature of each sector. This section examines major industries where occupational risks are high or rapidly evolving—construction, maritime and port operations, manufacturing, healthcare, and emerging service sectors. It highlights how global organizations address these risks, the innovations that sectors have adopted, and the lessons that can be generalized across industries.

3.1. Construction Industry: High-Risk Environments and Persistent Challenges

Construction is one of the most hazardous industries worldwide, (Winge et al., 2019) consistently ranking high in work-related injuries and fatalities. The combination of dynamic worksites, heavy machinery, working at heights, and shifting schedules creates a complex and unpredictable environment.

3.1.1 Nature of Risks in Global Construction

Common risk categories include:

- falls from height
- electrical hazards
- struck-by or caught-in-between incidents
- manual handling and back injuries
- exposure to dust, noise, and chemicals
- psychosocial pressures from time-bound projects (Sun et al., 2022)

Developing countries often experience additional issues such as informal labor practices, insufficient protective equipment, and minimal regulatory enforcement.

3.1.2 Global Best Practices and Innovations

Successful safety programs in construction emphasize a combination of engineering controls, administrative interventions, and behavioral measures. Internationally recognized approaches include:

- Building Information Modeling (BIM) for risk visualization and hazard identification
- prefabricated modular construction to reduce on-site risks
- wearable sensors to detect fatigue, motion patterns, and ergonomic strain
- drone-based inspections for hard-to-reach or hazardous areas
- real-time safety dashboards integrating data from IoT devices

Japan and Singapore have been leaders in integrating automation and robotics to reduce manual labor in high-risk environments. Australia and the United Kingdom emphasize training accreditation systems that mandate competency for specialized tasks.

3.1.3 Human and Organizational Considerations

Construction safety is deeply rooted in worker behavior and leadership involvement. Given the rotating and multicultural nature of construction workforces, communication plays a critical role. Multilingual safety signage, culturally sensitive training programs, and daily safety briefings (toolbox talks) are fundamental to building situational awareness.

3.2. Maritime and Port Operations: Complex Interactions Between Humans, Machines, and Environment (Ronza et al., 2006)

Maritime and port operations represent another sector with elevated OHS risks due to heavy cargo movement, large-scale machinery, environmental exposure, and busy traffic flows.

3.2.1 Key Occupational Risks in Ports

Workers in port terminals face hazards such as:

- musculoskeletal injuries from repetitive lifting
- slips, trips, and falls from uneven surfaces
- vibration and noise exposure
- collisions between equipment, vehicles, and cargo
- chemical exposure from hazardous materials
- extreme heat or cold stress

The integration of large vessels, automated cranes, trucks, loaders, and human workers creates significant complexity, where a single communication lapse can lead to disastrous outcomes.(Darbra & Casal, 2004)

3.2.2 Global Approaches to Enhancing Port Safety

International ports have adopted several advanced OHS solutions:

- **Automated and semi-automated cranes** to reduce human-machine interaction
- **Traffic management systems** to minimize collisions
- **Advanced ergonomics programs** targeting port-related musculoskeletal disorders
- **Fatigue management systems** integrating sensor technology and scheduling analytics
- **Smart PPE**, including vibration sensors and location-tracking tags

Northern European ports, such as those in the Netherlands and Denmark, integrate sustainability and worker well-being into their OHS frameworks, emphasizing psychosocial welfare alongside physical safety.

3.2.3 Organizational Systems in Maritime Safety

Ports function under structured safety management systems guided by maritime regulations such as the ISM (International Safety Management) Code. Effective ports emphasize:

- cross-department communication
- joint safety committees
- digital incident reporting
- simulation-based training for crane operators and vehicle drivers

Collaboration between port authorities, labor unions, and shipping companies is central to sustaining safety improvements.

3.3. Manufacturing and Heavy Industries: Transitioning Toward Automation and Safety 4.0. (Zanatelli et al., 2021)

Manufacturing sectors worldwide face hazards associated with machinery, chemicals, production lines, and repetitive work. However, these industries are rapidly transforming due to automation, robotics, and data analytics.

3.3.1 Traditional and Emerging Risks

Traditional risks include:

- cuts, burns, and crush injuries
- exposure to hazardous chemicals
- repetitive strain and ergonomic issues
- noise-induced hearing loss

As automation increases, new risks emerge related to:

- human-robot interaction
- cybersecurity breaches affecting safety systems
- worker deskilling and psychological stress
- increased cognitive load in monitoring automated systems

3.3.2 Implementing Safety 4.0 Principles

Safety 4.0 integrates advanced technologies with traditional OHS principles. (Guo et al., 2021) Global leaders adopt:

- machine vision systems to detect unsafe acts in real time
- collaborative robots (cobots) designed for safe human interaction
- AI-driven predictive analytics for equipment failures and incident prediction
- digital twins of factories to simulate hazards before physical changes
- automated chemical monitoring and ventilation systems

Germany, South Korea, and the United States lead advancements in safety automation, combining engineering solutions with strong regulatory oversight.

3.4. Healthcare Sector: High Risk but Often Overlooked

Healthcare workers face unique hazards that differ from industrial sectors. Their risks are not limited to physical injuries but extend to emotional strain, infectious diseases, and ethical stressors.

3.4.1 Core Occupational Hazards in Healthcare

Key risks include:

- exposure to infectious agents
- needlestick injuries
- musculoskeletal strain from patient handling
- violence and aggression from patients or visitors
- long working hours and shift work (Ahmadi et al., 2022)
- emotional exhaustion and compassion fatigue

The COVID-19 pandemic heightened awareness of respiratory risks, PPE shortages, and mental health issues among frontline workers.

3.4.2 Innovations and Safety Measures (Occupational Safety and Health Administration [OSHA], 2022)

Modern healthcare systems adopt:

- patient-lifting devices and ergonomic aids
- closed-system drug transfer devices for hazardous medications
- infection control technologies such as UV-C sterilization
- digital fatigue monitoring systems for staff
- specialized training on de-escalation and conflict management

Nordic countries and Canada lead in integrating psychosocial health into healthcare OHS policies.

3.5. Emerging Sectors: Service, Education, and IT Workplaces

Although traditionally viewed as "low risk," modern service-sector environments pose evolving safety concerns.

3.5.1 Digital Fatigue and Sedentary Risks

Prolonged screen exposure leads to:

- eye strain
- musculoskeletal discomfort
- mental overload
- stress from constant digital connectivity

3.5.2 Workplace Violence and Emotional Strain

Education and frontline service roles face incidents involving:

- verbal conflicts
- harassment
- emotional distress from student or customer interactions

These risks require preventive training, psychological support programs, and organizational justice measures.

3.5.3 Telework and Hybrid Work Challenges

Remote workers face:

- ergonomic issues from improvised home workstations
- blurred work-life boundaries
- increased isolation and stress

Organizations now adopt remote-work safety guidelines, digital wellness programs, and periodic virtual ergonomic assessments.

3.6. Cross-Sector Lessons and Global Insights

Across industries, several universal lessons emerge:

- strong leadership commitment drives long-term safety improvement
- integrated human-organizational-technical approaches outperform isolated interventions
- data-driven monitoring enhances hazard prediction
- worker participation and empowerment are consistent success factors

- psychosocial risks are rising across all sectors and must be treated as core OHS concerns

Industries that successfully combine regulatory frameworks, technological innovation, and supportive organizational culture demonstrate the highest resilience and safety performance.

4. THE TECHNOLOGICAL FRONTIER IN OCCUPATIONAL HEALTH AND SAFETY (BOURAHLA ET AL., 2024)

Technological innovation is reshaping the landscape of occupational health and safety worldwide. As industries modernize and digital transformation accelerates, organizations increasingly adopt advanced tools to predict risks, automate hazardous tasks, strengthen monitoring, and optimize decision-making. Technology is no longer an optional enhancement within OHS systems—it has become an essential component of effective prevention strategies and a driving force behind global safety improvements. This section explores how emerging technologies such as artificial intelligence, automation, wearable sensors, Internet of Things (IoT) systems, and data analytics contribute to safer workplaces across sectors.

4.1. Artificial Intelligence and Machine Learning in OHS

Artificial intelligence (AI) and machine learning (ML) are transforming how organizations identify hazards, detect unsafe behaviors, and manage risks. Their greatest strength lies in their ability to process vast quantities of data and generate insights that would be impossible to achieve through manual methods.

4.1.1 Predictive Analytics for Accident Prevention (Maheronnaghsh et al., 2023)

Traditional safety models depend on historical incident data and periodic risk assessments. AI-enabled systems, however, can analyze:

- near-miss reports
- environmental sensor data
- video observations
- worker behavior patterns
- equipment usage logs
- physiological measurements

By integrating these diverse data streams, predictive models forecast high-risk events before they occur. Many global companies now use real-time prediction systems to flag situations such as likely falls, fatigue-related errors, equipment failures, or hazardous environmental conditions.

4.1.2 Computer Vision for Hazard Detection

Computer vision systems—AI that analyzes images or video—monitor workplace activities to detect:

- unsafe body movements
- missing personal protective equipment (PPE)
- unauthorized access to restricted areas
- proximity to dangerous machines
- hazardous vehicle-person interactions

Factories, warehouses, and construction sites increasingly deploy camera-based systems that automatically trigger alerts, prevention protocols, or shut-down procedures.

4.1.3 Digital Incident Analysis and Reporting

Natural language processing (NLP) tools classify safety reports, identify recurring patterns, and generate insights for safety specialists. Global corporations use AI to streamline:

- incident classification
- root-cause analysis
- safety documentation
- regulatory compliance

This reduces administrative burden and ensures consistent reporting across international branches.

4.2. Internet of Things (IoT) and Connected Safety Systems

IoT refers to interconnected sensors, devices, and platforms that collect and transmit real-time data. In OHS, IoT has become a game-changer by enabling continuous monitoring of hazards and worker conditions.

4.2.1 Environmental Monitoring

Sensors detect:

- gas leaks
- temperature fluctuations
- noise and vibration levels
- particulate matter and dust
- radiation exposure
- humidity or heat index

These devices provide early warnings and trigger protective actions, especially in high-risk environments such as mining, chemical plants, construction tunnels, and confined spaces.

4.2.2 Worker Safety Monitoring

IoT-enabled wearables and smart PPE can track:

- heart rate
- body temperature
- posture
- movement patterns
- fatigue and micro-sleep episodes
- exposure to harmful substances

When thresholds are exceeded, alarms notify both workers and supervisors, allowing timely intervention. This is particularly valuable for lone workers, remote teams, and workers assigned to hazardous zones.

4.2.3 Smart Equipment and Machinery

Connected machines monitor their own operating conditions, detect anomalies, and communicate failures before accidents occur. Examples include:

- cranes with load stability sensors
- forklifts equipped with proximity warning systems
- automated shutdowns when guards are removed
- ventilation systems that adjust airflow based on contaminant levels

IoT enhances situational awareness and reduces reliance on manual inspection, enabling proactive safety management.

4.3. Robotics and Automation: Reducing Human Exposure to Hazards

Automation reduces the need for workers to perform dangerous, repetitive, or physically demanding tasks. Industries worldwide have adopted robots for operations that pose significant risks, including:

- high-temperature environments
- heavy material handling
- toxic chemical exposure
- tasks requiring repetitive bending or lifting
- deep-sea or underwater operations
- demolition and structural testing

4.3.1 Collaborative Robots (Cobots)

Cobots work alongside humans and are designed with built-in safety features, such as force-limiting sensors and emergency stop functions. They assist in assembly lines, packaging, and ergonomic tasks, reducing musculoskeletal strain and lowering injury rates.

4.3.2 Drones and Remote-Controlled Systems

Drones inspect:

- tall structures
- confined spaces
- unstable terrain
- high-voltage lines
- offshore facilities

They minimize worker exposure to falls, structural failures, or hazardous atmospheres.

4.3.3 Autonomous Vehicles and Material-Handling Equipment

Warehouses and ports increasingly deploy:

- autonomous guided vehicles (AGVs)
- robotic forklifts
- automated container handlers

These technologies reduce collision risks, improve efficiency, and streamline logistics operations.

4.4. Wearable Technology and Smart PPE

Wearable devices are becoming standard components of OHS programs, particularly in physically demanding industries. These devices support both real-time monitoring and long-term health surveillance.

4.4.1 Ergonomic and Posture Monitoring

Sensors embedded in belts, vests, or exoskeletons track back posture and lifting techniques. Alerts are issued when workers adopt high-risk positions. Wearable exoskeletons also reduce strain by supporting the spine and joints.

4.4.2 Physiological Monitoring for Fatigue and Heat Stress

Heat stress is a critical concern in many countries. Wearable systems measure:

- skin temperature
- hydration levels
- heart-rate variability

Real-time heat-stress algorithms notify supervisors when workers are approaching dangerous physiological limits.

4.4.3 Smart Helmets and Head-Mounted Displays

Advanced helmets integrate:

- impact sensors
- augmented reality (AR) instructions
- noise-reduction technology
- built-in communication devices

These tools support safer navigation, hands-free communication, and enhanced visibility in confined or hazardous settings.

4.5. Digital Twins and Virtual Simulation Technologies

Digital twins are virtual replicas of real-world environments, machines, or processes. They simulate hazards, test changes, and evaluate intervention strategies without exposing workers to danger.

4.5.1 Hazard Modeling and Scenario Planning

Industries use digital twins to:

- simulate fire outbreaks
- analyze evacuation routes
- test machinery modifications
- model chemical leaks or explosions
- predict structural failures

This supports risk-based decision-making and emergency preparedness.

4.5.2 Virtual Reality (VR) and Augmented Reality (AR) Training

VR immerses workers in realistic scenarios such as working at heights, handling chemicals, or responding to emergencies. AR overlays instructions during real tasks.

Benefits include:

- higher retention of safety knowledge
- safer practice for dangerous operations
- standardized training across global sites

Countries like Singapore, Japan, and the UAE increasingly incorporate VR training in national workforce programs.

4.6. Challenges and Ethical Considerations in Technology-Driven OHS

While technological advances offer significant benefits, they also introduce challenges that require thoughtful management.

4.6.1 Privacy and Worker Surveillance

Continuous monitoring raises concerns about:

- data privacy
- consent
- misuse of health information
- excessive surveillance pressure

Organizations must establish clear policies to protect workers' rights and maintain trust.

4.6.2 Over-reliance on Technology

Technology should assist—not replace—human judgment. Excessive dependence may lead to:

- complacency
- reduced situational awareness
- skill degradation

Balanced integration ensures humans remain central to safety systems.

4.6.3 Equity and Access

Low-income industries or developing economies may struggle with:

- cost of devices
- limited digital infrastructure
- insufficient technical training

Equitable global OHS advancement requires accessible, scalable technology solutions.

4.7. Strategic Role of Technology in the Future of OHS

The global move toward digitalization positions technology as a critical enabler of safer, more intelligent workplaces. As organizations integrate socio-technical systems, OHS becomes more predictive, personalized, and preventive.

Key future directions include:

- AI-driven personalization of safety interventions
- integration of OHS data with ESG (Environmental, Social, Governance) reporting
- automated regulatory compliance systems
- hybrid human-machine teams
- ethical digital safety governance frameworks

Technology enhances—not replaces—the essential human and organizational dimensions of safety, paving the way for more resilient and adaptive OHS systems worldwide.

5. INTERNATIONAL SUCCESS STORIES IN OCCUPATIONAL HEALTH AND SAFETY

Global success stories provide powerful evidence of how effective OHS strategies can transform workplace safety, productivity, and organizational culture. These examples offer practical lessons on leadership commitment, technological integration, worker participation, and system-wide innovation. This section examines how leading multinational corporations and national systems have implemented robust safety programs across diverse environments. It highlights transferable principles that can guide organizations and policymakers worldwide.

5.1. Toyota: Embedding Safety in a Culture of Continuous Improvement

Toyota is internationally recognized for its unparalleled safety culture and systematic approach to risk reduction. The company's OHS performance is deeply rooted in the Toyota Production System (TPS), which prioritizes efficiency, quality, and workplace well-being through continuous improvement, or *Kaizen*.

5.1.1 Key Pillars of Toyota's Approach

Toyota's safety model is founded on:

- **Respect for people** – ensuring every worker's health, dignity, and participation.
- **Standardized work processes** – minimizing variation that leads to accidents.
- **Root-cause analysis (5 Whys and fishbone diagrams)** – addressing underlying issues rather than symptoms.

- **Employee-driven problem-solving** – empowering workers to identify and improve unsafe conditions.
- **Visual management systems** – using clear signage and color coding to highlight hazards.

Workers participate actively in training, hazard assessments, and process redesigns. This high level of involvement fosters accountability and ownership.

5.1.2 Technological and Ergonomic Improvements

Toyota integrates advanced technologies such as:

- robotic automation for high-risk tasks
- wearable exoskeletons to prevent musculoskeletal strain
- digital monitoring systems for equipment safety
- ergonomic assessments embedded in workstation design

By combining human-centered design with engineered safety controls, Toyota significantly reduces both physical and psychosocial risks.

5.1.3 Global Impact

Toyota's methods have influenced manufacturing worldwide. Many organizations adopt the Toyota Safety Assessment Program and *Kaizen* principles to promote continuous learning, error-proofing, and shared responsibility.

5.2. Siemens: Integrating Digital Safety with Global Operations

Siemens is another global leader known for its comprehensive and technology-driven approach to OHS. Operating across energy, manufacturing, healthcare, and digital industries, the company manages diverse risks with a unified and proactive safety framework.

5.2.1 Safety-Integrated Systems

Siemens emphasizes a fully embedded safety approach where OHS is integrated into engineering design, operational workflows, and decision-making processes. The company uses the "Safety-Integrated" system, which ensures that safety functions are built directly into machines, processes, and automation architectures.

5.2.2 Digital Transformation in Safety

Siemens' technological innovations include:

- **digital twins** to simulate hazardous scenarios
- **AI-based safety analytics** for identifying risk patterns
- **IoT-enabled sensors** across production lines
- **smart helmets and AR tools** for maintenance staff
- **automated shutdown mechanisms** for safety-critical equipment

These digital developments support predictive safety, reduce reactive interventions, and strengthen global compliance mechanisms.

5.2.3 Human-Centered Safety Programs

Despite technological advancements, Siemens places equal importance on human factors. Programs such as "Zero Harm Culture" focus on behavioral training, leadership engagement, mental health support, and workforce participation.

Through this dual emphasis on technology and human systems, Siemens consistently achieves low accident rates across global operations.

5.3. Singapore's Tripartite Model: A National Success Story

Singapore's OHS framework serves as a benchmark for emerging economies seeking to build strong, systematic safety infrastructures. The country's success stems from collaboration among three major pillars: the government, industry associations, and labor unions.

5.3.1 Characteristics of the Tripartite Model

The model emphasizes:

- shared responsibility among stakeholders
- proactive enforcement and regular inspections
- competency-based training with national accreditation
- strong integration of migrant workers into safety programs
- transparent reporting across industries

This collaboration ensures that policies are practical while balancing organizational needs and worker welfare.

5.3.2 Institutional Strengths

Singapore established the Workplace Safety and Health (WSH) Council to promote safety culture across sectors. The national "Vision Zero" initiative champions the belief that all injuries are preventable and encourages organizations to adopt upstream risk management.

5.3.3 Outcomes and International Influence

Over the past decade, Singapore has significantly reduced workplace fatalities and injuries. Its tripartite model is now studied by international regulators and adopted in modified forms by countries in Asia and Europe.

5.4. The Nordic Model: Integrating Psychosocial Well-Being into OHS

Nordic countries—such as Sweden, Norway, Denmark, and Finland—are renowned for their holistic approach to OHS, emphasizing both physical safety and psychosocial well-being.

5.4.1 Principles of the Nordic Approach

Key features include:

- strong worker participation in OHS decision-making
- emphasis on work-life balance
- integration of mental health into safety laws
- collaborative workplace inspections
- universal access to occupational healthcare

5.4.2 Focus on Psychosocial Safety Climate

Nordic governments promote a proactive psychosocial safety climate (PSC), considering factors such as:

- job demands
- autonomy
- organizational fairness
- supervisor support

This national focus reduces workplace stress, improves engagement, and enhances overall health outcomes. (Leka et al., 2004)

5.4.3 Lessons for Global Application

The Nordic model demonstrates that mental well-being is inseparable from physical safety. Countries adopting similar approaches see positive impacts on productivity, retention, and morale.

5.5. Australia's Risk-Based Regulatory Model

Australia's OHS system is built on risk-based regulation, emphasizing preventive action rather than strict prescriptive rules.

5.5.1 Core Features

Australia's Work Health and Safety (WHS) laws promote:

- systematic hazard identification
- risk assessment and mitigation
- duty of care shared across all workplace parties
- worker consultation through formal committees
- competency-based training requirements

5.5.2 Industry Innovation and Implementation

Australian industries implement comprehensive safety systems supported by digital tools, behavioral programs, and standardized national training packages.

Mining, construction, and healthcare organizations use:

- fatigue monitoring systems
- remote operation centers
- resilience-building programs
- integrated safety dashboards

5.5.3 International Influence

Australia's risk-based regulatory framework is widely regarded as one of the most modern and practical systems, offering guidance for countries transitioning from prescriptive to performance-based legislation.

5.6. Cross-Learning from Global Success Stories

Despite differences in context, these international examples share several common principles:

5.6.1 Leadership Commitment and Vision

In all successful cases, leaders actively:

- participate in safety programs
- allocate resources
- reinforce safety values
- model desired behaviors

Leadership shapes the organizational climate and directly influences worker engagement. (Frick, 2019)

5.6.2 Worker Empowerment and Participation

Successful organizations encourage:

- reporting of near-misses
- involvement in audits and inspections
- joint problem-solving
- safety dialogues

Empowered workers serve as critical partners in implementing sustainable safety improvements.

5.6.3 Human-Centered and Technology-Enabled Integration

Global leaders balance engineering and technological innovations with human and organizational factors. This socio-technical integration produces more resilient and adaptable safety systems.

5.7. Strategic Value of Success Stories

These case studies demonstrate that effective OHS strategies produce benefits beyond compliance:

- improved productivity
- enhanced organizational reputation
- reduced costs and downtime
- higher worker satisfaction
- stronger resilience against disruptions

Organizations adopting lessons from global leaders can accelerate their safety maturity, foster innovation, and contribute to sustainable economic growth.

6. GLOBAL OHS FRAMEWORKS, STANDARDS, AND REGULATORY MODELS

Global occupational health and safety systems are shaped by a wide array of regulatory models, institutional mechanisms, and international standards. These frameworks not only guide national legislation but also influence industry practices, corporate governance, and cross-border safety management. As work environments evolve due to globalization, digital technology, and demographic change, the role of robust and adaptable OHS frameworks becomes increasingly important. This section analyzes major global standards, compares regulatory approaches, and highlights how countries integrate international guidelines into national systems.

6.1. International Standards Shaping Global OHS

Several organizations play crucial roles in developing universal OHS principles that guide global practice. These standards provide consistency, clarity, and measurable requirements for organizations to implement safe operations across borders.

6.1.1 ISO 45001: The Global Benchmark for OHS Management Systems

ISO 45001:2018 is the first international standard for occupational health and safety management systems. It provides a comprehensive framework that helps organizations:

- identify and manage OHS risks
- comply with legal and regulatory requirements
- implement worker participation and consultation
- evaluate performance through audits and monitoring
- ensure continuous improvement

The standard adopts a risk-based approach and encourages organizations to integrate safety management with business strategy. (International Organization for Standardization [ISO], 2018) Its High-Level Structure (HLS) ensures alignment with ISO 9001 (quality) and ISO 14001 (environment), promoting integrated management systems.

6.1.2 ISO 31000: Risk Management Guidelines

ISO 31000 provides a systematic approach to risk management applicable across all organizational processes. When applied in OHS contexts, it enhances:

- decision-making
- hazard identification
- risk assessment frameworks
- resilience in uncertain environments

Companies worldwide use ISO 31000 to strengthen their risk-based thinking and align OHS with enterprise-wide risk governance.

6.1.3 ILO-OSH 2001: A Foundational Framework

The International Labour Organization's ILO-OSH 2001 guidelines offer a flexible, non-prescriptive system that many developing countries adopt as a blueprint for national OHS policies. Key elements include:

- worker participation
- management leadership
- competency development
- continual improvement
- emergency preparedness

ILO-OSH remains influential, especially in regions where national regulatory structures are still emerging.

6.1.4 WHO and EU-OSHA Guidance

WHO supports occupational health through technical guidance, workplace health promotion, and recommendations on biological hazards. The European Agency for Safety and Health at Work (EU-OSHA) contributes through research, thematic campaigns, and directives shaping safety laws across the European Union.

Together, these bodies ensure global coherence in OHS policies and best practices.

6.2. Regulatory Models Across Countries: A Comparative Overview

Countries implement OHS regulation through diverse legal and enforcement systems reflecting their socio-economic context, political history, cultural norms, and institutional capacities. Comparative analysis reveals four dominant models.

6.2.1 Prescriptive (Rule-Based) Regulation

Countries with prescriptive systems specify detailed rules and technical requirements that organizations must follow. These models focus on compliance and enforcement.

Typical features include:

- strict regulations with clear specifications
- frequent inspections
- penalties for non-compliance
- limited flexibility for organizational innovation

Examples:

- Some Asian and Middle Eastern countries
- Historically, the United States before the shift to performance-based rules in some sectors

Strengths: clarity and uniformity
Limitations: reduced adaptability, potential bureaucracy

6.2.2 Performance-Based Regulation

Performance-based models emphasize outcomes rather than prescriptive rules. Organizations must demonstrate that their internal systems effectively manage risks.

Characteristics include:

- risk assessments tailored to operations
- documentation demonstrating control measures

- flexible approaches to meeting safety objectives

Examples:

- Australia
- New Zealand
- Norway and the United Kingdom (in many sectors)

Strengths: adaptability and innovation
Limitations: requires strong organizational competency and regulatory maturity

6.2.3 Cooperative and Tripartite Regulatory Systems

These systems emphasize collaboration between government, employers, and workers to jointly solve OHS challenges. Safety governance becomes a shared responsibility.
Examples:

- Singapore (tripartite model)
- Some Nordic countries
- Japan's labor-management cooperation systems

Strengths: high worker trust and involvement
Limitations: requires strong institutional capacity and social dialogue traditions

6.2.4 Hybrid Models

Many countries blend prescriptive rules with performance-based flexibility and cooperative structures. Hybrid systems are especially common in economies undergoing rapid industrial transitions.
Examples:

- India (with recent reforms toward risk-based models)
- South Korea
- Malaysia
- South Africa

Strengths: balanced approach suitable for varying workforce needs and industry conditions
Limitations: regulatory inconsistencies may arise during transitions

6.3. Implementation Challenges in Low- and Middle-Income Countries

Although international frameworks offer strong guidance, many low- and middle-income countries (LMICs) face barriers affecting implementation (Ahasan & Partanen, 2001)

6.3.1 Limited Institutional Capacity

Challenges include:

- insufficient number of trained inspectors
- lack of modern enforcement tools
- weak coordination across ministries
- difficulty implementing complex ISO-based systems

6.3.2 Informal and Migrant Labor Markets

Large informal sectors mean many workers lack:

- legal protections
- written contracts
- access to PPE
- training opportunities

Migrant workers often face cultural, linguistic, and economic barriers to participating in safety systems.

6.3.3 Financial Constraints

Small and medium enterprises (SMEs) may struggle to invest in:

- monitoring technology
- engineering controls
- training programs
- certification processes

6.3.4 Education and Awareness Gaps

Without adequate awareness, employers may perceive OHS as a cost rather than an investment. Education campaigns and incentives can address this perception.

6.4. Strengthening National OHS Frameworks: Pathways for Improvement

Countries seeking to strengthen OHS systems can adopt several strategic measures.

6.4.1 Capacity Building for Regulators

Investing in inspector training, digital enforcement tools, and research centers improves regulatory competence and credibility.

6.4.2 Incentive-Based Models

Governments can encourage compliance through:

- tax benefits
- certification support
- awards and recognition schemes
- financial subsidies for SMEs to adopt safer technologies

6.4.3 Harmonization with International Standards

Aligning national laws with ISO 45001, ILO-OSH 2001, and WHO recommendations ensures global consistency and improves export competitiveness.

6.4.4 Integrating OHS into National Development Goals

OHS contributes to:

- economic productivity
- public health
- human capital development
- sustainable growth

Countries that view safety as a strategic priority achieve long-term socio-economic gains.

6.5. Emerging Global OHS Regulations: New Areas of Focus

Modern workplaces face evolving risks requiring updated regulatory responses.

6.5.1 Mental Health and Psychosocial Safety

More countries now incorporate psychosocial risk assessments into legislation, recognizing mental health as a core component of OHS.

6.5.2 Digital Work and Remote Safety

Remote work policies increasingly mandate guidelines on:

- ergonomics
- digital fatigue
- data security
- communication demands

6.5.3 Climate-Related Health Risks

Regulators now consider:

- extreme heat policies
- air quality standards
- emergency plans for natural disasters

6.5.4 Ethical AI in Workplace Monitoring

New guidelines emphasize transparency, consent, and responsible use of digital surveillance tools.

6.6. Global Lessons from Regulatory Systems

Analysis of diverse national models reveals universal insights:

- safety improves when laws emphasize prevention rather than punishment
- strong worker participation enhances compliance and reporting
- technology strengthens enforcement but must be ethically governed
- collaboration between regulators, industry, and workers creates sustainable results

- harmonization with international standards supports global mobility and competitiveness.

7. EMERGING GLOBAL TRENDS AND FUTURE CHALLENGES IN OCCUPATIONAL HEALTH AND SAFETY

Occupational health and safety continues to evolve in response to rapid technological advancements, demographic transitions, environmental pressures, and the changing nature of work. These emerging global trends are reshaping how organizations approach risk management, worker well-being, and regulatory compliance. Understanding these developments is essential for designing OHS strategies that remain relevant and resilient in the decades ahead. This section examines the major global shifts influencing OHS today, forecasts future challenges, and presents a comprehensive overview of contemporary issues that must be addressed by policymakers, industries, and researchers.

7.1. Digital Transformation and the Expansion of Data-Driven Safety

Digitalization is transforming workplaces into interconnected ecosystems where data flows continuously across machines, workers, systems, and environments. This shift accelerates the adoption of advanced safety tools while also introducing new complexities.

7.1.1 Rise of Smart Safety Ecosystems

Organizations worldwide increasingly rely on:

- real-time dashboards integrating OHS metrics
- IoT sensors tracking environmental conditions
- wearable devices monitoring worker health
- AI-driven predictive analytics
- automated hazard identification systems

These ecosystems can forecast accidents, identify unsafe trends, and guide timely interventions. They support a proactive approach to safety, enabling organizations to transition from reactive compliance to continuous risk prediction and prevention. (Guo et al., 2021)

7.1.2 Growing Importance of Cybersecurity in OHS

As digital safety systems expand, cybersecurity risks directly affect physical safety. Compromised sensors, hacked equipment, or data manipulation could have catastrophic consequences in industries such as energy, healthcare, manufacturing, and transportation.

Thus, the integration of cybersecurity with OHS governance becomes essential in safeguarding both digital and physical infrastructures.

7.1.3 Ethical and Governance Concerns

Digital tools generate sensitive worker data related to physiological indicators, location, fatigue levels, and behavior. Ethical OHS governance requires:

- transparency on data usage
- clear consent mechanisms
- limitations on intrusive monitoring
- secure handling of biometric information

Future OHS systems must balance innovation with trust, fairness, and worker dignity.

7.2. Psychosocial Health, Mental Well-Being, and Organizational Justice

The global recognition of psychosocial risks represents one of the most significant evolutions in modern OHS. Organizations worldwide are acknowledging that psychological well-being significantly affects safety performance, productivity, and organizational resilience.

7.2.1 Growing Prevalence of Psychosocial Risks

Factors influencing psychosocial well-being include:

- job insecurity
- high workload and rapid pace
- poor work-life balance
- lack of autonomy
- organizational injustice
- harassment or discrimination

- excessive digital demands

These challenges manifest in stress, burnout, anxiety, depression, and reduced cognitive capacity—each contributing to heightened accident risk. (Rollin et al., 2022)

7.2.2 Integration of Mental Health into Safety Laws

Many countries now incorporate psychosocial hazards into legal frameworks. National policies mandate:

- mental health risk assessments
- monitoring of stress indicators
- anti-harassment protocols
- employee assistance programs
- training for supervisors in supportive leadership

This paradigm shift reflects a global understanding that mental health is inseparable from physical safety.

7.2.3 Organizational Justice and Climate

Fairness in performance evaluation, workload distribution, communication, and grievance handling strongly influences worker trust and safety participation. Organizations with high organizational justice experience:

- fewer accidents
- stronger reporting culture
- higher morale
- greater worker retention

Psychosocial safety climate (PSC) has become an emerging metric for evaluating how well organizations protect mental health and prevent stress-related issues.

7.3. Climate Change and Environmental Stressors

Climate change introduces new occupational hazards affecting workers globally, particularly those in outdoor or physically demanding roles.

7.3.1 Rising Temperatures and Heat Stress

Extreme heat impacts:

- construction workers
- agricultural laborers
- transportation and logistics staff
- emergency responders

Heat stress causes dehydration, fatigue, impaired judgment, reduced productivity, and increased accidents. Countries in Asia, the Middle East, and Africa experience growing burdens, requiring new safety protocols, hydration policies, and adaptive work schedules.

7.3.2 Air Pollution and Respiratory Risks

In industrialized and rapidly developing regions, poor air quality contributes to respiratory conditions, reduced lung function, and increased susceptibility to illness. Industries such as mining, construction, and manufacturing face heightened responsibilities for monitoring air quality and implementing respiratory protection programs.

7.3.3 Extreme Weather Events and Disaster Preparedness

Climate change increases the frequency of:

- hurricanes
- floods
- droughts
- wildfires

These events disrupt operations and create hazardous working environments. Organizations must invest in disaster preparedness, emergency response training, and climate-resilient infrastructure to ensure worker safety.

7.4. Demographic Transitions and Workforce Diversity

A diverse workforce brings new strengths and safety challenges that require inclusive and adaptable OHS strategies.

7.4.1 Aging Workforce

Older workers face:

- reduced physical endurance
- slower recovery from injury
- greater susceptibility to chronic conditions

Organizations must redesign tasks, improve ergonomics, and offer flexible work arrangements to accommodate aging workers without compromising safety.

7.4.2 Increasing Participation of Women in High-Risk Sectors

Women's participation is rising in traditionally male-dominated industries such as:

- construction
- maritime operations
- manufacturing
- emergency services

This shift requires attention to:

- gender-specific ergonomic design
- maternity protection
- PPE designed for women
- prevention of workplace harassment

Gender-responsive OHS policies are essential for inclusivity and safety equity.

7.4.3 Migrant and Contract Workers

Global supply chains rely heavily on migrant labor, often concentrated in high-risk jobs. Challenges include:

- language barriers
- unfamiliarity with safety regulations
- inadequate access to training
- unequal access to healthcare

Inclusive OHS practices must ensure that safety standards apply equally to all workers regardless of nationality or employment status.

7.5. The Transformation of Work: Remote, Hybrid, and Gig Economies

Work arrangements are changing rapidly, requiring modern OHS practices to extend beyond traditional physical workplaces.

7.5.1 Remote Work and Ergonomic Challenges

Home-based workers face risks such as:

- poor workstation ergonomics
- prolonged sedentary behavior
- musculoskeletal discomfort
- blurred work-life boundaries

Organizations now implement guidelines for home ergonomics, encourage regular breaks, and offer virtual ergonomic assessments.

7.5.2 The Gig Economy and Precarious Work

Gig workers in transportation, food delivery, and freelance digital labor often lack:

- employer-provided safety training
- workers' compensation
- access to OHS programs
- stable work environments

Modern OHS strategies must consider these workers' needs to promote equitable safety across diverse labor models.

7.5.3 Hybrid Work Complexities

Hybrid environments require rethinking:

- communication protocols
- emergency preparedness
- digital safety practices

- psychosocial support

Blended workspaces must maintain safety continuity across both physical and virtual domains.

7.6. The Need for Integrated, Holistic OHS Approaches

With complex and interconnected global trends, the future of OHS depends on integrating human, organizational, technological, and environmental dimensions.

7.6.1 Socio-Technical OHS Systems

Effective systems harmonize:

- human factors
- technological monitoring
- organizational culture
- regulatory requirements

Such integration improves adaptability, resilience, and predictive capacity.

7.6.2 Human-Centered Design

Future workplaces increasingly adopt design approaches that prioritize:

- ergonomics
- cognitive load reduction
- accessibility
- user-friendly interfaces
- inclusive PPE

Human-centered design ensures that safety technologies and systems align with workers' needs and capabilities.

7.6.3 Sustainability and OHS Convergence

Sustainability frameworks now view worker health as a core component of environmental and social governance (ESG). Organizations incorporate OHS into sustainability reporting, reflecting the growing understanding that safe work environments support long-term social and economic development.

7.7. Summary: Preparing for the Future of Workplace Safety

Emerging trends illustrate that OHS is entering a transformative era defined by digitalization, climate change, workforce diversity, and evolving work models. Responding to these challenges requires:

- proactive risk management
- ethical technological governance
- inclusive and equitable safety frameworks
- integration of mental and physical health
- strong global cooperation

Organizations that adapt to these shifts will achieve safer, healthier, and more resilient workplaces capable of thriving in dynamic global environments.

8. SECTOR-SPECIFIC OCCUPATIONAL HEALTH AND SAFETY CHALLENGES: COMPARATIVE INSIGHTS

Different industries present distinct occupational health and safety challenges shaped by their operational environments, technological dependencies, workforce demographics, and regulatory demands. Understanding sector-specific risk patterns provides valuable insights into how organizations can tailor their safety strategies, improve prevention programs, and implement targeted interventions. This section examines major high-risk and emerging sectors, illustrating how workplace conditions, human factors, and technological complexities shape occupational hazards.

8.1. Construction Industry: High Injury Burden and Complex Risk Dynamics

The construction sector remains one of the most hazardous industries globally (Winge et al., 2019) due to constantly changing worksites, tight project timelines, and the coexistence of multiple subcontractors. Hazards range from falls and struck-by incidents to musculoskeletal strain, noise exposure, and psychological stress.

8.1.1 Dynamic and Unpredictable Work Environments

Construction sites undergo continuous transformation as projects progress. Workers must navigate:

- elevated platforms
- heavy machinery
- power tools
- unstable ground conditions
- varying weather

These dynamic features create a constantly shifting risk landscape where hazard identification must be updated continuously.

8.1.2 Human Factors and Skill Variability

The workforce often includes individuals with differing skill levels, qualifications, and experience. Temporary, migrant, and contract workers may lack adequate training or familiarity with local safety regulations. Language barriers further complicate communication during critical operations, increasing the likelihood of errors and accidents.

8.1.3 Psychosocial and Organizational Pressures

Construction work involves long hours, physically demanding tasks, and deadline-driven environments. These factors contribute to:

- fatigue
- stress
- reduced attention
- impaired decision-making

When combined with hazardous physical environments, psychosocial risks significantly elevate accident probability.

8.1.4 Technological Innovations and Opportunities

Emerging technologies—such as drones for site inspections, AI-powered safety analytics, and exoskeletons for heavy lifting—offer promising avenues for accident prevention. Effective integration, however, requires training and organizational readiness to adapt traditional workflows.

8.2. Maritime and Port Operations: High-Risk Environments with Complex Logistics

Ports, shipping facilities, and maritime operations involve unique hazards due to large equipment, heavy cargo, confined spaces, and complex interactions between human and mechanical processes.

8.2.1 Physical Hazards in High-Motion Environments

Port workers are exposed to risks such as:

- falls from heights
- collisions with moving equipment
- entanglement in machinery
- shifts in cargo weight
- manual material handling injuries

Unpredictable weather and maritime conditions further complicate safe operations.

8.2.2 Ergonomic and Musculoskeletal Risks

Prolonged lifting, repetitive movements, and awkward postures contribute to high rates of musculoskeletal disorders. Back pain, shoulder strain, and joint injuries are widespread in cargo handling and container operations.

8.2.3 Technological Transition Toward Automation

Many ports are adopting automated cranes, robotic equipment, and AI-based logistics systems. While these technologies can reduce physical risks, they introduce:

- cybersecurity vulnerabilities
- new skills requirements
- potential job displacement concerns
- increased cognitive demands on operators

Balancing automation with human-centered safety strategies remains essential.

8.2.4 Mental Health and Shift Work Challenges

Port operations run continuously, requiring irregular shifts, night work, and extended hours. These conditions affect sleep patterns, cognitive functioning, and mood stability, which in turn influence accident risk and long-term health outcomes.

8.3. Manufacturing and Industrial Production: Evolving Risks in Modern Systems

Manufacturing environments vary widely—from small workshops to large automated factories—but they share common safety challenges involving machinery, chemicals, noise, and ergonomic strain.

8.3.1 Machinery-Related Hazards

Workers face risks from:

- rotating equipment
- conveyor belts
- cutting tools
- industrial robots
- press machines

Machine guarding, lockout-tagout (LOTO) systems, and automation significantly reduce these risks, but lapses in maintenance or procedural compliance can lead to severe injuries.

8.3.2 Chemical and Process Safety Issues

Industries such as pharmaceuticals, petrochemicals, and metallurgy require stringent control of hazardous substances, including:

- volatile chemicals
- toxic fumes
- corrosive materials
- flammable agents

Adherence to process safety management principles is vital to preventing catastrophic incidents.

8.3.3 Noise and Vibration Exposure

Continuous exposure to high noise levels and industrial vibrations contributes to hearing loss and long-term health effects. Engineering controls and PPE are essential, alongside health surveillance programs.

8.3.4 Industry 4.0 and Digital Manufacturing

Modern manufacturing integrates:

- cyber-physical systems
- real-time monitoring
- predictive maintenance
- autonomous robots

These technologies enhance safety but also create new demands for worker training, cybersecurity, and human-machine interface design.

8.4. Healthcare Sector: High-Stress Environments and Biological Risk Factors

Healthcare workers face some of the most diverse and intense risks of any occupational group due to close interaction with patients, exposure to biological agents, and demanding emotional labor.

8.4.1 Biological Hazards and Infectious Diseases

Healthcare environments expose workers to:

- airborne pathogens
- bloodborne viruses
- antimicrobial-resistant organisms
- contaminated materials

Pandemics have highlighted the need for robust infection control, PPE availability, and emergency preparedness.

8.4.2 Ergonomic and Physical Strain

Patient handling, repetitive movements, and long shifts lead to musculoskeletal disorders, fatigue, and reduced alertness—major contributors to procedural errors. (Ahmadi et al., 2022)

8.4.3 Psychological and Emotional Pressure

Healthcare workers experience:

- compassion fatigue
- trauma exposure
- emotional exhaustion
- moral distress

These psychosocial risks undermine both worker well-being and patient safety.

8.4.4 Technological and Workflow Complexity

Electronic health records, automated medication systems, and telemedicine platforms require continuous adaptation and training to prevent technological errors that may compromise safety.

8.5. Information Technology and Knowledge-Based Industries: Emerging White-Collar Risks

Although IT and service sectors are traditionally viewed as low-risk, evolving work patterns have introduced significant health challenges.

Digital Overload and Cognitive Strain

Prolonged screen time, multitasking, and high mental workload contribute to:

- eye strain
- headaches
- decreased concentration
- digital fatigue

Cognitive overload also increases the likelihood of errors in critical tasks.

8.5.2 Sedentary Work and Physical Health Concerns

Extended sitting correlates with musculoskeletal disorders, metabolic syndrome, and cardiovascular risks. Poor workstation ergonomics amplify these issues.

Psychosocial Stressors

Fast-paced project cycles, performance pressure, and remote work challenges lead to:

- burnout
- anxiety
- emotional exhaustion

Psychosocial safety programs and supportive leadership play key roles in addressing these issues.

8.6. Education Sector: Underrecognized but Growing OHS Concerns

Teachers and educational staff face a range of hazards that are often overlooked in safety discourse.

8.6.1 Physical and Ergonomic Issues

Regular classroom activities involve:

- prolonged standing
- repetitive writing
- poor seating ergonomics
- voice strain

These factors contribute to musculoskeletal pain and chronic fatigue.

8.6.2 Psychosocial Risks and Workload Pressure

Teaching roles often involve high emotional demands, dealing with difficult student behavior, and managing large workloads with limited resources.

8.6.3 Infection Exposure in School Environments

Schools are environments with high potential for infectious disease transmission, requiring improved ventilation and hygiene protocols.

8.7. Cross-Sector Learnings and Integrated Safety Improvements

Comparative analysis across industries demonstrates:

- construction and manufacturing highlight the importance of engineering controls
- healthcare and education underscore the need for psychosocial and emotional support systems
- IT and service industries illustrate the rising dominance of ergonomic and mental-health issues
- maritime and port sectors emphasize the importance of automation safety and logistical coordination
- all sectors show rising demand for human-centered design, inclusive ergonomics, and digital safety literacy

Sector-specific strategies are effective, but cross-sector collaboration provides stronger, more innovative solutions.

9. EMERGING GLOBAL TRENDS RESHAPING OCCUPATIONAL HEALTH AND SAFETY

Occupational Health and Safety (OHS) is experiencing a significant transformation driven by technological advancements, evolving workforce demographics, globalization, and rapid shifts in work environments. These developments are redefining traditional risk profiles and creating new opportunities—and challenges—for ensuring worker well-being. Understanding these emerging trends is essential for building proactive, resilient, and future-oriented safety systems.

9.1. Digital Transformation and the Rise of Smart OHS Systems

Digitalization is fundamentally reshaping the way organizations identify, assess, and manage workplace risks. Modern OHS programs increasingly rely on interconnected technologies that enhance real-time monitoring, decision-making, and predictive capabilities.

9.1.1 Internet of Things (IoT) for Continuous Monitoring

IoT-based safety systems employ sensors, wearables, and smart devices to monitor hazards such as:

- toxic gas exposure
- heat stress
- worker fatigue
- proximity to dangerous machinery
- structural vibration and instability

These devices provide real-time alerts, enabling faster emergency responses and reducing accident severity. In high-risk sectors like mining, manufacturing, and oil and gas, IoT has become integral to hazard detection.

9.1.2 Artificial Intelligence and Predictive Analytics

AI amplifies the value of collected data by identifying patterns that humans may overlook. Predictive models analyze:

- accident trends
- near-miss reports
- behavior patterns
- environmental conditions

to forecast potential incidents before they occur. This shift from reactive to proactive safety management enhances organizational resilience and resource allocation. (Guo et al., 2021)

9.1.3 Robotics and Automation

Robots increasingly perform high-risk tasks such as:

- material handling
- welding
- container stacking
- toxic waste disposal

- inspection of confined spaces

Automation reduces human exposure to danger but introduces new risks related to:

- human-machine interaction
- cybersecurity
- system reliability

Thus, organizations must also strengthen digital safety protocols and operator training.

9.2. The Growing Importance of Psychosocial Health and Emotional Well-Being

Psychosocial risks—stress, burnout, harassment, work overload, and emotional fatigue—have become dominant concerns in modern workplaces. The shift from physical labor to cognitive work, along with competitive work cultures, has intensified psychological pressures across sectors.

9.2.1 Burnout as a Global Occupational Hazard

Burnout has moved from being seen as a personal issue to a recognized organizational challenge. Contributing factors include:

- unrealistic productivity expectations
- excessive digital connectivity
- poor work-life boundaries
- role ambiguity

Chronic burnout affects not only worker well-being but also productivity, error rates, and retention. (Sun et al., 2022)

9.2.2 Organizational Justice and Supportive Work Cultures

Employees who feel valued, respected, and supported are more engaged and safer. Organizational justice—fairness in workload distribution, communication, and conflict management—reduces stress and enhances safety performance. Companies adopting participatory safety cultures experience:

- lower accident rates

- higher job satisfaction
- improved trust and cooperation

9.2.3 Mental Health Programs and Preventive Strategies

Progressive organizations are implementing:

- Employee Assistance Programs (EAPs)
- resilience-building workshops
- stress-management training
- workload regulation policies
- peer support networks
- psychological risk assessments

These are particularly crucial in healthcare, education, IT, and service industries where emotional labor and cognitive strain are high.

9.3. Climate Change and Environmental Stressors: The New Frontier of OHS

Climate change has become a significant occupational hazard, influencing work patterns, exposure conditions, and health outcomes.

9.3.1 Heat Stress and Outdoor Work Risks

Rising temperatures directly impact workers involved in:

- agriculture
- construction
- utilities
- transportation
- mining

Heat-related illnesses, dehydration, and reduced cognitive functioning are increasingly common. Organizations must adopt measures such as adaptive shift scheduling, hydration protocols, and wearables that monitor body temperature.

9.3.2 Air Quality and Environmental Toxins

Wildfires, industrial emissions, and climate-related pollution expose workers to hazardous particulates. Poor air quality affects respiratory health, especially for outdoor workers, emergency responders, and logistics personnel.

9.3.3 Extreme Weather and Disaster Preparedness

Floods, storms, and droughts disrupt workplace operations, creating new safety risks. Organizations must:

- incorporate climate resilience into safety planning
- use risk-mapping tools
- design emergency response frameworks
- strengthen infrastructure

9.3.4 Occupational Risks in the Green Economy

The transition to renewable energy introduces new hazards:

- high-voltage systems in solar farms
- chemical exposure in battery production
- fall hazards in wind turbine maintenance

Integrating safety into sustainability initiatives is essential for responsible green growth.

9.4. Post-Pandemic Workplace Evolution and Health Protocols

The COVID-19 pandemic permanently shifted global perspectives on workplace health, hygiene, and continuity planning.

9.4.1 Hybrid Work Models and Remote Work Challenges

Remote work brought benefits but also new risks:

- ergonomic strain due to poor home workstations
- blurred work-life boundaries
- increased screen time and digital fatigue
- social isolation and reduced peer support

Organizations now develop guidelines for safe remote work environments and monitor psychosocial well-being.

9.4.2 Infection Control and Workplace Hygiene

Workplaces increasingly implement:

- improved ventilation
- touchless technology
- health screening tools
- isolation protocols
- PPE reserves
- employee health monitoring systems

These measures extend beyond pandemics and enhance overall workplace hygiene.

9.4.3 Resilient Safety Governance

Organizations now recognize the need for:

- continuity planning
- rapid response systems
- multi-hazard preparedness
- decentralized decision-making

The post-pandemic era requires integrated health governance rather than isolated safety programs.

9.5. Workforce Demographic Shifts and Inclusive OHS Approaches

Workforces are becoming more diverse in terms of age, gender, nationality, and skill sets. Safety systems must evolve accordingly.

9.5.1 Ageing Workforce

Older workers may face:

- reduced physical capacity
- slower reaction times

- chronic health conditions

But they offer experience, stability, and high safety awareness. Ergonomic redesigns, flexible duties, and targeted well-being programs can optimize their participation.

9.5.2 Gender-Related Safety Considerations

Women encounter unique occupational risks related to:

- personal protective equipment not tailored to female anatomy
- reproductive health hazards
- harassment and psychosocial stress
- underrepresentation in safety leadership roles

Ensuring gender-responsive OHS policies enhances equity and protection.

9.5.3 Migrant and Contract Workers

Globalization has increased workforce mobility. Migrant workers often face challenges such as:

- language barriers
- unfamiliarity with local regulations
- limited access to training

Providing multilingual safety training, cultural adaptation programs, and fair employment policies improves inclusivity. (Sampson et al., 2014)

9.6. Ethical Governance and the Future of Responsible OHS Leadership

Ethical leadership is becoming central to OHS effectiveness. Transparent communication, fairness, and accountability strengthen safety cultures and trust.

9.6.1 Ethical Decision-Making in Safety Planning

Organizations must prioritize:

- risk transparency

- evidence-based decisions
- effective reporting systems
- whistleblower protection

9.6.2 Human-Centered Safety Innovation

Modern OHS emphasizes worker-centered strategies, valuing human experience and participation in designing solutions.

9.6.3 Global Harmonization of Standards

International frameworks such as ISO 45001 and WHO-ILO guidelines are encouraging unified safety approaches across borders. Harmonization supports multinational companies in establishing consistent safety programs worldwide.

9.7. Synthesis: The Future Landscape of OHS

The convergence of digital technologies, psychosocial awareness, climate adaptation, and inclusive policies is reshaping the global OHS landscape. Future-ready systems must be:

- **proactive**, using predictive tools to prevent harm
- **resilient**, capable of adapting to rapid change
- **integrated**, linking physical, mental, and organizational factors
- **inclusive**, addressing diverse worker needs
- **data-driven**, employing advanced analytics while protecting privacy

Organizations that embrace these evolving trends will be better equipped to safeguard worker health and achieve sustainable excellence.

10. INTEGRATING HUMAN, ORGANIZATIONAL, AND TECHNOLOGICAL DIMENSIONS IN GLOBAL OHS SYSTEMS

Creating effective occupational health and safety systems requires a balanced integration of human factors, organizational structures, and technological innovations. These three dimensions are deeply interconnected, and their alignment determines the overall success or failure of workplace safety strategies. A systems-based perspective recognizes that accidents and health issues rarely stem from isolated causes; rather, they emerge from interactions between worker behavior, organizational culture, technological complexity, and environmental conditions. This section explores how these dimensions converge and how organizations can harness their synergy to build robust, holistic, and future-ready OHS systems.

10.1. The Human Dimension: Behavior, Capacity, and Empowerment

Workers remain at the center of all OHS efforts. Regardless of how advanced technologies become, human judgment, decision-making, and behavioral responses continue to play crucial roles in preventing incidents and sustaining safe environments.

10.1.1 Worker Behavior and Safety Culture

Human behavior influences risk exposure, compliance with safety procedures, and the capacity to identify hazards. A strong safety culture encourages workers to internalize safety values and integrate them into everyday tasks. Central elements include:

- shared norms that prioritize safety
- open communication between supervisors and workers
- mutual trust and respect
- empowerment to report unsafe conditions without fear

Organizations with a strong safety culture experience fewer incidents, reduced absenteeism, and enhanced productivity. (Frick, 2019)

10.1.2 Competence, Skills, and Training

Human competency significantly affects safety performance. Training programs must be:

- accessible
- practical and hands-on
- tailored to worker backgrounds
- aligned with technological updates

Continuous learning ensures that workers remain aware of emerging risks, regulatory changes, and new equipment.

10.1.3 Physiological and Psychological Capacity

Workers' physical and mental health directly impact their ability to perform tasks safely. Fatigue, stress, poor ergonomics, and emotional strain reduce attentiveness and reaction time. OHS programs must therefore focus on:

- reasonable workload distribution
- ergonomic redesign
- fatigue management
- supportive work environments
- mental health resources

By enhancing human capacity, organizations create more resilient and reliable safety systems.

10.2. The Organizational Dimension: Systems, Leadership, and Governance

Organizational structures determine how safety principles are operationalized. Even well-trained workers and sophisticated technologies cannot compensate for weak leadership or poorly designed safety systems. (Quinlan, 2015)

10.2.1 Leadership Commitment and Accountability

Strong OHS performance begins with leadership that demonstrates commitment to safety. Leaders influence safety outcomes by:

- allocating adequate resources
- modeling safe behavior
- establishing clear accountability mechanisms
- integrating OHS goals into strategic planning

When leaders treat safety as a core value rather than a compliance obligation, it becomes embedded in organizational identity.

10.2.2 Policy Frameworks and Standardization

Formal safety policies and procedures guide organizational behavior. Effective OHS systems include:

- hazard identification processes
- risk assessment protocols
- emergency preparedness plans
- reporting mechanisms
- performance metrics

Global standards such as ISO 45001 provide frameworks for consistent implementation across multinational operations.

10.2.3 Work Organization and Communication

Communication is essential for seamless coordination among teams, especially in high-risk industries. Efficient communication reduces misunderstandings, minimizes delays, and ensures that workers receive timely safety information.

10.2.4 Organizational Justice and Employee Engagement

Fairness in management practices enhances worker engagement and compliance. Employees who feel valued and respected are more likely to contribute to safety initiatives. Organizational justice encompasses:

- transparency in decision-making
- equitable workload distribution
- conflict management
- recognition of contributions

These elements strengthen morale and foster shared responsibility for safety outcomes.

10.3. The Technological Dimension: Tools, Automation, and Data-Driven Solutions

Technology enhances OHS by enabling precise monitoring, predictive insights, and the automation of hazardous tasks. However, it must be integrated thoughtfully to avoid new risks and inequalities.

10.3.1 Safety Automation and Robotics

Robotics are increasingly used to perform high-risk tasks, reducing human exposure to danger. Applications include:

- automated cranes in ports
- robotic welding in manufacturing
- drones for inspection of inaccessible areas
- exoskeletons to support lifting tasks

Automation must be paired with proper interface design, operator training, and cybersecurity measures.

10.3.2 Digital Systems and Real-Time Monitoring

Digital platforms support:

- centralized safety data management
- incident reporting
- remotely monitored operations
- communication during emergencies

Wearables track worker movement, environmental conditions, and physiological indicators, helping prevent accidents in real time.

10.3.3 Predictive Analytics and Machine Learning

Machine learning algorithms analyze vast datasets to identify early signs of risk. They can:

- detect patterns in near-miss data
- predict equipment failure
- forecast environmental hazards
- evaluate compliance gaps

Predictive analytics help shift organizations from reactive responses to anticipatory safety management.

10.3.4 Challenges in Technology Adoption

Despite numerous advantages, technological integration can raise new concerns:

- privacy issues related to biometric data
- digital fatigue from constant monitoring
- skill gaps among workers

- increased cognitive demands
- dependence on automated systems

These challenges require careful ethical, educational, and regulatory attention.

10.4. Convergence of Human, Organizational, and Technological Factors

Holistic OHS systems recognize that no single dimension operates independently. (Guo et al., 2021) Instead, safety outcomes emerge from the interplay of workers, organizational structures, and technologies.

10.4.1 Human-Technology Interaction

The effectiveness of technology depends on how workers interact with it. Poor interface design or inadequate training can override safety benefits. Human-centered design principles ensure that technologies:

- support natural human behavior
- minimize cognitive load
- enhance task efficiency
- align with user capabilities

10.4.2 Organizational Structures Supporting Technology and People

Organizations must create structures that facilitate synergy between humans and technology. This includes:

- aligning policies with technological capabilities
- ensuring equitable access to training
- allocating resources for maintenance and upgrades
- integrating technology feedback into policy revision

10.4.3 Resilience Through Integration

Resilient OHS systems are adaptable, learning-oriented, and capable of responding to disruptions. When human, organizational, and technological systems work together, organizations can:

- rapidly adjust to workforce changes

- respond effectively to emerging risks
- maintain continuity during crises
- cultivate a culture of learning and innovation

10.5. Global Implications for OHS System Development

Countries differ in regulatory structures, economic capacity, and cultural practices. However, the integrated model is universally applicable because it emphasizes:

- adaptability
- systemic thinking
- shared responsibility
- evidence-based improvement

This integration supports global harmonization of OHS systems and helps nations progress from compliance-based approaches to strategic safety leadership.

11. ANALYTICAL ASSESSMENT AND INTEGRATIVE PERSPECTIVE

Occupational health and safety has evolved into a complex, multidimensional domain that extends far beyond regulatory compliance. Across the global evidence synthesized in this chapter—from human factors and industry-specific risks to advanced technological solutions—it is clear that OHS now functions as a strategic driver of productivity, economic resilience, and worker well-being. My assessment of the reviewed literature, international models, and sectoral case studies reflects a central realization: effective OHS systems emerge from the dynamic interplay between people, organizational systems, and intelligent technologies, (Man et al., 2025) and no single dimension can produce sustainable safety outcomes in isolation. The purpose of this chapter has been to understand how these elements converge, diverge, and mutually reinforce each other across different cultural, economic, and industrial contexts. In reflecting on these diverse strands, several overarching analytical themes become evident.

11.1. Integrating Global Insights: What the Evidence Reveals About OHS

From a human-centered perspective, the studies consistently show that workers remain the pivotal element in OHS performance. Whether examining stress among construction workers, musculoskeletal strain in port operations, or the psychosocial impact of job demands, a common thread emerges: safety outcomes cannot be improved unless worker well-being, competence, and engagement are prioritized. Technological tools can detect risks, and organizational policies can prescribe procedures, but ultimately, it is human behavior—motivation, communication, situational awareness, decision-making—that determines whether safety measures succeed in daily practice.

Organizationally, the most successful OHS systems examined in this chapter share three defining attributes: leadership commitment, safety culture maturity, and governance coherence. Countries and companies with the best outcomes have leaders who allocate resources, model safe behaviors, and establish accountability mechanisms. Their systems are structured around continuous learning, transparent reporting, and open communication. Conversely, where leadership is fragmented or reactive, even the most advanced technologies or comprehensive policies fail to translate into meaningful safety improvements.

In terms of technology, the global transition toward AI, IoT-enabled monitoring, automation, wearables, and predictive analytics represents a transformative shift in how risks are identified and managed. These tools enhance precision, provide real-time insights, and enable predictive decision-making. Yet, the literature also highlights a tension: technology amplifies human capability but cannot replace human judgment, adaptability, and ethical reasoning. Thus, technology must be integrated thoughtfully, ensuring usability, privacy, and equitable access across diverse workforce segments.

The sectoral analyses—including construction, ports, green industries, and technologically advanced manufacturing—further emphasize that OHS is shaped by contextual realities. High-risk sectors require specialized interventions, while low-risk sectors need holistic well-being strategies. In developing economies, resource limitations challenge the implementation of international models, while developed economies grapple with emerging risks such as automation-induced stress and aging workforces. This diversity reinforces the need for adaptable and context-sensitive OHS frameworks.

11.2. Cross-Cutting Synthesis: The Interdependence of Human, Organizational, and Technological Factors

The most important insight from Parts 1 to 10 is the strong interdependence among human, organizational, and technological factors. No single domain operates independently, and weaknesses in one dimension inevitably undermine the others. (Fabiano et al., 2010)

For example:

- Human factors influence how technology is used and how policies are interpreted.
- Organizational structures determine resource allocation, training quality, and technology governance.
- Technological systems generate data that organizations must analyze and workers must act upon.

This triadic relationship forms the basis of an integrated OHS model. When one dimension is neglected—such as lack of training for new technologies, weak leadership support, or unaddressed psychosocial risks—the entire system becomes fragile. But when all three dimensions reinforce each other, the outcome is a resilient, continuously improving OHS ecosystem.

Across all reviewed studies and examples, I observe five universal success factors:

1. Leadership commitment that is visible, consistent, and proactive
2. Worker participation and empowerment through communication, engagement, and shared responsibility
3. Data-driven safety intelligence, leveraging analytics for prevention rather than reaction
4. Adaptive policy frameworks that evolve with technological and industrial changes
5. Human-centered design, ensuring that technologies and policies align with worker needs

These factors collectively shape how effectively OHS systems function in practice.

11.3. Lessons from International Models: What Works and Why

The international comparisons show that high-performing OHS systems—whether in Japan, the EU, or multinational corporations—share several characteristics:

- Institutional maturity supported by regulatory consistency
- Emphasis on training and skills as fundamental components of safety
- Systems thinking, ensuring hazards are addressed holistically
- Collaborative governance, involving unions, regulators, and industries
- Investment in technology, not as a substitute for human labor but as an enabler of safer work

In contrast, developing economies often display fragmented systems due to resource constraints, informal labor sectors, or inconsistent enforcement. Yet, successful cases from emerging economies demonstrate that progress is possible when leadership commitment, education, and incremental technological adoption occur together.

A key analytical insight is that international success stories are not merely the result of advanced technologies or strict laws—they are the product of integrated approaches that align human values, organizational priorities, and technological capabilities.

11.4. Future Outlook: Strategic Imperatives for the Next Generation of OHS

The evolving landscape of occupational health and safety demands forward-looking strategies. Based on the collective evidence, several strategic imperatives emerge:

1. Address psychosocial risks with the same rigor as physical hazards

Stress, burnout, moral injury, job instability, and remote-work fatigue require structured, evidence-based interventions.

2. Adopt predictive and preventive OHS models using AI and data analytics

Prevention must replace reaction, especially in high-risk and rapidly changing industries.

3. Prepare for climate-related occupational risks

Heat stress, air quality concerns, extreme weather disruptions, and vector-borne diseases will increasingly shape OHS policies.

4. Create gender-responsive and inclusive safety systems

Women entering traditionally male-dominated sectors require tailored interventions addressing unique ergonomic, psychosocial, and health risks.

5. Strengthen capacity-building in developing economies

International collaboration, knowledge sharing, and affordable digital tools will be essential for equitable global safety outcomes.

6. Ensure ethical governance of digital safety technologies

Worker privacy, data transparency, consent, and digital fairness must become core principles.

7. Promote lifelong learning and adaptive training

Workforce capability must evolve alongside automation and smart technologies.
The future of OHS depends on how well organizations, industries, and nations anticipate these challenges and redesign their systems proactively.

11.5. Final Integrative Insight

My overall assessment is that occupational health and safety is fundamentally a systems-driven, human-centered discipline. The global evidence consistently shows that successful OHS outcomes do not arise from isolated interventions or compliance checklists. (Enshassi et al., 2015) They emerge when:

- workers feel valued and psychologically supported,
- leaders demonstrate genuine commitment,
- organizations embed safety into daily practice, and
- technologies are used intelligently to augment—not replace—human capability.

When these dimensions act in harmony, OHS becomes a strategic asset that enhances productivity, strengthens morale, and contributes to national development. The future of OHS lies in integration: integration of global insights, integration of human and machine intelligence, integration of policy and practice, and integration of prevention and innovation.

CONCLUSION

Occupational health and safety is undergoing a profound transformation driven by globalization, technological advancement, and evolving workforce expectations. Across all industries and national contexts, the evidence demonstrates that OHS has moved beyond its traditional regulatory boundaries and now serves as an essential component of organizational performance, national development, and human well-being. This chapter synthesized diverse international models, empirical studies, and sector-specific insights to highlight the multidimensional nature of modern OHS systems.

A clear conclusion emerging from this synthesis is that effective OHS cannot be achieved through isolated efforts. Safe workplaces arise when human, organizational, and technological elements function in alignment. Workers require supportive environments, meaningful participation, adequate training, and protection from physical and psychosocial hazards. Organizations must demonstrate leadership commitment, develop mature safety cultures, and implement coherent policy frameworks. Technologies must be deployed thoughtfully, ethically, and in ways that enhance rather than overwhelm human capability.

The global case studies and comparative models discussed in this chapter illustrate that success is not defined by economic wealth or technological sophistication alone. Rather, it is shaped by how effectively systems are integrated, how deeply safety is embedded into organizational identity, and how consistently learning and adaptation occur. Nations and industries that achieve strong OHS outcomes do so by treating safety as a strategic priority, investing in capacity-building, and fostering collaboration among regulators, employers, and workers.

As the risk landscape continues to evolve—with climate change, digitalization, remote work, and demographic shifts introducing new pressures—OHS systems must become more resilient and forward-looking. The future of global OHS depends on embracing holistic, evidence-based, and human-centered strategies that promote prevention rather than reaction. Such approaches will ensure not only the protection of workers but also the long-term sustainability and competitiveness of organizations and economies.

Ultimately, OHS is an investment in people, productivity, and societal well-being. Its success reflects the value placed on human dignity and the commitment to creating work environments where safety, health, and performance reinforce each other. By integrating global insights, applying systems thinking, and embracing innovation, organizations can build OHS strategies that are not only effective today but also adaptable to the challenges of tomorrow.

REFERENCES

Ahasan, M. R., & Partanen, T. (2001). Occupational health and safety in the least developed countries—A simple case of neglect. *Journal of Epidemiology*, *11*(2), 74–80. DOI: 10.2188/jea.11.74 PMID: 11388496

Ahmadi, M., Choobineh, A., & Ghaem, H. (2022). Physical and psychological workloads and their association with occupational fatigue among hospital service personnel. *BMC Health Services Research*, *22*(1), 1163. DOI: 10.1186/s12913-022-08530-0 PMID: 36096773

Bourahla, A., Fernandes, G., & Ferreira, L. M. D. F. (2024). Managing occupational health and safety risks in construction projects to achieve social sustainability: A review of literature. *Procedia Computer Science*, *239*, 1–8. DOI: 10.1016/j.procs.2024.06.269

Darbra, R. M., & Casal, J. (2004). Historical analysis of accidents in seaports. *Safety Science*, *42*(2), 85–98. DOI: 10.1016/S0925-7535(03)00002-X

Enshassi, A., Mohamed, S., & Mayer, P. E. (2015). Job stress, job burnout and safety performance in the Palestinian construction industry. *Journal of Financial Management of Property and Construction*, *20*(2), 170–187. DOI: 10.1108/JFMPC-01-2015-0004

Fabiano, B., Currò, F., Reverberi, A. P., & Pastorino, R. (2010). Port safety and the container revolution: A statistical study on human factor and occupational accidents over the long period. *Safety Science*, *48*(8), 980–990. DOI: 10.1016/j.ssci.2009.08.007

Frick, K. (2019). Worker influence on voluntary OHS management systems: A review of its ends and means. *Safety Science*, *113*, 356–364. DOI: 10.1016/j.ssci.2018.12.006

Fukuzawa, A., Sakaki, M., & Imada, T. (2020). Relationship between the internal locus of control and collective action: A comparison of East Asian and Western countries. *Asian Journal of Social Psychology*, *23*(4), 453–463. DOI: 10.1111/ajsp.12406

Guo, B. H., Yiu, T. W., & González, V. A. (2021). Computer vision technologies for safety science and management in construction: A critical review and future research directions. *Safety Science*, *135*, 105130. DOI: 10.1016/j.ssci.2020.105130

International Organization for Standardization. (2018). *Occupational health and safety management systems—Requirements with guidance for use (ISO 45001:2018)*. https://www.iso.org/standard/63787.html

Leka, S., Griffiths, A., & Cox, T. (2004). *Work organization and stress: Systematic problem approaches for employers, managers and trade union representatives.* World Health Organization. https://www.who.int/publications/i/item/9241590475

Liang, Q., Leung, M., & Cooper, C. (2022). Unveiling the mechanism of construction workers' unsafe behaviors from an occupational stress perspective: A stress–cognition–safety model. *Safety Science, 148*, 105647. DOI: 10.1016/j.ssci.2021.105486

Maheronnaghsh, S., Zolfagharnasab, H., Gorgich, M., & Duarte, J. (2023). Machine learning in occupational safety and health: A systematic review. *International Journal of Occupational and Environmental Safety*, *7*(1), 14–32. DOI: 10.24840/2184-0954_007-001_001586

Man, S. S., Wang, D., Tsang, S. N. H., Liu, L., & Chan, A. H. S. (2025). Relationships between occupational stress and occupational safety and health outcomes amongst construction workers: A meta-analysis of evidence from the past twenty years. *Safety Science*, *191*, 107207. DOI: 10.1016/j.ssci.2025.106939

Occupational Safety and Health Administration. (2022). *OSHA at a glance.* U.S. Department of Labor. https://www.osha.gov/aboutosha

Quinlan, M. (2015). *Ten pathways to death and disaster: Learning from fatal incidents in mines and other high hazard workplaces*. Federation Press.

Rollin, L., Dutheil, F., & Naughton, G. (2022). Occupational stressors in healthcare workers in France. *Revue d'Epidemiologie et de Sante Publique*, *70*(2, Suppl. 1), S95–S102. DOI: 10.1016/j.respe.2022.02.002 PMID: 35337699

Ronza, A., Carol, S., Espejo, V., Vílchez, J. A., & Arnaldos, J. (2006). A quantitative risk analysis approach to port hydrocarbon logistics. *Journal of Hazardous Materials*, *128*(1), 10–24. DOI: 10.1016/j.jhazmat.2005.07.032 PMID: 16125840

Sampson, J. M., DeArmond, S., & Chen, P. Y. (2014). Role of safety stressors and social support on safety performance. *Safety Science*, *64*, 137–145. DOI: 10.1016/j.ssci.2013.11.025

Sun, C., Zhang, J., & Wang, X. (2022). The relationship between psychosocial hazards and mental health in the construction industry: A meta-analysis. *Safety Science*, *150*, 105707. DOI: 10.1016/j.ssci.2021.105485

Wachter, J. K., & Yorio, P. L. (2014). A system of safety management practices and worker engagement for reducing and preventing accidents: An empirical and theoretical investigation. *Accident; Analysis and Prevention*, *68*, 117–130. DOI: 10.1016/j.aap.2013.07.029 PMID: 23993683

Winge, S., Albrechtsen, E., & Mostue, B. A. (2019). Causal factors and connections in construction accidents. *Safety Science*, *112*, 130–141. DOI: 10.1016/j.ssci.2018.10.015

World Health Organization. (2021). *Occupational safety and health in public health emergencies: A manual for protecting health workers.* https://www.who.int/publications/i/item/9789240033116

Zanatelli, M. M., Guimarães, A. V., Storte, G. R., Velloso, N., Emidio, M. V., Peruzzetto, M. C., & Bastos, P. A. S. (2021). Prevalence of low back pain in Port of Santos workers. *Revista Brasileira de Medicina do Trabalho : Publicacao Oficial da Associacao Nacional de Medicina do Trabalho-ANAMT*, *19*(2), 173–180. DOI: 10.47626/1679-4435-2021-580 PMID: 34603413

Chapter 3
Occupational Health and Safety Strategies of the European Union, United States, Japan, Australia–New Zealand, Canada, Nordic Countries, and Singapore

S. Hemalatha
http://orcid.org/0009-0005-0397-8738
Panimalar Engineering College, India

Karedla Nandhini
Nalla Narasimha Reddy Group of Institutions Integrated Campus, India

Lakshmi Banavath
Nalla Narasimha Reddy Group of Institutions Integrated Campus, India

D. Naga Malleswari
Koneru Lakshmaiah Education Foundation, India

Jyoti D. Shendage
Sri Balaji University, India

S. Krithick Ajay
Panimalar Engineering College, India

ABSTRACT

Occupational Health and Safety (OHS) systems across the world have evolved in diverse ways, shaped by historical, socio-economic, regulatory, and cultural contexts. Understanding these models is essential for policymakers, industries, and global organizations seeking to benchmark practices, improve workplace safety, and foster

DOI: 10.4018/979-8-3373-6370-7.ch003

sustainable work environments. This chapter presents a comparative overview of major international OHS models—European Union, United States, Japan, Australia–New Zealand, Canada, Nordic countries, and Singapore. For each model, the chapter outlines key strategies, regulatory systems, structural features, and notable success stories. The analysis concludes by highlighting global best practices and strategic insights that can guide countries and organizations in enhancing workforce safety, resilience, and well-being.

1. INTRODUCTION

Occupational Health and Safety (OHS) (Kajiki et al., 2020) has evolved into one of the most critical pillars of modern industrial societies. Regardless of a nation's economic standing or regulatory philosophy, ensuring safe and healthy workplaces has become an essential requirement for economic growth, social justice, workforce well-being, and organizational sustainability. At its core, OHS encompasses the systems, institutions, regulations, cultural practices, and scientific principles designed to protect workers from injury, illness, and fatality. Yet, while the objectives of OHS are universal, the pathways chosen by countries vary dramatically based on history, culture, industry structure, and levels of development.

The past century has witnessed profound transformations in global labor markets (Zanko & Dawson, 2012). The transition from agrarian to industrial economies, the rise of manufacturing, the growth of extractive industries, mass urbanization, and the digital revolution have all brought new categories of workplace hazards. Early OHS efforts in most countries were crisis-driven, emerging as reactions to disasters, mass injuries, or public outcry. Over time, the focus shifted toward systematic prevention, enforcement-based regulation, and public health approaches. In the 21st century, the emphasis is increasingly on risk management, data-driven monitoring, psychosocial health, and the integration of technology for predictive safety.

Studying international models is essential for several reasons. First, no single country has perfected OHS; each model embodies strengths and limitations shaped by social values and regulatory contexts. Second, globalized supply chains mean that workplace risks transcend national boundaries—unsafe labor practices in one region can disrupt production elsewhere or create reputational and legal risks for multinational corporations. Third, emerging hazards such as climate-related risks, automation, artificial intelligence, digital fatigue, and global pandemics demand collaborative learning from diverse OHS cultures. Finally, international benchmarking allows governments and organizations to adopt best practices suited to their unique contexts.

This chapter examines seven major global OHS models that represent distinct regulatory philosophies and cultural foundations: the European Union, the United States, Japan, Australia–New Zealand, Canada, the Nordic countries, and Singapore. These models were selected because they illustrate diverse approaches: regulation-driven harmonization, compliance-oriented enforcement, continuous improvement cultures, worker–management collaboration, welfare-state integration, and technologically advanced systems thinking.

The European Union exemplifies a harmonized regulatory system (da Silva & Amaral, 2019) underpinned by common directives, mandatory risk assessment, and participatory governance. The United States follows a compliance-heavy model built around legal enforcement, penalties, and industry standards driven by OSHA and supported by scientific research from NIOSH. Japan represents a unique human-centered culture grounded in Kaizen, quality systems, and holistic attention to mental health and overwork. Australia and New Zealand prioritize risk-based regulation and shared responsibility supported by competency-based training frameworks. Canada's Joint Responsibility System (JRS) illustrates a highly collaborative environment where workers and employers co-manage safety decisions. The Nordic countries, powered by robust welfare states, showcase global leadership in psychosocial health, worker equity, and prevention-first thinking. Singapore represents a technologically sophisticated model integrating predictive analytics, public–private partnerships, and a national Vision Zero movement.

Although these systems differ, several common principles emerge among high-performing OHS regimes. Effective models tend to balance regulation with organizational autonomy, emphasize prevention rather than reaction, foster a culture of safety led by top management, and encourage worker participation(Ashari, 2022). They also establish long-term institutional structures, invest in research, develop robust training systems, and utilize data for continuous improvement. Countries that excel in OHS consistently demonstrate political will, sufficient budget allocation, industry collaboration, and mechanisms for ensuring compliance.

However, international comparisons are not meant to imply that systems are interchangeable. Regulatory philosophies arise from deep cultural values: for example, Nordic welfare states prioritize social equality and universal public services; the U.S. emphasizes legal enforcement and individual responsibility; Japan emphasizes harmony, continuous improvement, and collective accountability. Understanding these philosophical foundations is essential before attempting to adapt policies across borders(Zimolong & Elke, 2006).

Moreover, OHS challenges today extend beyond traditional hazards such as falls, machinery accidents, chemical exposure, or noise. New forms of risk psychosocial stress, burnout, technology-induced pressure, gig-worker precarity, climate-induced heat stress, and work performed in digital and remote environments require novel global learning. Also, rapid industrialization in emerging countries raises the urgency for effective systems, yet these nations often lack the institutional maturity found in long-established industrial economies. International models can therefore guide capacity-building and governance reforms.

This chapter will analyze each international model in terms of key strategies, system features, enforcement mechanisms, cultural attributes, success stories, and global contributions. It will also highlight distinctive lessons each system offers the world. For policymakers, employers, workers, researchers, and health professionals, understanding how successful OHS frameworks operate globally can inform inclusive, evidence-driven reforms. As workplaces become increasingly interconnected, and as global challenges from pandemics to automation create shared vulnerabilities, the need for cross-national learning becomes stronger (Chen et al., 2020).

Ultimately, this chapter argues that world-class OHS performance emerges not from isolated regulations but from the integration of law, culture, governance, worker empowerment, and continuous improvement. High-performing nations demonstrate that when safety becomes a core societal value, workplaces become healthier, industries more productive, and communities more resilient. Global OHS models therefore serve not only as regulatory frameworks but also as reflections of national priorities, moral philosophies, and visions for the dignity and well-being of labor(Nastasia & Rives, 2024). Beyond reviewing international OHS models, this chapter advances the argument that student safety represents a critical and under-addressed dimension of occupational health and safety in higher education. The challenges identified in this chapter are grouped into three categories: (i) legal and policy gaps, (ii) physical and research-related risks, and (iii) human and organizational challenges. Each strategic recommendation presented later is explicitly mapped to these challenges

2. EUROPEAN UNION MODEL: REGULATION-DRIVEN HARMONIZATION

The European Union (EU) represents one of the most sophisticated examples of a harmonized, regulation-driven occupational health and safety system. Unlike countries where OHS evolves in a fragmented or sector-specific manner, the EU strives for coherence across its member states through framework directives, shared legal standards, and enforcement mechanisms that ensure consistency while allowing national adaptation.

2.1 Historical Context and Evolution

The EU's approach to OHS has evolved alongside industrialization, urbanization, and the expansion of the European labor market. Initially, workplace safety regulations were highly localized, reflecting national labor traditions and industrial priorities. In the post-World War II era, with increased cross-border trade and labor mobility(Kale et al., 2022), the EU recognized the need to harmonize safety standards to protect workers and create a level playing field for industries.

The Framework Directive 89/391/EEC, adopted in 1989, marked a turning point in European OHS policy. It established the fundamental obligation for employers to ensure the health and safety of workers, introduced systematic risk assessment, and promoted worker participation. Subsequent directives addressed specific hazards, including chemical agents, machinery, noise, and asbestos, creating a comprehensive regulatory ecosystem(Kim et al., 2016).

2.2 Key Strategies

Risk Assessment and Prevention

At the core of the EU model is the principle of prevention through risk assessment. Employers are required to identify potential hazards, evaluate risks, and implement control measures before incidents occur. Risk assessment is not limited to physical hazards but extends to psychosocial risks, such as stress, harassment, and fatigue, reflecting the EU's holistic approach(Madsen et al., 2020).

Worker Participation and Representation

Worker involvement is another cornerstone. The EU mandates the creation of health and safety committees, where employee representatives actively participate in risk evaluation, policy formulation, and the monitoring of safety practices. This participatory framework fosters ownership, compliance, and cultural reinforcement of safety norms.

Training and Education

Continuous education is a legal and cultural requirement. All employees, especially those in high-risk roles, undergo mandatory OHS training that covers hazard identification, safe work procedures, emergency preparedness, and rights and responsibilities. Training is reinforced through periodic audits, ensuring knowledge retention and proactive engagement(Köper et al., 2009).

Integration with National Health Systems

OHS in the EU is tightly integrated with national health services. Occupational health physicians and safety specialists monitor worker health, conduct regular medical check-ups, and contribute to risk analysis. This integration ensures early identification of work-related illnesses and timely interventions.

2.3 System Features

National Inspectorates

Each member state operates national OHS authorities responsible for enforcing EU directives(Pryor et al., 2019):

- United Kingdom – Health and Safety Executive (HSE): Focuses on risk-based inspections, data-driven enforcement, and sector-specific guidance.
- France – Institut National de Recherche et de Sécurité (INRS): Provides scientific support, training programs, and safety research.
- Germany – Berufsgenossenschaften (BGs): Sector-specific insurance funds that combine regulatory oversight, accident prevention, and worker compensation.

These inspectorates ensure that directives are translated into operational practices, balancing regulatory compliance with flexibility for organizational implementation.

Emphasis on Psychosocial Risks

European countries have increasingly recognized mental health hazards. Regulations address workplace stress, harassment, and occupational burnout, highlighting a comprehensive understanding of worker well-being beyond physical safety. Countries like Finland and Denmark have developed national frameworks for psychosocial risk management, including mandatory reporting and risk mitigation plans.

Data-Driven Safety and Monitoring

Advanced EU member states employ digital reporting systems to track workplace incidents, near-misses, and hazards. Data analytics support predictive prevention, policy refinement, and benchmarking across industries and regions.

2.4 Sectoral Adaptation

EU OHS directives are flexible enough to address sector-specific challenges:

- Construction: Mandatory scaffolding inspections, fall protection measures, and high visibility reporting systems.
- Manufacturing: Machine safety protocols, ergonomic assessments, and chemical hazard management.
- Healthcare: Infection control, ergonomic procedures, and psychosocial support for high-stress environments.
- Transportation: Driver fatigue management, load safety standards, and exposure monitoring.

This sectoral customization ensures that core principles are maintained while accommodating industrial diversity.

2.5 Success Stories

Sweden: Vision Zero

Sweden's Vision Zero policy, originally applied to traffic safety, has been extended to workplace safety. Its fundamental principle is that no workplace injury or fatality is acceptable, emphasizing design, prevention, and systemic risk elimination. Sweden reports some of the lowest industrial accident and fatality rates in Europe, demonstrating the power of a values-driven safety culture combined with regulatory enforcement(Ozobu et al., 2023).

United Kingdom: Risk-Based Regulation

The UK has adopted a risk-based approach, focusing regulatory attention on high-hazard industries while allowing lower-risk sectors to self-regulate within defined frameworks. Over the past 30 years, the UK has achieved a greater than

70% reduction in workplace fatalities, underscoring the efficacy of data-driven enforcement, worker participation, and continuous improvement.

France: Integration of Health and Research

France's INRS combines inspection, research, and training. Programs like the "Well-being at Work" initiative integrate physical and mental health, ergonomics, and psychosocial risk mitigation, positioning France as a leader in holistic OHS practices(Gallagher et al., 2003).

2.6 Lessons and Global Implications

The EU model demonstrates that harmonized regulation, combined with national enforcement flexibility, can yield:

- Predictable and consistent safety standards across borders
- Worker-centered approaches that emphasize participation and mental well-being
- Risk prevention rather than reactive enforcement
- Continuous improvement supported by research, training, and data analytics

Global industries, particularly those operating across multiple EU countries, benefit from these harmonized standards, which reduce operational complexity, enhance workforce safety, and facilitate cross-border compliance.

3. UNITED STATES MODEL: COMPLIANCE AND ENFORCEMENT THROUGH OSHA

The United States offers a contrasting approach to occupational health and safety (OHS) compared to the European Union. While the EU emphasizes harmonization and worker participation, the U.S. model is compliance- and enforcement-driven, prioritizing legal accountability, penalties, and scientifically informed standards. The backbone of this approach is the Occupational Safety and Health Act of 1970 (OSH Act), supported by the research and guidance provided by the National Institute for Occupational Safety and Health (NIOSH).

3.1 Historical Context and Evolution

Before the 1970s, U.S. OHS practices were largely voluntary and fragmented across states and industries. Industrial expansion, coupled with a rise in workplace accidents, galvanized public demand for federal intervention. The OSH Act was enacted in 1970 to establish comprehensive, uniform standards for workplace safety across the country. The act created OSHA to enforce compliance and NIOSH to conduct research, training, and recommendations for preventive strategies(Marhavilas et al., 2022).

Since its inception, the U.S. model has evolved to include industry-specific standards, technological monitoring, worker protection mechanisms, and proactive compliance programs, while maintaining its core philosophy of legal enforcement coupled with scientific guidance.

3.2 Key Strategies

Federal Regulation and Industry Standards

The OSH Act provides a legal framework for safety compliance across all sectors. OSHA sets mandatory standards for construction, mining, chemical handling, healthcare, and general industry. Standards cover hazardous materials, machine safety, personal protective equipment (PPE), ergonomics, fall protection, and workplace violence prevention.

Inspection and Enforcement

Compliance is ensured through scheduled and unscheduled inspections. OSHA inspectors have the authority to assess workplaces, issue citations, and levy fines for violations. Inspections focus primarily on high-risk industries, but can also be triggered by employee complaints, incident reports, or high injury rates. Enforcement mechanisms are designed to deter negligence and maintain a culture of accountability.

Worker Protection and Rights

U.S. law protects whistleblowers, ensuring that employees can report unsafe conditions without fear of retaliation. OSHA requires employers to maintain safety records, incident reports, and injury logs, which are publicly accessible for transparency and accountability. Employees also have the right to refuse unsafe work, providing an additional safeguard.

Scientific Guidance through NIOSH

NIOSH, part of the Centers for Disease Control and Prevention (CDC), conducts research on occupational hazards, provides recommendations for safe practices, and develops educational resources. This separation of enforcement (OSHA) and research (NIOSH) ensures that safety standards are scientifically grounded, evidence-based, and periodically updated.

3.3 System Features

Sector-Specific Standards

- Construction: Fall prevention, scaffolding safety, PPE usage, heavy machinery protocols.
- Mining: Ventilation, hazard monitoring, emergency evacuation procedures, dust and chemical exposure control.
- Healthcare: Bloodborne pathogen control, ergonomics, infection prevention, patient handling.
- Chemical and Manufacturing: Hazard communication, chemical exposure monitoring, emergency response procedures.

Data-Driven Enforcement

OSHA maintains databases of workplace injuries, fatalities, and citations, which inform policy, inspections, and targeted interventions. High-risk industries and repeat violators receive increased scrutiny, optimizing resource allocation.

Training and Certification Programs

OSHA supports programs such as the OSHA Outreach Training Program, which provides 10-hour and 30-hour safety training modules tailored to construction and general industry. These programs empower employees and supervisors with knowledge to recognize, prevent, and respond to workplace hazards.

Voluntary Compliance Programs

To complement enforcement, OSHA implements Voluntary Protection Programs (VPP). VPP recognizes workplaces that demonstrate excellent safety management, proactive hazard control, and employee engagement. Participants often achieve lower injury and fatality rates compared to industry averages, demonstrating the value of combining regulatory oversight with self-motivation and corporate safety culture.

3.4 Case Studies and Success Stories

Reduction in Workplace Fatalities

Since the establishment of OSHA, the United States has experienced a dramatic reduction in workplace injuries and fatalities, particularly in manufacturing, construction, and mining. For example, the fatal injury rate in the construction sector decreased by more than 60% over the past four decades, illustrating the effectiveness of rigorous enforcement, standardized protocols, and safety education.

Voluntary Protection Programs (VPP)

VPP-certified sites demonstrate injury rates significantly below national averages, showing that worker empowerment, corporate commitment, and proactive hazard management can complement formal legal enforcement. Industries that adopt VPP principles often integrate continuous improvement cycles, regular audits, and employee-driven reporting systems, creating a safety culture beyond mere compliance.

Technological Innovations

Many U.S. industries have leveraged technology for safety compliance:

- Wearable sensors to detect fatigue and monitor hazardous exposure.
- Digital reporting systems for near-miss events, accident tracking, and predictive risk analytics.
- Simulation and virtual reality training for high-risk tasks such as scaffolding, confined-space entry, and emergency evacuation.

These innovations illustrate how the U.S. model has evolved to incorporate data-driven safety management without losing its foundational compliance focus.

3.5 Challenges and Lessons Learned

While highly effective, the U.S. OHS model faces several challenges:

1. Resource Constraints: OSHA staffing levels and funding often limit inspection frequency, particularly in small and medium enterprises.
2. Fragmentation Across States: Some states operate their own OSHA-approved programs, leading to variations in enforcement rigor.
3. High-Risk Industries: Despite overall improvement, sectors such as construction, mining, and agriculture still report disproportionately high injury rates.
4. Emerging Hazards: Psychosocial risks, automation, and digital fatigue are not yet fully codified in regulatory standards.

Nonetheless, the U.S. experience demonstrates the efficacy of compliance-driven systems, the importance of scientific research integration, and the value of voluntary programs in reinforcing safety culture.

3.6 Global Influence

The U.S. model has influenced developing countries, multinational corporations, and international standard-setting bodies. OSHA standards often serve as benchmarks for industrial safety, chemical handling, and construction protocols worldwide. The model illustrates how legal enforcement, coupled with research-driven guidance, can substantially reduce occupational hazards, especially in high-risk industrial settings.

4. JAPAN MODEL: HUMAN-CENTRIC SYSTEMS AND CONTINUOUS IMPROVEMENT

Japan offers a unique perspective on occupational health and safety (OHS) that emphasizes human-centric approaches, continuous improvement, and integration with organizational quality systems. Unlike the EU's regulatory harmonization or the U.S.'s compliance-driven enforcement, the Japanese model is deeply rooted in cultural philosophies, corporate ethos, and holistic employee welfare, which have collectively shaped some of the safest industrial practices globally.

4.1 Historical Context and Evolution

Post-World War II industrialization brought rapid economic growth in Japan, accompanied by increased workplace hazards. The Ministry of Health, Labour and

Welfare (MHLW) gradually formalized occupational safety laws, complemented by industry-led initiatives. A cultural emphasis on collective responsibility, discipline, and lifelong employment facilitated the development of proactive safety measures, with companies adopting internal monitoring and participatory risk management.

By the late 20th century, Japan began integrating Kaizen principles—the philosophy of continuous improvement—into safety management, creating a system where OHS is not merely regulatory compliance but an ongoing organizational goal. Concurrently, concerns about karoshi (death from overwork) prompted the government to adopt regulations on working hours, mental health, and psychosocial risk management.

4.2 Key Strategies

Kaizen and Continuous Improvement

Kaizen, a cornerstone of Japanese management philosophy, extends beyond productivity to workplace safety. Companies encourage employees at all levels to identify hazards, propose improvements, and implement solutions, ensuring safety becomes part of daily operations. This continuous improvement culture fosters ownership, accountability, and innovation in risk management.

Integration with Quality and Productivity Systems

Japan's OHS approach is closely linked to quality management frameworks, such as:

- 5S (Sort, Set in order, Shine, Standardize, Sustain): Organizes workspaces to eliminate hazards and inefficiencies.
- Total Quality Management (TQM): Integrates safety, quality, and productivity metrics into corporate strategy.

This integration ensures that safety is not treated as an isolated function but as an embedded organizational value.

Psychosocial and Mental Health Measures

Recognizing the link between work-related stress and accidents, Japan mandates stress checks and mental health monitoring in workplaces above a certain size. Programs are designed to detect early signs of burnout, anxiety, or depression, providing counseling and workplace adjustments to mitigate risks.

Work-Hour Regulations

In response to karoshi, the Japanese government introduced strict limits on overtime and mandatory rest periods, ensuring employees have sufficient recovery time. Policies are complemented by organizational initiatives promoting work-life balance and health monitoring.

4.3 System Features

Shared Responsibility

In Japanese workplaces, both employers and employees share responsibility for safety. Workers participate in risk assessments, emergency drills, and reporting of hazards. Employers provide resources, training, and structural safeguards, creating a mutually accountable environment.

Safety Committees and Risk Management

Most medium and large enterprises have safety committees that regularly evaluate workplace risks, review incidents, and plan preventive measures. These committees often include cross-functional representation from production, maintenance, HR, and management, reinforcing collaboration and awareness.

Regular Safety Drills and Preparedness

Japanese companies prioritize emergency preparedness, including earthquake drills, fire evacuations, chemical spill response, and machinery shutdown procedures. Continuous training ensures that workers can respond effectively in crises, minimizing injury and operational disruptions.

Integration of Technology

Japan has leveraged technology to enhance OHS:

- Automated monitoring systems for machinery and process hazards.
- Wearable devices to track fatigue and ergonomics.
- Data analytics to detect patterns in near-misses and incidents, enabling predictive interventions.

Focus on Manufacturing and Heavy Industry

Industries such as automotive manufacturing, electronics, and chemical processing exemplify Japan's approach. Companies like Toyota integrate OHS into production lines, ensuring that employee safety is intertwined with operational efficiency.

4.4 Success Stories

Low Fatal Accident Rates

Japan maintains some of the lowest fatality rates in the manufacturing sector globally. Systematic integration of safety into production, combined with worker participation and continuous improvement, contributes to this achievement.

Toyota Production System (TPS)

The TPS exemplifies the synergy of quality, productivity, and safety. Safety checkpoints are built into workflows, hazards are eliminated through design, and employees are empowered to halt processes when unsafe conditions are detected. TPS has influenced global manufacturing standards, highlighting Japan's export of safety culture.

Mental Health Programs

Nationwide stress-check programs have reduced work-related psychological illness and improved early detection of burnout. Companies that adopt these programs report higher productivity, lower absenteeism, and reduced accident rates, illustrating the connection between psychosocial health and physical safety.

Regulation-Industry Collaboration

The MHLW and labor unions collaborate to monitor workplace conditions, disseminate best practices, and provide guidance on emerging hazards. This tripartite collaboration ensures legislation, industry adoption, and worker engagement are aligned.

4.5 Lessons and Global Implications

Japan's model offers several valuable lessons:

1. Culture Matters: Embedding safety into corporate culture yields sustainable improvements beyond mere compliance.
2. Continuous Improvement: Kaizen encourages proactive identification of hazards and adaptive solutions.
3. Integration with Productivity: Linking safety to quality and efficiency fosters executive buy-in and resource allocation.
4. Mental Health Integration: Addressing psychosocial risks complements traditional hazard control, ensuring holistic well-being.
5. Worker Participation: Empowering employees to actively contribute to safety strengthens accountability and early hazard detection.

For global organizations, the Japanese model demonstrates that regulatory compliance alone is insufficient. Sustainable OHS requires alignment of culture, leadership, continuous learning, and employee engagement, complemented by technological innovation and government support.

5 . AUSTRALIA & NEW ZEALAND MODEL: RISK-BASED AND WORKER-CENTERED

Australia and New Zealand exemplify a risk-based, worker-centered approach to occupational health and safety (OHS) that balances regulatory enforcement with proactive employee engagement. This model emphasizes shared responsibility, consultation, and continuous learning while leveraging harmonized legislative frameworks across regions.

5.1 Historical Context and Evolution

Both Australia and New Zealand transitioned from fragmented, state- or territory-specific safety regulations to cohesive, harmonized Work Health and Safety (WHS) legislation in the 21st century. In 2011, Australia implemented the Model WHS Act, serving as a blueprint for national consistency across states and territories. Similarly, New Zealand strengthened its Health and Safety at Work Act (2015), emphasizing duty of care, worker consultation, and risk management. These frameworks emerged from growing recognition that workplace accidents not only cause human suffering but also economic losses and societal costs.

Historically, mining, construction, and heavy manufacturing sectors in these countries faced high accident rates, prompting rigorous legal reform, enhanced training, and proactive hazard identification programs.

5.2 Key Strategies

Risk-Based Regulation

The cornerstone of the Australia–New Zealand model is risk-based regulation, which focuses regulatory attention proportionally to workplace hazards. Rather than enforcing uniform inspections across all sectors, authorities prioritize industries with high accident potential, using data-driven risk assessment to allocate resources efficiently.

This approach encourages employers to implement internal risk management systems, with legal oversight serving as a safety net rather than the primary driver of compliance. Organizations are incentivized to proactively identify hazards, assess risks, and implement preventive measures before incidents occur.

Worker-Centered Safety and Consultation

A defining feature of this model is shared responsibility between employers and workers. Workers are actively consulted in safety planning, incident investigations, and policy development. Regulations mandate the establishment of health and safety representatives (HSRs) in workplaces above a certain size, providing a formal channel for employee input and feedback.

Union and Stakeholder Involvement

Labor unions play a significant role in workplace safety, collaborating with employers and government regulators. Regular safety audits, joint inspections, and participatory committees foster a culture of shared accountability, ensuring hazards are addressed promptly and effectively.

Emphasis on Competency-Based Training

Australia and New Zealand have developed advanced OHS training frameworks, requiring workers in high-risk roles to attain formal competency certifications. Training is continuously updated to reflect emerging hazards, technological changes, and best practices, supporting both compliance and continuous improvement.

5.3 System Features

Harmonized WHS Laws

The Model WHS Act harmonizes regulations across Australian states and territories, creating uniform standards for duty of care, consultation, and risk management. Similarly, New Zealand's legislation ensures legal clarity, enforcement mechanisms, and worker empowerment. Harmonization reduces confusion for multinational companies operating across regions and supports consistent safety performance.

Proactive Inspection and Enforcement

Regulators in both countries emphasize inspections guided by risk intelligence, focusing on workplaces where hazards are high, previous violations exist, or incidents have occurred. Enforcement actions include cautions, improvement notices, and penalties, but are supplemented by education, guidance, and collaborative problem-solving.

Integration of Psychosocial Safety

Both countries recognize psychosocial hazards, including stress, fatigue, workplace bullying, and mental health risks. Legislation requires employers to identify, assess, and mitigate these risks, often incorporating employee assistance programs (EAPs), mental health policies, and training in psychological safety.

Sectoral Emphasis

- Construction: High priority due to fall risks, scaffolding hazards, and heavy machinery operation.
- Mining: Stricter regulations, mandatory emergency preparedness, and fatigue management programs.
- Healthcare: Infection control, manual handling protocols, and mental health interventions.
- Manufacturing: Ergonomics, machinery safety, chemical handling, and incident reporting.

Technology and Data Utilization

Advanced OHS data systems track incident reports, hazard alerts, and compliance trends, enabling regulators and companies to identify patterns and implement preventive interventions. Digital dashboards facilitate monitoring of high-risk activities and predictive analytics for resource allocation.

5.4 Success Stories

Construction Safety Improvements

Australia and New Zealand have witnessed substantial reductions in construction accidents over the past two decades. Risk-based inspections, mandatory training, and participatory safety committees have significantly lowered fatality rates while improving compliance and worker engagement.

Mining Sector Performance

Through proactive regulation and technology adoption, mining fatalities and serious injuries have declined steadily. Fatigue management systems, real-time monitoring, and joint safety committees contribute to this success.

Recognition of Psychosocial Hazards

These countries were early adopters in recognizing psychosocial risks as part of OHS, ahead of many industrialized nations. Workplace stress reduction programs and bullying prevention measures demonstrate a holistic view of worker well-being.

Cultural Integration of Safety

The focus on worker consultation, empowerment, and shared responsibility has led to a robust safety culture, where employees are active participants rather than passive recipients of regulations. Surveys indicate higher satisfaction, engagement, and compliance across industries, reinforcing the efficacy of the model.

5.5 Lessons and Global Implications

The Australia–New Zealand model highlights several critical lessons for global OHS practice:

1. Risk-Based Allocation: Prioritizing high-hazard sectors enhances regulatory efficiency and impact.
2. Worker Participation: Consultation and engagement foster accountability and early hazard detection.
3. Training and Competency: Continuous skill development strengthens safety culture and operational compliance.
4. Psychosocial Integration: Addressing mental health alongside physical hazards improves overall well-being.
5. Harmonized Legislation: Uniform standards across jurisdictions facilitate compliance, particularly for multinational enterprises.

Overall, the Australia–New Zealand approach demonstrates that risk-based regulation combined with worker-centered practices produces effective, sustainable OHS outcomes. Its balance of enforcement, consultation, and continuous learning offers a replicable framework for emerging economies and multinational corporations seeking to enhance workplace safety.

6. CANADA MODEL: JOINT RESPONSIBILITY SYSTEM (JRS)

Canada's approach to occupational health and safety (OHS) emphasizes collaboration between employers, employees, and unions, creating a framework known as the Joint Responsibility System (JRS). This model integrates shared accountability, participatory governance, and provincial-federal regulatory oversight to enhance safety outcomes across diverse industries.

6.1 Historical Context and Evolution

Canada's OHS landscape evolved through a combination of federal legislation and provincial regulations. While some sectors, such as mining, aviation, and federal public services, fall under the Canada Labour Code Part II, provinces have authority over workplaces within their jurisdiction. Over time, high-profile industrial incidents and public advocacy led to the adoption of joint responsibility principles, wherein both employers and employees are expected to proactively manage safety risks.

The development of JRS was influenced by strong labor unions, a cooperative industrial culture, and evidence-based research. By embedding worker participation in safety committees and decision-making, Canada established a co-managed, inclusive safety model.

6.2 Key Strategies

Shared Responsibility

Under JRS, employers and employees share primary responsibility for safety. Employers provide safe workplaces, training, and resources, while employees actively participate in risk identification, reporting, and preventive measures. This dual responsibility fosters a sense of ownership, ensuring hazards are addressed promptly.

Union and Worker Representation

Unions play a central role in workplace safety, particularly in industries like mining, forestry, and manufacturing. Safety committees must include union representatives, ensuring that worker concerns are heard, analyzed, and acted upon. Collective bargaining agreements often incorporate OHS commitments, reinforcing the culture of shared responsibility.

Embedded Occupational Health Services

Occupational health services, including medical examinations, ergonomic assessments, and health surveillance, are embedded within provincial OHS laws. These services allow early detection of work-related illnesses and ergonomic hazards, enabling timely intervention.

Training and Competency

Mandatory training is an essential component of JRS. Both employers and employees receive instruction on hazard recognition, emergency response, and safe work procedures. Training programs are updated to reflect technological advancements and emerging risks, ensuring workforce preparedness.

6.3 System Features

Federal and Provincial Oversight

- Federally regulated industries adhere to the Canada Labour Code Part II, which sets minimum safety standards, enforcement procedures, and reporting requirements.
- Provincial authorities supplement federal regulations with industry-specific guidelines, inspections, and educational programs.

Comprehensive Guidelines

Canada's OHS guidelines cover a wide array of hazards, including:

- Ergonomic risks: Manual handling, repetitive strain, workstation design.
- Violence prevention: Workplace violence policies, incident reporting, and risk assessment.
- Chemical and environmental hazards: Exposure limits, PPE, and emergency response protocols.
- Psychosocial safety: Stress management, bullying prevention, and mental health support.

Joint Safety Committees

Safety committees serve as a central mechanism for collaboration, responsible for hazard identification, incident investigation, risk assessment, and policy recommendations. Committees meet regularly, producing actionable reports that guide organizational safety strategies.

Worker Engagement Metrics

Many organizations monitor employee participation rates, hazard reporting frequency, and near-miss analyses to evaluate the effectiveness of joint responsibility mechanisms. High engagement correlates with reduced incidents and improved safety culture.

6.4 Success Stories

British Columbia – Forestry and Mining

In British Columbia, joint responsibility frameworks have significantly reduced injuries in the forestry and mining sectors. Collaborative safety committees, union involvement, and competency-based training have led to measurable reductions in lost-time injuries and fatalities.

Ontario – Manufacturing

Ontario's adoption of JRS in manufacturing has resulted in improved incident reporting, hazard mitigation, and employee engagement. Companies report enhanced safety awareness, reduced downtime, and increased productivity as a result of co-managed safety initiatives.

Integrated Health Services

Provinces that integrate occupational health services within regulatory frameworks have demonstrated better outcomes in terms of early detection of work-related illnesses, reduced absenteeism, and higher employee satisfaction.

6.5 Lessons and Global Implications

The Canadian JRS model provides several valuable insights for global OHS practice:

1. Shared Accountability: Combining employer authority with employee participation ensures proactive hazard management.
2. Union Collaboration: Formal union involvement strengthens enforcement, training, and policy adherence.
3. Integrated Health Services: Embedding occupational health within regulatory frameworks promotes early intervention and long-term worker well-being.
4. Data-Driven Monitoring: Tracking engagement, incident rates, and near misses supports continuous improvement.
5. Flexible Yet Structured: The federal-provincial dual system allows adaptation to local conditions while maintaining national safety standards.

By fostering cooperative governance, participatory decision-making, and preventive measures, Canada's model demonstrates that high OHS performance can be achieved when all stakeholders are empowered and accountable. Organizations adopting similar frameworks globally benefit from reduced workplace hazards, enhanced productivity, and sustainable employee engagement.

7. NORDIC COUNTRIES MODEL: WELFARE-BASED OHS LEADERSHIP

The Nordic countries, including Denmark, Norway, Finland, and Iceland, present a distinctive model of occupational health and safety (OHS) that emphasizes welfare-based policies, proactive prevention, and integration of worker well-being into national health frameworks. Unlike regulatory-heavy or compliance-driven systems, the Nordic approach blends social welfare principles, employer accountability, and participatory governance, producing some of the safest work environments globally.

7.1 Historical Context and Evolution

Industrialization in the Nordic countries during the late 19th and early 20th centuries prompted the creation of state-backed worker protection programs, initially focusing on industrial accidents and basic labor rights. Over time, a comprehensive welfare state emerged, emphasizing social equity, healthcare access, and worker safety.

By the late 20th century, the Nordic nations had integrated occupational health into broader national welfare policies, embedding OHS within public healthcare, labor law, and education. The result is a system where workplace safety is a shared societal responsibility, with both private and public sectors actively contributing to preventive measures.

7.2 Key Strategies

Proactive and Preventive Policies

Nordic countries prioritize prevention over reaction. Employers are required to conduct systematic workplace assessments, risk evaluations, and preventive measures before accidents occur. The legal frameworks encourage continuous monitoring and adjustment, ensuring hazards are addressed before incidents escalate.

Integration of Welfare and Safety

OHS in these countries is fully integrated with national healthcare and social welfare systems, providing workers with easy access to medical assessments, rehabilitation, and mental health support. This integration ensures that workplace safety extends beyond physical protection to holistic well-being, including psychological, social, and occupational dimensions.

Participatory Governance

The model emphasizes worker participation through elected safety representatives, union involvement, and collaborative committees. Employees are actively engaged in risk assessments, incident investigations, and policy development, promoting a sense of shared ownership and accountability.

Investment in Worker Well-Being

Significant government investment supports mental health initiatives, ergonomics programs, and workplace wellness. Nordic OHS policies recognize that healthy, motivated, and well-supported employees are less likely to experience accidents, creating a virtuous cycle of safety and productivity.

7.3 System Features

Legally Mandated Workplace Assessments

Employers must conduct regular safety audits that include physical, chemical, and psychosocial risk evaluations. In Finland and Denmark, psychosocial risk management is embedded in legislation, ensuring that factors like stress, fatigue, and harassment are systematically addressed.

OHS Integration with National Healthcare

Occupational health services are often state-funded, providing preventive care, monitoring, and rehabilitation. This integration enables early identification of health risks, reduces absenteeism, and supports reintegration after illness or injury.

Government-Funded Research and Innovation

The Nordic countries invest heavily in occupational health research, studying injury patterns, ergonomic improvements, and psychosocial interventions. Research findings inform legislation, training programs, and corporate policies, ensuring evidence-based OHS management.

Technology and Data Utilization

Advanced digital monitoring systems track workplace incidents, near misses, and compliance metrics. Predictive analytics help employers anticipate risks, allocate resources effectively, and implement proactive measures.

Sectoral Applications

- Oil and Gas (Norway): Offshore safety standards exceed global benchmarks, incorporating advanced hazard monitoring and emergency response.
- Forestry and Manufacturing (Finland): Low accident rates achieved through automated safety systems, participatory safety committees, and ergonomic interventions.
- Public Services: Focus on psychosocial risks, workplace stress, and mental health programs across education, healthcare, and municipal services.

7.4 Success Stories

Finland: National OHS Program

Finland implemented a nationwide OHS program that integrates occupational health with productivity and public health initiatives. Results include significant reductions in injury rates, improved worker satisfaction, and enhanced productivity.

Norway: Offshore Oil and Gas Safety

Norway's offshore industry demonstrates world-leading safety performance. Mandatory risk assessments, integrated emergency systems, and continuous worker training have minimized fatalities despite the inherently high-risk environment.

Denmark and Iceland: Mental Health and Well-Being

These countries have pioneered psychosocial safety initiatives, including mental health screenings, counseling, and stress reduction programs. Surveys indicate higher employee well-being and lower absenteeism, confirming the effectiveness of holistic OHS strategies.

Participatory Culture

Across Nordic countries, worker engagement in safety governance is deeply embedded, leading to rapid identification of hazards, innovative solutions, and sustained cultural adherence to safe practices. Companies consistently report lower incident rates and higher operational efficiency compared to global benchmarks.

7.5 Lessons and Global Implications

The Nordic welfare-based model highlights several principles relevant to international OHS practice:

1. Holistic Worker Well-Being: Safety extends beyond physical hazards to mental, social, and occupational health.
2. Proactive Prevention: Continuous risk assessments and early interventions reduce accidents before they occur.
3. Integrated Systems: Linking OHS with national healthcare ensures comprehensive monitoring and early treatment.
4. Participatory Governance: Employee involvement strengthens accountability, transparency, and safety culture.
5. Research-Driven Policies: Evidence-based legislation and practices improve effectiveness and adaptability.

This model illustrates that societal commitment, government investment, and cultural integration can produce exceptional OHS outcomes. For organizations globally, adopting elements of the Nordic approach—particularly holistic well-being, participatory governance, and proactive prevention—can enhance employee health, organizational efficiency, and social responsibility.

8. SINGAPORE MODEL: VISION ZERO AND SYSTEMS THINKING

Singapore has emerged as a global leader in occupational health and safety (OHS) by implementing a Vision Zero philosophy combined with systems-thinking approaches. The country's strategy emphasizes elimination of fatalities, data-driven decision-making, and collaboration between government, industry, and training institutions. Despite rapid industrial growth and a high-density workforce, Singapore maintains remarkably low workplace accident rates, offering a model of excellence for emerging economies and urbanized industrial hubs.

8.1 Historical Context and Evolution

Singapore's OHS evolution is closely linked to its industrialization strategy. During the 1970s and 1980s, rapid expansion in manufacturing, construction, and offshore oil and gas exposed workers to significant hazards. Recognizing the societal and economic impact of workplace accidents, the government introduced comprehensive legislation under the Workplace Safety and Health (WSH) Act of 2006, consolidating previous fragmented safety regulations.

The WSH Act is grounded in Vision Zero principles, which advocate that all workplace injuries and fatalities are preventable through proactive planning, risk management, and safety culture. Over time, the government invested in systems-oriented approaches, integrating technology, analytics, and training to enhance enforcement and compliance.

8.2 Key Strategies

Vision Zero Philosophy

Singapore's Vision Zero campaign sets the aspirational goal of zero workplace fatalities. This philosophy underpins all OHS initiatives and emphasizes:

- Prevention over reaction: Identifying and mitigating hazards before incidents occur.
- Continuous improvement: Regular assessment of processes and policies to reduce risks.
- Shared accountability: Involving employers, workers, regulators, and other stakeholders in maintaining safe workplaces.

Systems Thinking Approach

Singapore employs systems thinking in OHS management, viewing workplace safety as an interconnected system rather than a collection of isolated hazards. This approach incorporates:

- Risk mapping: Understanding relationships between hazards, processes, and human behavior.
- Predictive analytics: Using incident data to anticipate potential accidents.
- Feedback loops: Implementing continuous monitoring to refine safety measures and policies.

Regulatory Enforcement

The WSH Act provides comprehensive legal authority for inspections, citations, and penalties. Enforcement prioritizes high-risk industries, including construction, oil and gas, and chemical manufacturing. Companies are required to implement internal safety management systems, conduct risk assessments, and maintain incident reporting mechanisms.

Mandatory Training and Certification

All employees in high-risk sectors must undergo WSH training, including hazard recognition, emergency response, and safe work procedures. Supervisors and managers receive advanced training in safety leadership and risk management, ensuring that safety is embedded at all organizational levels.

8.3 System Features

Data-Driven Safety Management

Singapore leverages technology to create analytics-driven inspection and monitoring systems. Incident reports, near misses, and hazard alerts are systematically analyzed to identify trends and prevent recurrence. Predictive risk models allow regulators to focus resources on high-risk areas.

Government-Industry Collaboration

The government works closely with industry associations, training institutions, and safety consultants to promote best practices and continuous improvement. Collaborative initiatives include:

- National WSH campaigns: Public awareness, safety awards, and benchmarking programs.
- Shared learning platforms: Dissemination of case studies and hazard alerts.
- Industry partnerships: Development of sector-specific safety guidelines and training materials.

High-Risk Sector Focus

Singapore identifies sectors with the highest injury and fatality potential, such as construction, maritime, and manufacturing. Regulatory and preventive efforts are prioritized based on hazard potential and historical incident data.

Culture of Continuous Improvement

Companies are encouraged to adopt self-regulation alongside enforcement. Safety management systems integrate incident review, risk assessment, and employee feedback, fostering a culture of learning and proactive hazard mitigation.

8.4 Success Stories

Reduction in Workplace Fatalities

Singapore has achieved a rapid decline in workplace fatalities despite industrial growth. For example, the construction industry, historically high-risk, saw fatality rates drop significantly over a decade due to Vision Zero campaigns, mandatory training, and rigorous enforcement.

Technology-Enabled Safety

Digital systems for real-time incident reporting, hazard monitoring, and predictive analytics have helped identify potential risks before accidents occur. Automated dashboards enable regulators and companies to monitor safety trends and intervene proactively.

International Recognition

Singapore's construction safety framework has been recognized globally as a model for urbanized, high-density work environments. The country's approach combines regulatory rigor, proactive management, and culture-driven safety, demonstrating that Vision Zero is achievable with commitment and systems integration.

Workplace Safety Culture

Organizations adopting Singapore's model report higher safety awareness, improved compliance, and enhanced employee engagement. Workers are empowered to report hazards, participate in safety planning, and contribute to continuous improvement, creating a collaborative environment that sustains long-term safety gains.

8.5 Lessons and Global Implications

The Singapore model highlights several best practices relevant to international OHS:

1. Vision Zero as a guiding philosophy: Setting ambitious but achievable safety goals fosters a proactive safety mindset.
2. Systems thinking: Viewing hazards, processes, and human behavior as interconnected enables predictive and preventive measures.
3. Data-driven decision-making: Analytics and monitoring systems enhance regulatory efficiency and targeted interventions.
4. Collaborative governance: Partnerships among government, industry, and training institutions promote shared accountability and continuous improvement.
5. Integration of culture and enforcement: Embedding safety into corporate culture reinforces compliance and sustainability.

Singapore's experience demonstrates that ambitious safety goals, underpinned by strong regulatory frameworks and advanced technology, can dramatically improve workplace safety, even in high-risk industries.

9. GLOBAL BEST PRACTICES AND CONCLUSION: SYNTHESIZING LESSONS FROM INTERNATIONAL MODELS

International occupational health and safety (OHS) models, spanning regions such as the European Union, United States, Japan, Australia, New Zealand, Canada, Nordic countries, and Singapore, offer diverse strategies and practices. While each country or region is shaped by its economic development, cultural values, regulatory philosophy, and industrial composition, common themes emerge that collectively define global best practices. Synthesizing these models provides organizations and policymakers with guidelines for effective OHS systems, adaptable across industries and geographies.

9.1 Comparative Overview of International Models

The comparative analysis of OHS frameworks highlights seven major dimensions:

1. Regulatory Philosophy: Models vary from strict directive-based systems (EU) to risk-based regulation (Australia/New Zealand) and shared responsibility (Canada, Japan). Enforcement mechanisms, penalties, and compliance expectations differ, but all aim to ensure safe workplaces.
2. Worker Participation: Nordic countries, Canada, and Japan demonstrate high levels of worker engagement through safety committees, unions, and participatory governance, while the U.S. emphasizes compliance with OSHA standards but encourages voluntary safety programs.
3. Integration with Broader Systems: Nordic nations and Japan integrate OHS into healthcare, productivity, and quality management systems, promoting holistic well-being. Singapore combines OHS with data-driven governance, predictive analytics, and Vision Zero philosophy.
4. Risk Management Approaches: Risk-based strategies are central in Australia, New Zealand, and Singapore, where regulatory focus is proportional to hazard potential, optimizing resource allocation.
5. Psychosocial Safety: EU, Nordic countries, Japan, and Australia/New Zealand recognize mental health, stress, and psychosocial risks as integral to OHS, integrating these concerns into legislation, workplace programs, and training.
6. Training and Competency Development: All regions emphasize continuous education and competency certification, with high-risk sectors receiving specialized training to ensure preparedness and compliance.
7. Technology and Data Utilization: Advanced technologies—including wearable devices, digital monitoring, predictive analytics, and automated reporting systems—are increasingly used to enhance hazard detection, trend analysis, and proactive safety measures.

By comparing these dimensions, it is evident that effective OHS requires a combination of regulatory rigor, organizational culture, worker participation, and technological innovation.

9.2 Key Global Best Practices

Drawing from international experience, the following best practices emerge as central to successful OHS implementation:

1. Risk-Based Regulatory Oversight

Adopting risk-prioritized regulation, as seen in Australia, New Zealand, and Singapore, allows regulators to focus resources on high-hazard sectors while promoting internal safety management. Risk-based oversight enhances efficiency, reduces administrative burden, and motivates organizations to develop proactive internal systems.

2. Shared Responsibility and Worker Engagement

Models like Canada's JRS, Nordic participatory governance, and Japan's Kaizen culture demonstrate that employee involvement improves hazard identification, fosters ownership, and strengthens compliance. Establishing formal safety committees, union representation, and consultative frameworks ensures inclusive decision-making and effective hazard mitigation.

3. Integration of Safety with Organizational Systems

The Japanese model shows the benefits of linking OHS with quality management, continuous improvement, and productivity metrics, creating safety-conscious organizational cultures. Similarly, Nordic integration with national healthcare and welfare systems ensures holistic well-being, encompassing physical, mental, and social health.

4. Proactive Mental Health and Psychosocial Risk Management

Workplace safety extends beyond physical hazards. Countries like Finland, Denmark, Australia, and Japan have implemented stress-check programs, counseling, and psychosocial risk mitigation strategies. Organizations adopting holistic safety programs report lower absenteeism, reduced incidents, and higher employee satisfaction.

5. Continuous Training and Competency Development

OHS effectiveness depends on knowledgeable and trained personnel. High-risk sectors in the U.S., Singapore, and Australia emphasize formal training, certification, and ongoing professional development, ensuring that employees understand hazards, safety procedures, and emergency response.

6. Technology-Enabled Safety Systems

Advancements in automation, predictive analytics, wearable monitoring, and digital reporting are increasingly integrated into OHS systems. Singapore and Nordic countries exemplify data-driven safety management, enabling real-time hazard detection, incident tracking, and preventive action.

7. Cultural and Organizational Integration

Embedding safety into corporate culture is a hallmark of the Japanese Kaizen approach and Nordic participatory systems. When safety is part of organizational ethos rather than a regulatory obligation, adherence becomes voluntary, continuous, and self-reinforcing.

8. Holistic, Multi-Level Governance

Effective OHS frameworks integrate multiple layers of governance, including:

- Government legislation and enforcement (EU, Singapore, U.S.)
- Industry standards and codes of practice (Japan, Australia)
- Organizational policies and internal systems (Nordic, Canada)
- Worker participation and union involvement (Canada, Nordic)

This multi-level structure ensures redundancy, accountability, and sustainability.

9.3 Cross-Regional Lessons

Balancing Enforcement and Culture

While stringent enforcement ensures compliance (U.S., EU), cultural integration (Japan, Nordic countries) fosters voluntary adherence and continuous improvement. Organizations benefit from combining regulatory rigor with cultural embedding to sustain safety outcomes.

Customization and Adaptability

Global best practices are not "one-size-fits-all." Regulatory frameworks and cultural contexts differ, but key principles—risk assessment, worker engagement, continuous improvement, and technology adoption—are universally applicable. Organizations should adapt models to local legal, industrial, and cultural environments.

Emphasis on Data and Continuous Improvement

Modern OHS models increasingly rely on evidence, data analytics, and performance metrics. Predictive modeling, trend analysis, and benchmarking allow organizations to anticipate hazards, optimize interventions, and enhance safety culture.

Incorporating Psychosocial Safety

As work environments evolve, mental health and psychosocial risks are becoming integral to OHS. Addressing stress, burnout, and harassment enhances productivity, worker satisfaction, and overall safety.

International Collaboration and Knowledge Exchange

Many countries benefit from learning from global leaders. For instance, Singapore adopts Vision Zero lessons from Sweden, while Nordic ergonomic innovations influence Japanese manufacturing practices. Cross-border knowledge exchange strengthens policy development, training, and research.

9.4 Integrated OHS Framework for Organizations

Based on the synthesis of international models, organizations can develop an integrated OHS framework comprising:

1. Risk-Based Regulation and Internal Management: Prioritize high-risk areas, enforce compliance, and establish internal hazard management systems.
2. Worker Participation: Form safety committees, involve employees in decision-making, and encourage hazard reporting.
3. Holistic Well-Being: Address physical, mental, and social health, integrating occupational health services and psychosocial risk management.
4. Continuous Training: Ensure employees and supervisors are competent, updated on best practices, and certified where required.
5. Technology Utilization: Implement monitoring systems, predictive analytics, and digital dashboards to support proactive safety interventions.
6. Cultural Integration: Embed safety into organizational ethos, linking it with quality, productivity, and continuous improvement.
7. Performance Measurement and Benchmarking: Regularly assess safety metrics, incident reports, and near-miss trends, benchmarking against global best practices.
8. Collaborative Governance: Align internal policies with national legislation, industry standards, and international best practices.

9.5 Conclusion

International OHS models demonstrate that safety excellence is multifaceted, encompassing regulatory enforcement, organizational culture, worker participation, technology, and holistic well-being. Key takeaways include:

- Regulatory frameworks establish baseline safety standards but must be complemented by organizational initiatives and cultural integration.
- Worker engagement and participatory governance enhance hazard detection, accountability, and sustainable safety culture.
- Integration with broader systems, such as healthcare, productivity, and quality management, ensures a holistic approach to workplace safety.
- Data-driven decision-making and technology adoption allow predictive hazard management and proactive interventions.
- Global knowledge exchange promotes innovation, benchmarking, and continuous improvement.

By synthesizing lessons from the EU, U.S., Japan, Australia, New Zealand, Canada, Nordic countries, and Singapore, organizations worldwide can design context-sensitive, effective, and sustainable OHS programs. Adopting a blend of regulatory rigor, worker-centric practices, proactive risk management, and continuous improvement not only reduces workplace accidents and fatalities but also enhances employee well-being, operational efficiency, and organizational reputation. The collective experience of international OHS models underscores that safety is both a moral and strategic imperative, essential for sustainable development in today's globalized and industrially diverse economy.

REFERENCES

Ashari, M. R. (2022). Occupational disease prevention strategies through ohs interventions: Global evaluation and challenges. *Journal of Health Literacy and Qualitative Research*, *2*(2), 56–70. DOI: 10.61194/jhlqr.v2i2.531

Chen, H., Hou, C., Zhang, L., & Li, S. (2020). Comparative study on the strands of research on the governance model of international occupational safety and health issues. *Safety Science*, *122*, 104513. DOI: 10.1016/j.ssci.2019.104513

da Silva, S. L. C., & Amaral, F. G. (2019). Critical factors of success and barriers to the implementation of occupational health and safety management systems: A systematic review of literature. *Safety Science*, *117*, 123–132. DOI: 10.1016/j.ssci.2019.03.026

. Gallagher, C., Underhill, E., & Rimmer, M. (2003). Occupational safety and health management systems in Australia: barriers to success. *Policy and practice in health and safety, 1*(2), 67-81.

Kajiki, S., Mori, K., Kobayashi, Y., Hiraoka, K., Fukai, N., Uehara, M., Adi, N. P., & Nakanishi, S. (2020). Developing a global occupational health and safety management system model for Japanese companies. *Journal of Occupational Health*, *62*(1), e12081. DOI: 10.1002/1348-9585.12081 PMID: 31386257

Kale, S. R., Kota, S., Jasti, N. V. K., Soni, G., & Prakash, S. (2022). An occupational health and safety management system framework for lean process industries: An interpretive structural modelling approach. *International Journal of Lean Six Sigma*, *13*(6), 1367–1394. DOI: 10.1108/IJLSS-11-2020-0185

Kim, Y., Park, J., & Park, M. (2016). Creating a culture of prevention in occupational safety and health practice. *Safety and Health at Work*, *7*(2), 89–96. DOI: 10.1016/j.shaw.2016.02.002 PMID: 27340594

Köper, B., Möller, K., & Zwetsloot, G. (2009). The occupational safety and health scorecard-a business case example for strategic management. *Scandinavian Journal of Work, Environment & Health*, *35*(6), 413–420. DOI: 10.5271/sjweh.1361 PMID: 19806280

Madsen, C. U., Kirkegaard, M. L., Dyreborg, J., & Hasle, P. (2020). Making occupational health and safety management systems 'work': A realist review of the OHSAS 18001 standard. *Safety Science*, *129*, 104843. DOI: 10.1016/j.ssci.2020.104843

Marhavilas, P. K., Pliaki, F., & Koulouriotis, D. (2022). International management system standards related to occupational safety and health: An updated literature survey. *Sustainability (Basel)*, *14*(20), 13282. DOI: 10.3390/su142013282

Nastasia, I., & Rives, R. (2024). Successful strategies for occupational health and safety in small and medium enterprises: Insights for a sustainable return to work. *Journal of Occupational Rehabilitation*, •••, 1–16. PMID: 39616561

Ozobu, C. O., Onyekwe, F. O., Adikwu, F. E., Odujobi, O., & Nwulu, E. O. (2023). Developing a national strategy for integrating wellness programs into occupational safety and health management systems in Nigeria: A conceptual framework. *International Journal of Multidisciplinary Research and Growth Evaluation*, *4*(1), 914–927. DOI: 10.54660/.IJMRGE.2023.4.1.914-927

Pryor, P., Hale, A., & Hudson, D. (2019). Development of a global framework for OHS professional practice. *Safety Science*, *117*, 404–416. DOI: 10.1016/j.ssci.2019.04.033

Zanko, M., & Dawson, P. (2012). Occupational health and safety management in organizations: A review. *International Journal of Management Reviews*, *14*(3), 328–344. DOI: 10.1111/j.1468-2370.2011.00319.x

Zimolong, B., & Elke, G. (2006). Occupational health and safety management. *Handbook of human factors and ergonomics*, *10*(0470048204), 673-707.

Chapter 4
Learning From Success:
International Best Practices in Reducing Occupational Accidents and Diseases

Kaan Koçali
http://orcid.org/0000-0002-1329-6176
Istanbul Gelisim University, Turkey

ABSTRACT

This chapter contributes to the book by providing a structured, comparative analysis of successful international occupational health and safety models, moving beyond regulatory descriptions to focus on why and how certain systems effectively reduce occupational accidents and diseases. Unlike chapters that concentrate on single-country cases or specific regulations, this contribution synthesizes diverse regional experiences—developed and developing alike—to extract transferable success factors and policy-relevant lessons. The chapter aligns closely with the book's core themes of strategies, systems, and success stories, offering both analytical depth and practical implications. Its emphasis on prevention, learning from success, and adaptability ensures relevance for academics, policymakers, and practitioners seeking evidence-based pathways to improve OHS performance across different institutional and socio-economic contexts.

1. INTRODUCTION

Workplace accidents and occupational diseases continue to be one of the most significant occupational health and safety issues on a global scale today. According to international estimates, millions of workers are injured in workplace accidents

DOI: 10.4018/979-8-3373-6370-7.ch004

each year, while hundreds of thousands lose their lives due to occupational diseases or causes related to workplace accidents. This situation is not limited to individual health losses; it also has multidimensional consequences such as loss of productivity, increased social security expenditures, and weakened economic sustainability. The fact that occupational diseases are largely unreported shows that the real scale of the problem goes beyond official statistics (International Labour Organization [ILO], 2019; Takala et al., 2014).

Traditional occupational health and safety approaches have been shaped primarily within a reactive framework for many years, focusing on investigating causes and compensating for consequences after accidents have occurred. However, the limited effectiveness of this approach has become apparent, particularly in high-risk sectors where accidents and occupational diseases persist. This situation has accelerated the shift towards preventive, systematic, and learning organization-based models in occupational health and safety management. This new approach, which emphasizes elements such as risk-based management, worker participation, leadership, and continuous improvement, treats accidents not as "inevitable" outcomes but as preventable events (Reason, 2000; Robson et al., 2007).

In this transformation process, not only the existence of regulatory legislation but also the question of how these regulations are implemented and under what conditions they achieve success has become critically important. In this context, the experiences of countries and sectors that have achieved meaningful and sustainable reductions in occupational accidents and diseases provide an important source of learning on a global scale. The "learning from success stories" approach aims to go beyond abstract policy goals in the field of occupational health and safety and reveal the mechanisms through which concrete practices produce effective results. This approach requires not only identifying good practices but also analyzing why they work (EU-OSHA, 2019; Walters et al., 2016).

This section aims to examine occupational health and safety practices that have been successful in reducing work accidents and occupational diseases in different countries and contexts, identifying their common features and distinguishing characteristics. The section's original contribution is that it does not merely describe successful examples; it evaluates the institutional structure, management approach, risk management approach, and safety culture components behind these successes within a holistic framework. Thus, it aims to provide transferable policy and practice lessons for countries with different socio-economic and institutional structures. The fundamental assumption is clear: Accidents are not inevitable; they can be largely prevented through the right systems, effective practices, and learning organizations (Hämäläinen, Takala, & Saarela, 2017; Rasmussen, 1997).

2. CONCEPTUAL FRAMEWORK: WHAT DOES 'SUCCESS' MEAN IN OCCUPATIONAL HEALTH AND SAFETY?

In the field of occupational health and safety, the concept of "success" has long been evaluated solely based on a decrease in the number of workplace accidents. However, this narrow perspective fails to adequately reflect the multidimensional nature of workplace accidents and, in particular, the delayed and often invisible effects of occupational diseases. Current approaches consider success not only in terms of outcome indicators but also within a broader framework that includes preventive capacity, organizational learning, and safety culture. In this context, success is associated with the ability of a business or country to systematically identify risks, control them, and maintain a cycle of continuous improvement (Rasmussen, 1997; Hale, 2003).

Work accident frequency and severity rates are among the most commonly used classic (reactive) indicators for measuring occupational health and safety performance. The accident frequency rate refers to the ratio of work accidents occurring within a specific time period to the number of employees or total working hours, while the severity rate is based on the number of work days lost as a result of accidents. While these indicators are important for monitoring safety performance, particularly in high-risk sectors, they provide limited information about preventive capacity when used alone. Indeed, in many cases, temporary declines in accident rates may reflect changes in reporting or inspection practices rather than the elimination of actual risks (Hinze, Thurman, & Wehle, 2013; ILO, 2013).

The decrease in occupational disease incidence is another dimension that is often overlooked but critically important in assessing occupational health and safety success. Occupational diseases cannot be easily measured using short-term performance indicators due to long exposure periods and delayed effects. Nevertheless, establishing early diagnosis systems, effectively implementing health surveillance programs, and controlling exposures at source can significantly reduce the burden of occupational diseases in the long term. Therefore, success should be evaluated not only by the decrease in the number of reported cases, but also by the increase in diagnostic capacity and the institutionalization of preventive health approaches (Hämäläinen, Takala, & Saarela, 2017; Leigh, 2011).

The maturity level of safety culture has increasingly become a central concept in explaining success in occupational health and safety. Safety culture reflects the values, beliefs, and behavioral patterns that employees and managers attribute to safety. In organizations with a mature safety culture, safety is not only a legal requirement but is also treated as an integral part of corporate decision-making processes. In such structures, employee participation is high, learning from mistakes is encouraged, and reporting unsafe conditions is seen as a developmental tool rather than a punitive one. Research has revealed a strong relationship between safety culture maturity and low accident rates (Reason, 2000; Cooper, 2000).

In recent years, proactive performance indicators have gained increasing importance in occupational health and safety literature for success evaluation. Indicators such as near-miss incident reports, behavior-based safety observations, risk assessment updates, and employee participation levels in safe behaviors provide early warning signals about how well the system is functioning before accidents occur. These indicators strengthen organizations' learning capacity and preventive reflexes, enabling them to move beyond reactive approaches. Successful occupational health and safety systems stand out as those that regularly monitor such proactive indicators and integrate them into managerial decision-making processes (Hopkins, 2006; Guldenmund, 2010).

Within this holistic framework, success in occupational health and safety should be defined not only as a reduction in accidents, but also as the ability to anticipate risks, protect worker health, foster a mature safety culture, and sustainably operate a learning system. This approach provides a solid analytical foundation for evaluating international best practices and drawing lessons that can be adapted to different contexts.

Table 1. Indicators of Success in Occupational Health and Safety

Indicator Category	Key Indicators	Description
Reactive Indicators	Accident frequency rate, accident severity rate	Measures outcomes after incidents occur
Occupational Health Indicators	Occupational disease incidence, health surveillance coverage	Reflects long-term worker health protection
Safety Culture Indicators	Safety culture maturity, leadership commitment, employee participation	Indicates organizational attitudes and values toward safety
Proactive Indicators	Near-miss reporting, behavior-based safety observations, risk assessment updates	Provides early warning signals and prevention capacity

3. METHODOLOGICAL APPROACH

This section adopts a hybrid methodological approach aimed at systematically identifying and comparatively analyzing practices that have achieved success in reducing occupational accidents and diseases on an international scale. The study combines the transparency provided by systematic reviews based on strict exclusion and inclusion criteria with the contextual depth offered by narrative-based reviews. This approach allows for an understanding not only of quantitative results but also of the managerial and cultural mechanisms behind the practices in a multidisciplinary and institutionally sensitive field such as occupational health and safety (Grant & Booth, 2009; Snyder, 2019).

The literature review covers not only studies published in peer-reviewed scientific journals but also policy reports and evaluation documents published by international organizations. A significant portion of global trends and best practices in occupational health and safety are addressed in detail in such institutional reports rather than directly in academic publications. Therefore, the literature review process aims to ensure both scientific validity and policy and practice relevance by evaluating academic sources and international organization reports together (ILO, 2019; EU-OSHA, 2019; OECD, 2021).

Case studies were structured based on specific selection criteria from among the examples identified through literature and report analysis. First, it was determined as a basic condition that the selected countries or sectors should have shown measurable and documented improvement in the frequency and severity rates of occupational accidents and the incidence of occupational diseases. This criterion ensures that the concept of success is based on traceable and comparable indicators rather than subjective assessments (Hämäläinen, Takala, & Saarela, 2017).

The second selection criterion is that the practices must be institutionally sustainable. Rather than short-term campaigns or temporary arrangements, occupational health and safety practices that have been integrated into legal and administrative structures and have continued for many years have been preferred. This approach makes it possible to evaluate success through institutionalized management systems and continuous improvement processes rather than through random or periodic results (Hale, 2003; Walters et al., 2016).

Finally, the criterion of transferability was used in the selection of cases. Although the socio-economic and cultural conditions of each country differ, the aim was to select examples that could be adapted to other contexts in terms of basic principles and management approaches. Accordingly, not only advanced practices in high-income countries but also flexible and adaptable models that have been successful in countries with different levels of development were included in the scope of the study. Thus, rather than offering universal "one-size-fits-all" solutions, the chapter aims to present learnable and replicable lessons that can be applied in different contexts (Rasmussen, 1997; Hopkins, 2006).

4. INTERNATIONAL BEST PRACTICE CASE STUDIES

4.1 Northern Europe: Preventive and Participatory Models

Northern European countries attract international attention with their models that focus on preventive and participatory approaches in occupational health and safety. The success achieved in reducing occupational accidents and diseases in these countries is based not only on the existence of legal regulations but also on the fact that these regulations are supported by strong institutional structures, effective monitoring mechanisms, and the active participation of social partners. In this context, occupational health and safety is addressed on the basis of social consensus and institutional cooperation rather than individual responsibilities (EU-OSHA, 2019; Walters & Nichols, 2007).

One of the most distinctive features of the Northern European model is the operation of strong inspection systems in conjunction with social dialogue mechanisms. Inspection activities serve a guiding and developmental function rather than being a punitive tool. The continuous communication established between labor inspectors, employers, and employee representatives enables the early detection of risks and the timely implementation of corrective measures. This approach ensures that occupational health and safety is integrated into corporate decision-making processes, going beyond legal compliance (Frick, 2011; ILO, 2013).

Employee participation and strong union structures are among the key determinants of occupational health and safety success in Northern European countries. Employees are actively involved in risk assessment processes, workplace inspections, and safety committees; they are encouraged to report unsafe conditions. Trade unions are not limited to the role of defending rights, but are positioned as constructive partners in the development and implementation of occupational health and safety policies. This participatory structure ensures that the safety culture is spread throughout the organization and strengthens employees' sense of ownership of safety (Walters et al., 2016; Quinlan, 2014).

Another distinguishing feature of the Northern European model is the sophistication of early diagnosis and health surveillance systems for the prevention of occupational diseases. Pre-employment and periodic health examinations are treated not only as a legal obligation but also as a strategic tool for protecting worker health. Systematic monitoring of occupational exposures enables the development of targeted interventions in high-risk sectors and allows for the detection of occupational diseases before clinical symptoms appear. This approach significantly reduces both individual health losses and societal costs in the long term (Hämäläinen, Takala, & Saarela, 2017; Leigh, 2011).

In general, the preventive and participatory occupational health and safety models in Northern European countries demonstrate that success is not only achievable through technical measures, but also through the integrated implementation of strong institutional structures, social dialogue, and worker participation. While this model cannot be directly replicated in different socio-economic contexts, its fundamental principles offer significant potential for learning and adaptation in other countries.

4.2 Anglo-Saxon Model: Risk-Based and Data-Driven Approaches

The approach adopted in Anglo-Saxon countries in the field of occupational health and safety is largely based on a risk-based and data-driven management philosophy. In this model, the reduction of workplace accidents and occupational diseases relies on the systematic identification, prioritization, and management of risks rather than on strict sets of rules. The role of the state is not so much to provide detailed technical instructions, but rather to establish frameworks that enable employers to manage risks effectively and to oversee this process. This approach treats occupational health and safety as a dynamic and context-sensitive management activity (Hale & Hovden, 1998; Gunningham & Sinclair, 2017).

The risk-based inspection approach is one of the distinguishing features of the Anglo-Saxon model. Instead of being distributed equally across all workplaces, inspection resources are directed towards high-risk sectors, businesses, and activities. This allows limited public resources to be used more effectively and increases the preventive impact of inspections. Audit processes cover not only the verification of legal compliance but also the quality of risk assessment, the effectiveness of control measures, and the functioning of management systems in businesses. Research shows that risk-based audit practices make significant contributions, particularly in reducing serious and fatal accidents (Black, 2005; Hutter, 2011).

In Anglo-Saxon countries, occupational health and safety performance is evaluated through measurable and trackable performance indicators. In addition to accident frequency and severity rates, indicators such as near-miss reports, audit results, risk assessment updates, and training participation rates are regularly monitored. These indicators make companies' safety performance comparable and provide a data-driven basis for managerial decision-making processes. The central role of performance measurement removes occupational health and safety from being an abstract goal and makes it an accountable area of management (Hinze, Thurman, & Wehle, 2013; Hopkins, 2009).

Employer responsibility and a strong reporting culture are another fundamental component of the Anglo-Saxon model. Employers bear the non-transferable responsibility of protecting the health and safety of their employees and are directly accountable for identifying and controlling risks. Mandatory reporting systems ensure that serious accidents and hazardous incidents are reported transparently; this data is used in national policy development and prioritization processes. This reporting culture encourages errors to be treated as opportunities for learning and improvement rather than being concealed (Reason, 2000; HSE, 2019).

In general, the Anglo-Saxon risk-based and data-driven approach demonstrates that success in occupational health and safety can be built on the principles of flexibility, accountability, and measurability. This model may not have the same impact in every context due to its requirement for a robust data infrastructure and organizational capacity; however, it offers significant learning potential, particularly for sectors with complex and variable risk profiles.

4.3 East Asia: Technology-Enhanced Safety Systems

East Asian countries stand out for their widespread and comprehensive implementation of technology-supported preventive systems in the field of occupational health and safety. Rapid industrialization, high production intensity, and complex supply chains have made the management of occupational accidents and diseases a strategic priority in these countries. In this context, success has been achieved not only through regulatory frameworks but also through the systematic integration of digital technologies, automation solutions, and engineering-based design approaches into occupational health and safety management (Park, Kim, & Cho, 2019; ILO, 2021).

Digital monitoring systems are one of the most prominent components of the East Asian model. Thanks to sensor-based monitoring, wearable technologies, and real-time data collection applications, workers' exposures, environmental conditions, and machine performance can be continuously tracked. These systems enable the detection of risks before they turn into accidents, ensuring that preventive interventions are carried out in a timely manner. Digital monitoring applications have made significant contributions to reducing occupational diseases and serious accidents, particularly in the chemical, electronics, and heavy industry sectors (Liu et al., 2020; EU-OSHA, 2020).

Automation and ergonomics-focused design approaches are another fundamental element of technology-supported occupational health and safety systems in East Asia. Eliminating human intervention from hazardous, repetitive, or physically demanding tasks through automation has been effective in preventing musculoskeletal disorders and serious injuries. Furthermore, redesigning human-machine interaction in line with ergonomic principles has increased worker performance while reducing health risks associated with workload. This approach positions occupational health and safety not as an element that conflicts with production efficiency, but as a strategic investment that supports efficiency (Dul & Neumann, 2009; Karwowski, 2012).

Large-scale industrial applications have played a significant role in the widespread technological transformation in the field of occupational health and safety in East Asian countries. Multinational production facilities and large industrial clusters have provided a suitable ground for pilot applications of advanced technologies; successful applications have quickly become standard across the sector. In this process, government policies, incentive mechanisms, and national digitalization strategies have emerged as important factors accelerating companies' technology investments. As a result, technology-supported occupational health and safety systems have shown the potential to spread not only in large enterprises but also to smaller-scale businesses throughout the supply chain (Kim & Park, 2020; OECD, 2019).

Overall, the East Asian model demonstrates how success in occupational health and safety can be made sustainable through technological capacity, engineering solutions, and system integration. Although this approach requires high investment costs and technical expertise, it offers a robust and scalable solution framework, particularly for high-risk and high-volume production environments.

4.4 Emerging Economies: Targeted and Adaptive Interventions

Occupational health and safety practices in developing economies are often shaped by structural challenges such as limited institutional capacity, informal employment, the predominance of high-risk sectors, and insufficient oversight resources. In this context, success is achieved not by directly replicating the comprehensive and costly systems of developed countries, but rather through targeted, adaptable, and context-sensitive interventions. International experience shows that flexible and prioritized strategies produce more effective results in reducing occupational accidents and diseases in these countries (ILO, 2019; World Bank, 2020).

Sector-based interventions are among the most common and effective practices in occupational health and safety in developing economies. Regulations and inspection programs focusing on high-risk sectors such as construction, mining, agriculture, and textiles allow limited resources to be concentrated in areas of greatest risk. Compared to general regulations covering all sectors, this approach delivers more measurable improvements in the short and medium term. Pilot sector programs supported by international organizations demonstrate that such targeted interventions can lead to significant reductions in accident rates (Takala et al., 2014; EU-OSHA, 2018).

Education-focused transformation is one of the cornerstones of sustainable occupational health and safety success in developing countries. Training programs aimed at increasing risk awareness among workers and managers play a critical role, especially in environments with limited technical infrastructure. In these countries, training is not limited to the transfer of regulatory knowledge; it is used as a tool to encourage safe behavior, teach basic risk control measures, and develop a safety culture. Research shows that low-cost but continuous training interventions make significant contributions to reducing workplace accidents (Robson et al., 2007; Burke et al., 2011).

Building institutional capacity is a long-term factor that determines success in occupational safety and health in developing economies. Establishing an effective legal framework, strengthening regulatory bodies, and developing data collection and reporting systems are key components of this process. International cooperation projects and technical support programs have played an important role in enhancing the institutional competencies of occupational health and safety agencies in many countries. Although such capacity-building efforts produce limited results in the short term, they lay the groundwork for the establishment of sustainable and self-sustaining systems in the long term (Walters et al., 2016; OECD, 2021).

In general, success in developing economies is shaped by flexible, prioritizing, and learning-oriented institutional structures. Targeted and adaptable interventions implemented in these countries demonstrate that contextually designed strategies are more effective than universal, one-size-fits-all solutions in the field of occupational safety and health. This approach offers important policy lessons for countries at different levels of development.

Table 2. Comparison of International Best Practices in Reducing Occupational Accidents and Diseases

Region / Model	Core Approach	Key Instruments	Main Strengths	Transferable Lessons
Northern Europe	Preventive & participatory	Social dialogue, strong inspection, worker representation	High safety culture maturity, early disease detection	Worker participation, trust-based governance
Anglo-Saxon Countries	Risk-based & data-driven	Risk-based inspections, performance indicators, reporting systems	Accountability, measurable performance	Risk prioritization, evidence-based regulation
East Asia	Technology-enhanced systems	Digital monitoring, automation, ergonomic design	Scalable prevention, real-time risk control	Technology integration, engineering controls
Emerging Economies	Targeted & adaptive interventions	Sector-based programs, training, capacity building	Cost-effective improvements, contextual adaptability	Prioritization, education-focused prevention

5. KEY SUCCESS FACTORS ACROSS COUNTRIES

A comparative analysis of international best practices shows that, despite different socio-economic and institutional contexts, there are common success factors in reducing occupational accidents and diseases. These factors are directly related to how occupational health and safety systems are designed, implemented, and maintained, rather than being specific to a particular country or model. The elements that determine success are not limited to technical measures; they exhibit a multi-layered structure that includes legal, managerial, cultural, and cognitive dimensions (Hale, 2003; Walters et al., 2016).

A strong legal framework and its effective implementation are among the key success factors that stand out in cross-country comparisons. However, findings reveal that the existence of legislation alone is not sufficient; it must be accompanied by inspection capacity, deterrent sanctions, and guidance practices. In successful countries, legal regulations are not merely an obligation for employers and employees, but provide a clear and understandable roadmap for managing risks. This prevents legislation from remaining "on paper" and ensures real transformation at the implementation level (Black, 2005; ILO, 2019).

The preventive risk management approach forms the common denominator of all successful models examined. The systematic identification, assessment, and control of risks enable intervention before accidents occur. This approach goes beyond reactive accident investigations, reflecting a holistic perspective that acknowledges the dynamic and variable nature of risks. In systems where preventive risk management is institutionalized, the reduction in occupational accidents and diseases is not coincidental but becomes a predictable outcome (Rasmussen, 1997; Hale & Hovden, 1998).

Leadership and safety culture emerge as critical multipliers that determine the impact of technical and legal regulations. Top management's emphasis on occupational health and safety is reflected throughout the entire organization through resource allocation, prioritization, and role modeling behaviors. In systems with a mature safety culture, employees perceive safety not as an individual responsibility but as a shared value; reporting unsafe conditions is encouraged. This cultural transformation forms the basis for sustainable safety performance in the long term (Reason, 2000; Cooper, 2000).

Education and continuous learning are another important factor explaining differences in success between countries. In successful systems, education is approached not as a one-time activity but as a continuous and cyclical process. Training programs for employees and managers aim not only to convey regulatory knowledge but also to develop risk awareness, reinforce safe behaviors, and support a learning organization culture. The continuous learning approach ensures that systems remain up-to-date in the face of changing technologies and emerging risks (Robson et al., 2007; Burke et al., 2011).

Data-driven decision-making has become an indispensable element of success in modern occupational health and safety management. The systematic collection and analysis of data on accidents, near misses, occupational diseases, and inspection results enables policies and practices to be shaped based on evidence. In successful countries, data is considered a strategic tool not only for reporting purposes but also for prioritizing risks and utilizing resources effectively (Hopkins, 2009; OECD, 2019).

These shared success factors offer adaptable policy lessons rather than models that should be directly replicated in Turkey and similar countries. In particular, increasing the enforceability of legislation, strengthening preventive risk management, investing in leadership and safety culture, and developing a data-driven management approach stand out as key areas that can deliver meaningful improvements in different contexts. This framework provides a solid analytical foundation for building an occupational health and safety strategy that is sensitive to national conditions but informed by international experience.

6. DISCUSSION: TRANSFERABILITY OF SUCCESSFUL OHS PRACTICES

One of the most fundamental questions in examining international best practices is, “Does every model fit every country?” This question highlights the limitations of seeking universal solutions in the field of occupational health and safety. Comparative analyses show that models that achieve high success in certain countries do not automatically produce the same effect in other contexts. The fundamental reason for this is that occupational health and safety systems are strongly influenced not only by technical and legal regulations, but also by contextual factors such as social values, managerial traditions, and institutional capacity. Therefore, transferability should be assessed through the adaptation of principles and mechanisms rather than direct replication (Rasmussen, 1997; Hale, 2003).

Cultural factors play a decisive role in transferring successful occupational health and safety practices to different countries. Cultural dimensions such as the perception of safety, approach to authority, employee-employer relations, and risk tolerance have a direct impact on the acceptance and effectiveness of these practices. For example, models based on employee participation are more easily adopted in countries with a strong culture of social dialogue, while such approaches may have limited impact in contexts where hierarchical structures are dominant. This explains why policy transfers that disregard cultural adaptation can fail (Hofstede, 2001; Hopkins, 2006).

Economic and institutional constraints are also key factors shaping the transferability of successful practices. Technology-intensive or data-driven occupational health and safety systems, common in developed countries, are not equally applicable in every country due to high investment costs and technical expertise requirements. Similarly, strong monitoring and guidance mechanisms may not have the expected impact in countries that lack sufficient human resources and institutional capacity. In this context, in addition to the question of "what can be done" in policy transfer, the question of "what is sustainable" must also be considered (World Bank, 2020; OECD, 2019).

In light of these limitations, adaptation strategies have become central to the transfer of successful occupational health and safety practices. Adaptation refers to the redesign of good practices in a manner appropriate to local conditions, while preserving their fundamental principles. Sector-based prioritization, phased implementation, pilot projects, and capacity development programs are among the effective tools in this process. This approach enables countries to achieve long-term and sustainable improvements while remaining within their existing institutional and economic constraints (Dolowitz & Marsh, 2000; Walters et al., 2016).

From a policy perspective, the transfer of successful occupational health and safety practices is not merely about importing technical solutions. The real value lies in integrating the underlying governance logic, risk management understanding, and learning culture of these practices into local systems. Therefore, the fundamental question for policymakers should not be "which model should we adopt?" but rather "which principles should we adapt, with which tools, and at what pace?" This approach enables different countries to develop occupational health and safety systems that are informed by international experience but owned locally, within their own unique conditions (ILO, 2019; Gunningham & Sinclair, 2017).

In conclusion, the transferability of successful occupational health and safety practices is directly related to flexibility, contextual awareness, and organizational learning capacity. International experiences show that adaptable and phased strategies produce more lasting results than one-size-fits-all solutions. This finding emphasizes the importance of establishing a balanced bridge between global knowledge and local realities in the design of occupational health and safety policies.

7. IMPLICATIONS FOR POLICY AND PRACTICE

This section presents policy and practice implications applicable to different actors in the field of occupational health and safety, based on findings from previous comparative analyses. International success stories demonstrate that effective occupational health and safety systems depend not only on well-designed legislation, but also on how this legislation is interpreted and implemented by actors in the field.

7.1 Implications for Policymakers

The most fundamental conclusion for policymakers is that occupational health and safety legislation must be designed not only comprehensively but also in a way that is both enforceable and preventive. International best practices demonstrate that a risk-based regulatory approach enables more effective use of limited public resources and enhances the preventive impact of inspections. In this context, it is important for policymakers to develop flexible but accountable frameworks that guide employers to systematically manage risks, rather than creating detailed sets of rules (Black, 2005; ILO, 2019).

Furthermore, strengthening data-driven decision-making mechanisms in the design of occupational health and safety policies is a critical requirement. The reliable collection and analysis of accident, near-miss, and occupational disease data facilitates the determination of policy priorities. International experience shows that transparent reporting systems not only increase monitoring capacity but also support social awareness and organizational learning processes (OECD, 2019; HSE, 2019).

7.2 Implications for Organizations and Employers

The key takeaway for businesses and employers is that occupational health and safety should be viewed not merely as a legal obligation, but as a strategic investment in terms of corporate performance and sustainability. Successful examples demonstrate that when occupational health and safety management is embraced at the top management level and integrated into corporate decision-making processes, it leads to a sustained reduction in accident and illness rates. A lack of leadership, however, can render even the best-designed systems ineffective in the field (Hale, 2003; Cooper, 2000).

Another important conclusion for employers is the institutionalization of a preventive risk management approach. Risk assessments must be removed from being static documents and regularly updated according to changing working conditions; they must be supported by near-miss reports and employee feedback. This approach makes it possible to control risks before they arise, rather than intervening after accidents occur (Rasmussen, 1997; Robson et al., 2007).

7.3 Implications for OHS Professionals

The most important message that international best practices offer for occupational health and safety professionals is that their roles are not limited to technical expertise alone. In successful systems, OHS professionals are positioned as consultants, facilitators, and agents of change rather than in a supervisory or certifying role. This role involves building bridges between management and employees, promoting safe behaviors, and supporting a learning organization culture (Hopkins, 2006; Guldenmund, 2010).

However, it is becoming increasingly important for OSH professionals to develop their data literacy and analytical skills. The accurate interpretation of data generated through digital monitoring systems, performance indicators, and reporting tools is emerging as a critical area of competence in terms of prioritizing risks and providing meaningful feedback to management. Continuous professional development and interdisciplinary learning are indispensable elements of this transformation process (Hinze, Thurman, & Wehle, 2013; EU-OSHA, 2020).

8. FUTURE RESEARCH DIRECTIONS

The comparative analyses discussed in this section demonstrate significant progress in the field of occupational health and safety, while also highlighting unanswered questions and new research needs in the existing literature. Future studies should not be limited to merely describing successful practices; they should provide more in-depth analyses explaining under what conditions, through what mechanisms, and to what extent these practices are effective.

Comparative quantitative studies are still limited in the occupational health and safety literature. It is important that future research focuses on analyses based on longitudinal data sets covering variables such as accident rates, occupational disease incidence, and safety culture indicators across different countries and sectors. Such studies will contribute to a stronger demonstration of the causal relationships between policy interventions and concrete outcomes (Hämäläinen, Takala, & Saarela, 2017; Robson et al., 2007).

The integration of the Safety 4.0 approach into occupational health and safety systems constitutes an important focus area for future research. How artificial intelligence, big data analytics, the Internet of Things, and digital monitoring technologies can be used in risk prediction and preventive interventions should be addressed from both technical and ethical perspectives. In particular, the applicability and effectiveness of these technologies in countries with different levels of development should be evaluated through comparative studies (EU-OSHA, 2020; OECD, 2019).

The invisibility of occupational diseases remains one of the most important research gaps in the field of occupational health and safety. In many countries, occupational diseases are not sufficiently recognized or reported. Future studies should focus on the early diagnosis of occupational diseases, the development of recording systems, and the effectiveness of exposure-based monitoring approaches. This area is critical not only in terms of improving health outcomes but also in terms of assessing the long-term effects of occupational health and safety policies (Leigh, 2011; ILO, 2013).

Finally, the relationship between occupational health and safety and ESG (Environmental, Social, Governance) and sustainability is increasingly becoming a research agenda. The effects of occupational health and safety performance on corporate sustainability, social responsibility, and investor perception should be examined through interdisciplinary approaches. Future research should provide empirical evidence demonstrating that occupational health and safety is not merely an area of compliance but also a fundamental component of sustainable development (Gunningham, Kagan, & Thornton, 2004; United Nations, 2015).

9. CONCLUSION

This section examines occupational health and safety practices that have achieved success in reducing workplace accidents and occupational diseases in different countries and contexts, revealing the common dynamics behind these successes. The participatory and preventive models of Northern Europe, the risk-based and data-driven approaches of Anglo-Saxon countries, the technology-supported systems of East Asia, and the targeted interventions of developing economies demonstrate that the paths to success are not uniform, but are shaped around certain fundamental principles. The findings reveal that sustainable improvement in occupational health and safety is possible not only through technical measures but also through strong governance structures and learning systems (Hale, 2003; Walters et al., 2016).

One of the key messages of this study is that success is not accidental, but rather learnable and replicable. The examples examined show that occupational accidents and diseases should not be viewed as “inevitable” outcomes, but rather as events that can be largely prevented through the right strategies and systematic interventions. Countries that have achieved success have structures that identify risks early, involve workers in the process, use data-driven decision-making mechanisms, and establish a safety culture as an institutional value. Although these characteristics cannot be copied verbatim into different contexts, they can be adapted to other countries at a fundamental level (Reason, 2000; Rasmussen, 1997).

From a global occupational health and safety management perspective, these findings provide a strategic roadmap for the future. This roadmap includes strengthening preventive risk management, investing in leadership and safety culture, institutionalizing training and continuous learning processes, and promoting a data-driven management approach. Furthermore, it is critically important for countries to develop adaptable policies that take into account their own socio-economic and institutional conditions in order to maximize the benefits of international experience. This approach emphasizes that occupational health and safety is not only a legal obligation but also a fundamental component of sustainable development and work with human dignity (ILO, 2019; OECD, 2019).

In conclusion, the comparative analyses presented in this section highlight the need for flexible and holistic approaches based on global learning in the field of occupational health and safety. Lessons learned from success stories provide a strong reference framework that enables countries to develop solutions appropriate to their own contexts, rather than offering ready-made prescriptions. This framework contributes to advancing global efforts to reduce occupational accidents and diseases on a more effective, equitable, and sustainable basis.

REFERENCES

Black, J. (2005). The emergence of risk-based regulation and the new public risk management in the United Kingdom. *Public Law*, *2005*(3), 512–549.

Burke, M. J., Sarpy, S. A., Smith-Crowe, K., Chan-Serafin, S., Salvador, R. O., & Islam, G. (2011). Relative effectiveness of worker safety and health training methods. *American Journal of Public Health*, *101*(2), 315–324. DOI: 10.2105/AJPH.2004.059840 PMID: 16380566

Cooper, M. D. (2000). Towards a model of safety culture. *Safety Science*, *36*(2), 111–136. DOI: 10.1016/S0925-7535(00)00035-7

Dolowitz, D. P., & Marsh, D. (2000). Learning from abroad: The role of policy transfer in contemporary policy-making. *Governance: An International Journal of Policy, Administration and Institutions*, *13*(1), 5–24. DOI: 10.1111/0952-1895.00121

Dul, J., & Neumann, W. P. (2009). Ergonomics contributions to company strategies. *Applied Ergonomics*, *40*(4), 745–752. DOI: 10.1016/j.apergo.2008.07.001 PMID: 18775532

EU-OSHA. (2018). *Improving occupational safety and health in micro and small enterprises*. European Agency for Safety and Health at Work.

EU-OSHA. (2019). *International comparison of occupational safety and health strategies*. European Agency for Safety and Health at Work.

EU-OSHA. (2020). *Smart digital monitoring systems for occupational safety and health*. European Agency for Safety and Health at Work.

Frick, K. (2011). Worker influence on voluntary OHS management systems: A review of its ends and means. *Safety Science*, *49*(7), 974–987. DOI: 10.1016/j.ssci.2011.04.007

Grant, M. J., & Booth, A. (2009). A typology of reviews: An analysis of 14 review types and associated methodologies. *Health Information and Libraries Journal*, *26*(2), 91–108. DOI: 10.1111/j.1471-1842.2009.00848.x PMID: 19490148

Guldenmund, F. W. (2010). (Mis)understanding safety culture and its relationship to safety management. *Risk Analysis*, *30*(10), 1466–1480. DOI: 10.1111/j.1539-6924.2010.01452.x PMID: 20626685

Gunningham, N., Kagan, R. A., & Thornton, D. (2004). Social license and environmental protection: Why businesses go beyond compliance. *Law & Social Inquiry*, *29*(2), 307–341. DOI: 10.1111/j.1747-4469.2004.tb00338.x

Gunningham, N., & Sinclair, D. (2017). *Smart regulation*. Oxford University Press.

Hale, A. (2003). Safety management in production. *Human Factors and Ergonomics in Manufacturing, 13*(3), 185–201. DOI: 10.1002/hfm.10040

Hale, A., & Hovden, J. (1998). Management and culture: The third age of safety. *Safety Science, 34*(1–3), 129–165. DOI: 10.1016/S0925-7535(00)00020-5

Hämäläinen, P., Takala, J., & Saarela, K. L. (2017). Global estimates of occupational accidents and work-related illnesses 2014. *Journal of Occupational and Environmental Hygiene, 14*(7), 581–600. DOI: 10.1080/15459624.2017.1304579

Health and Safety Executive (HSE). (2019). *Managing for health and safety (HSG65)*. UK Health and Safety Executive.

Hinze, J., Thurman, S., & Wehle, A. (2013). Leading indicators of construction safety performance. *Safety Science, 51*(1), 23–28. DOI: 10.1016/j.ssci.2012.05.016

Hofstede, G. (2001). *Culture's consequences: Comparing values, behaviors, institutions, and organizations across nations* (2nd ed.). Sage Publications.

Hopkins, A. (2006). Studying organisational cultures and their effects on safety. *Safety Science, 44*(10), 875–889. DOI: 10.1016/j.ssci.2006.05.005

Hopkins, A. (2009). *Learning from high reliability organisations*. CCH Australia.

Hutter, B. M. (2011). *Managing food safety and hygiene: Governance and regulation as risk management*. Edward Elgar Publishing. DOI: 10.4337/9780857935717

International Labour Organization. (2013). *Prevention of occupational diseases*. ILO.

International Labour Organization. (2019). *Safety and health at the heart of the future of work*. ILO.

International Labour Organization. (2021). Safety and health through supply chains. ILO.

Karwowski, W. (2012). A review of human factors challenges of complex adaptive systems. *Human Factors, 54*(6), 983–995. DOI: 10.1177/0018720812467459 PMID: 23397808

Kim, Y., & Park, J. (2020). Industrial safety policies and digital transformation in East Asia. *Safety Science, 124*, 104587. DOI: 10.1016/j.ssci.2019.104587

Leigh, J. P. (2011). Economic burden of occupational injury and illness in the United States. *The Milbank Quarterly, 89*(4), 728–772. DOI: 10.1111/j.1468-0009.2011.00648.x PMID: 22188353

Liu, X., Wang, Z., Zhang, L., & Chen, Y. (2020). Real-time risk monitoring in industrial workplaces using IoT-based systems. *Journal of Cleaner Production, 258*, 120742. DOI: 10.1016/j.jclepro.2020.120742

OECD. (2019). *The future of work: Occupational safety and health.* Organisation for Economic Co-operation and Development.

OECD. (2021). *Reducing the precarity of work.* Organisation for Economic Co-operation and Development.

Park, J., Kim, S., & Cho, Y. (2019). Technology-driven occupational safety management systems in manufacturing industries. *Safety Science, 118*, 112–121. DOI: 10.1016/j.ssci.2019.05.012

Quinlan, M. (2014). *Ten pathways to death and disaster: Learning from fatal incidents in mines and other high hazard workplaces.* Federation Press.

Rasmussen, J. (1997). Risk management in a dynamic society: A modelling problem. *Safety Science, 27*(2–3), 183–213. DOI: 10.1016/S0925-7535(97)00052-0

Reason, J. (2000). Human error: Models and management. *BMJ (Clinical Research Ed.), 320*(7237), 768–770. DOI: 10.1136/bmj.320.7237.768 PMID: 10720363

Robson, L. S., Clarke, J. A., Cullen, K., Bielecky, A., Severin, C., Bigelow, P. L., Irvin, E., Culyer, A., & Mahood, Q. (2007). The effectiveness of occupational health and safety management system interventions: A systematic review. *Safety Science, 45*(3), 329–353. DOI: 10.1016/j.ssci.2006.07.003

Snyder, H. (2019). Literature review as a research methodology: An overview and guidelines. *Journal of Business Research, 104*, 333–339. DOI: 10.1016/j.jbusres.2019.07.039

Takala, J., Hämäläinen, P., Saarela, K. L., Yun, L. Y., Manickam, K., Jin, T. W., Heng, P., Tjong, C., Kheng, L. G., Lim, S., & Lin, G. S. (2014). Global estimates of the burden of injury and illness at work in 2012. *Journal of Occupational and Environmental Hygiene, 11*(5), 326–337. DOI: 10.1080/15459624.2013.863131 PMID: 24219404

United Nations. (2015). Transforming our world: The 2030 Agenda for Sustainable Development. UN.

Walters, D., & Nichols, T. (2007). Worker representation and workplace health and safety. *Industrial Relations Journal, 38*(4), 302–319. DOI: 10.1111/j.1468-2338.2007.00441.x

Walters, D., Wadsworth, E., & Quinlan, M. (2016). *Analysis of the determinants of workplace occupational safety and health practice in a selection of EU member states*. EU-OSHA.

World Bank. (2020). *Protecting workers from COVID-19 and other occupational risks*. World Bank Group.

Chapter 5
Risk Assessment and Hazard Prevention Techniques Across Industries

Shubhashree Sahoo
http://orcid.org/0000-0002-7859-2638
Sphoorthy Engineering College, India

T. C. Manjunath
http://orcid.org/0000-0003-2545-9160
Rajarajeswari College of Engineering, Bengaluru, India

P. Selvakumar
http://orcid.org/0000-0002-3650-4548
Department of Science and Humanities, Nehru Institute of Technology, Coimbatore, India

Subramaniam Seshan Iyer
http://orcid.org/0000-0002-3843-9399
Datta Meghe Institute of Management Studies, India

R. Shobarani
http://orcid.org/0000-0003-0183-3867
Dr. M.G.R. Educational and Research Institute, India

Rishabh Patel
Datta Meghe Institute of Management Studies, India

ABSTRACT

Risk assessment is a cornerstone of public health, clinical medicine, and pharmaceutical development, serving as a structured approach for identifying, evaluating, and managing the potential hazards associated with drugs, chemicals, or environmental exposures. In the context of drug safety and efficacy, risk assessment enables stakeholders—regulators, healthcare providers, manufacturers, and researchers—to make informed decisions by quantifying potential harms and weighing them against anticipated therapeutic benefits. The science of risk assessment has evolved significantly, driven by advancements in toxicology, pharmacology, epidemiology, and

DOI: 10.4018/979-8-3373-6370-7.ch005

computational modeling. This comprehensive exploration outlines the fundamental principles, methodologies, and practical applications of risk assessment, emphasizing its central role in safeguarding public health and promoting responsible innovation in medicine. Risk assessment refers to the systematic process of evaluating the likelihood and severity of adverse outcomes resulting from exposure to a particular agent under defined conditions.

UNDERSTANDING RISK ASSESSMENT: PRINCIPLES AND PRACTICES

Risk assessment is a cornerstone of public health, clinical medicine, and pharmaceutical development, serving as a structured approach for identifying, evaluating, and managing the potential hazards associated with drugs, chemicals, or environmental exposures. In the context of drug safety and efficacy, risk assessment enables stakeholders—regulators, healthcare providers, manufacturers, and researchers—to make informed decisions by quantifying potential harms and weighing them against anticipated therapeutic benefits. The science of risk assessment has evolved significantly, driven by advancements in toxicology, pharmacology, epidemiology, and computational modeling. This comprehensive exploration outlines the fundamental principles, methodologies, and practical applications of risk assessment, emphasizing its central role in safeguarding public health and promoting responsible innovation in medicine. Definition and Scope of Risk Assessment: Risk assessment refers to the systematic process of evaluating the likelihood and severity of adverse outcomes resulting from exposure to a particular agent under defined conditions. In the realm of pharmacology and healthcare, the term typically applies to evaluating the safety profile of drug substances, biologics, medical devices, or treatment protocols. The assessment often spans preclinical and clinical stages of development, as well as post-marketing surveillance. It encompasses a range of considerations, including toxicological effects, pharmacokinetics, pharmacodynamics, drug-drug interactions, demographic variability, and comorbidities. Risk assessments are used not only for regulatory approvals but also in guiding prescribing practices, formulating risk mitigation strategies, and informing patient education. Core Components of Risk Assessment: Risk assessment is generally structured around four interrelated components: hazard identification, dose-response assessment, exposure assessment, and risk characterization.

This involves determining whether a substance or intervention is inherently capable of causing adverse effects. This is typically based on data from animal studies, in vitro assays, human case reports, and mechanistic research. Dose-Response Assessment: This step quantifies the relationship between the magnitude of exposure and the probability or severity of the adverse effect. It helps establish thresholds such as the No Observed Adverse Effect Level (NOAEL) or Lowest Observed Adverse Effect Level (LOAEL). Exposure Assessment: This evaluates the extent, frequency, and duration of exposure to the agent under specific conditions. It includes factors such as route of administration, dosage, patient demographics, and environmental considerations. Risk Characterization: This final stage integrates data from the previous components to estimate the overall risk and its uncertainty. It provides a basis for making regulatory decisions or designing risk management strategies. Qualitative and Quantitative Approaches: Risk assessments can be qualitative, semi-quantitative, or fully quantitative, depending on the data availability and context. Qualitative risk assessment relies on expert judgment and categorical descriptors (e.g., "low", "moderate", "high" risk), often used when empirical data is limited.

Quantitative risk assessment (QRA) applies mathematical models to estimate numerical values of risk, such as incidence rates, risk ratios, or probability distributions. Modern pharmacovigilance and regulatory science increasingly favor data-driven, probabilistic models that provide transparent and reproducible risk estimates. Quantitative methods allow for scenario analyses, uncertainty modeling, and sensitivity testing, which are particularly valuable in benefit-risk decision-making.4. Risk Assessment vs. Risk Management: Although closely related, risk assessment and risk management serve different functions. Risk assessment is the scientific process of evaluating risks, while risk management involves implementing policies or actions to minimize those risks. For example, once a drug's hepatotoxicity risk is identified, risk management strategies might include liver function monitoring, contraindications, restricted distribution programs, or black box warnings. Regulatory authorities often mandate risk management plans (RMPs) or Risk Evaluation and Mitigation Strategies (REMS) as part of drug approval and marketing authorization. Thus, effective risk assessment lays the groundwork for evidence-based, targeted risk mitigation. Understanding risk assessment is essential for ensuring the safety and effectiveness of therapeutic interventions. By combining scientific rigor, structured methodologies, and ethical principles, risk assessment allows for evidence-based decision-making across the healthcare ecosystem. As medical science advances and healthcare becomes increasingly personalized, the practice of risk assessment must evolve accordingly—integrating new technologies, real-world data, and interdisciplinary perspectives. Ultimately, robust risk assessment supports the dual goals of innovation and patient safety, enabling societies to harness the full potential of medical advances while safeguarding public health.

THE IMPORTANCE OF RISK ASSESSMENT IN VARIOUS INDUSTRIES

Risk assessment is a foundational process across multiple industries, designed to identify, evaluate, and prioritize potential threats to human health, safety, organizational assets, the environment, and operational continuity. Whether applied to pharmaceuticals, manufacturing, finance, construction, energy, information technology, or public health, the role of risk assessment is universally critical (Nobanee et al., 2021). It enables decision-makers to proactively mitigate hazards, allocate resources efficiently, comply with regulations, and protect stakeholders from unforeseen consequences. As global operations become increasingly complex and interconnected, the demand for robust, systematic, and dynamic risk assessment practices continues to rise. This comprehensive exploration examines the importance of risk assessment across key industries, the specific methodologies they adopt, and the broader implications for economic sustainability, innovation, safety, and regulatory compliance. General Principles of Risk Assessment: At its core, risk assessment is a structured approach for identifying potential hazards, evaluating the likelihood and severity of adverse outcomes, and determining strategies to mitigate or manage these risks. It typically consists of four key steps: Hazard Identification: Recognizing potential threats or events that could cause harm. Risk Analysis: Evaluating the probability and impact of the hazard materializing. Risk Evaluation: Comparing the estimated risk against benchmarks or acceptable levels. Risk Control or Treatment: Implementing preventive or corrective actions to reduce risk to an acceptable level.

These principles are applied universally but tailored according to the industry's unique requirements, operational context, and stakeholder concerns (Huy et al., 2021). Pharmaceutical and Healthcare Industry: In the pharmaceutical and healthcare sectors, risk assessment is indispensable in drug development, clinical trials, pharmacovigilance, and patient safety. The consequences of overlooking risks can result in adverse drug reactions, product recalls, or harm to patients. Preclinical and Clinical Phases: Risk assessments evaluate toxicity, dosage limits, and side effects. Post-Marketing Surveillance: Real-world evidence and pharmacovigilance systems assess long-term safety. Regulatory Compliance: Risk assessments support submissions to agencies like the FDA, EMA, and WHO, and underpin Risk Management Plans (RMPs) or Risk Evaluation and Mitigation Strategies (REMS). Effective risk assessment in healthcare not only ensures the safety of therapeutic interventions but also fosters trust among patients and providers. 3. Construction and Infrastructure: Construction is one of the most hazardous industries, where risk assessment helps prevent accidents, ensure regulatory compliance, and protect worker welfare. Site Risk Assessments: Identify physical dangers (e.g., falls, structural collapse, electrical hazards) (Alvand et al, 2023). Environmental Impact Assessments (EIA): Evaluate risks related to land use, water resources, and ecosystem disruption. Operational Risks: Include supply chain issues, labor shortages, or equipment failures. By integrating safety management systems with risk assessments, construction firms reduce incidents, improve project timelines, and enhance stakeholder confidence.

In the financial sector, risk assessment is fundamental for evaluating market volatility, creditworthiness, operational integrity, and compliance. Market Risk: Assesses exposure to fluctuations in interest rates, exchange rates, and asset prices. Credit Risk: Analyzes the likelihood of a borrower defaulting on their obligations (Naim, 2022). Operational Risk: Involves evaluating internal processes, IT systems, and human error. Regulatory Risk: Ensures compliance with laws like Basel III, Dodd-Frank, or AML guidelines. Risk modeling, stress testing, and scenario analysis are commonly employed to guide investment decisions, pricing models, and policy frameworks. Energy and Utilities: Energy production—particularly in oil, gas, and nuclear sectors—presents considerable risks due to the potential for environmental damage, catastrophic failures, and worker injuries. Process Hazard Analysis (PHA): Identifies risks in chemical or mechanical operations. Failure Modes and Effects Analysis (FMEA): Evaluates how equipment failures could lead to system breakdowns. Environmental Risk Assessment: Examines emissions, spills, and ecological impacts. Risk assessment ensures that energy companies meet environmental standards, avoid costly disruptions, and maintain a social license to operate. The importance of risk assessment across various industries cannot be overstated. It is a foundational tool for ensuring safety, supporting innovation, enhancing resilience, and achieving compliance with regulatory and ethical standards (Kunnathully et al., 2025). Whether preventing workplace accidents, safeguarding public health, or protecting digital infrastructure, robust risk assessment empowers organizations to navigate complexity and uncertainty with confidence. As industries evolve and confront emerging global challenges—from climate change to cyber warfare—the role of risk assessment will only become more central. Proactive, interdisciplinary, and technologically advanced approaches to risk assessment are vital for building a safer, more sustainable, and future-ready world.

HAZARD IDENTIFICATION: METHODS AND TOOLS

Hazard identification is a fundamental component of risk management processes across industries and disciplines. It involves the systematic process of recognizing potential sources of harm or adverse effects to people, property, processes, or the environment. The accuracy and completeness of hazard identification directly influence the quality of subsequent risk analysis and control strategies. Effective hazard identification not only reduces the likelihood of accidents and system failures but also enhances organizational resilience, regulatory compliance, and stakeholder confidence (Wahab et al., 2021). Across domains—such as healthcare, manufacturing, construction, energy, transportation, environmental management, and information technology—hazard identification serves as the first and often the most crucial step in ensuring operational safety and sustainability. This comprehensive overview explores the core principles of hazard identification, widely used methodologies and tools, applications in various sectors, integration with modern technologies, and emerging trends shaping its future. 1. Fundamentals of Hazard Identification: A *hazard* is any source of potential damage, harm, or adverse health effects on something or someone under certain conditions. Hazard identification aims to determine what hazards exist or may potentially arise in a system or environment. Hazards may be categorized into several types: Physical Hazards: Such as noise, heat, radiation, or vibration. Chemical Hazards: Exposure to toxic, reactive, flammable, or explosive substances. Biological Hazards: Bacteria, viruses, fungi, or biological toxins (Panchal et al., 2022). Ergonomic Hazards: Poor workstation design, repetitive motions, or improper lifting. Psychosocial Hazards: Workplace stress, harassment, or violence. Mechanical Hazards: Involving equipment, machines, or falling objects. Environmental Hazards: Natural disasters, pollution, or ecological degradation. The purpose of hazard identification is to recognize these threats before they result in accidents or harm, enabling effective mitigation.

Common Methods of Hazard Identification: Numerous systematic approaches and methodologies are employed across industries for identifying hazards. The choice of method depends on the complexity of the system, the nature of the operations, and the level of risk tolerance. a. Preliminary Hazard Analysis (PHA): PHA is a broad, qualitative technique used early in project development. It identifies potential hazards and evaluates the possible effects of failure or system malfunction. Pros: Rapid, cost-effective, and suitable for initial project phases. Cons: Less detailed; may overlook rare but high-consequence hazards. b. Hazard and Operability Study (HAZOP): HAZOP is a structured, team-based method commonly used in chemical, petrochemical, and process industries. It examines potential deviations from design intent using guidewords (e.g., “more,” “less,” “none”). Strengths: Highly detailed; encourages thorough exploration. Limitations: Time-consuming and requires experienced facilitators. c. Failure Mode and Effects Analysis (FMEA): FMEA identifies all possible failure modes within a system and assesses their effects and causes. It is widely used in manufacturing, aerospace, and healthcare (EFSA Scientific Committee et al., 2021). Application: Focuses on reliability and prioritizes risks using Risk Priority Numbers (RPNs). Advantage: Quantifies risk based on severity, occurrence, and detectability. d. Fault Tree Analysis (FTA): FTA is a top-down deductive approach that begins with a potential undesirable event and analyzes the system's failures leading to it. Use Case: Suitable for analyzing complex systems with interdependencies. Benefit: Graphical representation makes it intuitive and traceable. e. What-If Analysis: A brainstorming method in which experts ask "what if" questions to explore different failure scenarios. Example: "What if the power supply fails?" or "What if the operator forgets to close a valve?" Usefulness: Encourages creative thinking and scenario development

Tools and Technologies Supporting Hazard Identification: Modern hazard identification is increasingly supported by advanced tools and digital platforms that enhance precision, efficiency, and real-time monitoring. a. Digital Twins: A digital twin is a virtual model of a physical process, asset, or system. It allows simulation and real-time monitoring to detect deviations and emerging hazards (Scannell et al., 2021). b. Artificial Intelligence (AI) and Machine Learning: AI can process vast amounts of data to detect patterns and predict potential hazards. Machine learning algorithms improve over time by learning from new incidents and safety reports. c. Sensor Technology and Internet of Things (IoT): Connected sensors monitor environmental conditions (e.g., gas leaks, temperature spikes, structural stress), enabling continuous hazard detection. d. Drones and Remote Imaging: Used in inaccessible or hazardous environments (like mines or offshore rigs), drones can identify physical and structural hazards safely. e. Augmented Reality (AR) and Virtual Reality (VR): AR and VR enable immersive training for hazard recognition and emergency response planning. f. Hazard Identification Software: Specialized software platforms (e.g., BowTieXP, PHA-Pro, RiskWatch) provide structured frameworks for documenting, visualizing, and analyzing hazards. 4. Challenges in Hazard Identification: Human Factors: Biases, oversights, or lack of expertise can lead to missed hazards. Evolving Systems: Rapid technological change introduces new hazards not yet well-understood. Interconnected Risks: Complex systems may hide cascading failures that are hard to detect. Data Gaps: Incomplete or outdated information hampers accurate identification. Overcoming these challenges requires interdisciplinary collaboration, continuous training, and adaptive hazard identification models. Hazard identification is the cornerstone of any robust risk management system (Kotte et al., 2025). It empowers organizations to detect threats early, design appropriate controls, and safeguard human life, infrastructure, and the environment. The methods and tools used in hazard identification continue to evolve, incorporating advanced technologies, interdisciplinary frameworks, and dynamic analytics. By investing in comprehensive hazard identification strategies—tailored to their operational context—industries can not only prevent harm but also drive innovation, ensure compliance, and build resilient systems that stand the test of time. As risks become more interconnected and unpredictable in a globalized world, the importance of proactive, intelligent, and collaborative hazard identification will only continue to grow.

RISK ANALYSIS: QUALITATIVE AND QUANTITATIVE APPROACHES

Risk analysis is a critical function in decision-making processes across various industries and disciplines, enabling organizations to evaluate potential threats and uncertainties and determine appropriate actions for mitigation or acceptance. It provides the structured assessment necessary for identifying, understanding, and managing risks that may affect project goals, safety standards, regulatory compliance, financial performance, or operational continuity (Fjeld et al., 2023). Risk analysis is generally categorized into two major approaches: qualitative and quantitative. Each approach has its unique methodologies, tools, advantages, and limitations, and is often applied depending on the complexity of the system, the availability of data, and the required level of precision. This discussion provides a comprehensive overview of qualitative and quantitative risk analysis approaches, comparing their characteristics, exploring their use in real-world scenarios, and emphasizing their complementary role in comprehensive risk management systems. Understanding Risk Analysis: Risk analysis involves a systematic process of identifying potential hazards, assessing the likelihood of their occurrence, and evaluating their potential consequences. It supports informed decision-making by enabling stakeholders to understand: What can go wrong? How likely is it to happen? What will be the impact if it does happen The outcome of risk analysis serves as the basis for prioritizing risks and implementing mitigation strategies, thereby reducing uncertainty and optimizing resource allocation. 2Qualitative Risk Analysis Definition: Qualitative risk analysis is a subjective assessment method that relies on descriptive data, expert judgment, and categorical scales (e.g., low, medium, high) to evaluate risks (Crotty, 2022) . Key Features: Does not require extensive numerical data. Useful in the early stages of risk management. Provides a rapid and cost-effective overview of risk exposure. Often used to rank and prioritize risks. Methods and Tools: a. Risk Matrices: A risk matrix (or risk heat map) visually represents the probability and impact of risks in a two-dimensional grid. It helps categorize risks into various zones (e.g., acceptable, tolerable, unacceptable). b. SWIFT (Structured What-If Technique): This brainstorming tool is used to anticipate potential risk scenarios through structured questions like "What if this system fails?" It is especially useful for identifying process and operational risks. c. Delphi Technique: A structured communication method where a panel of experts anonymously provides risk assessments over several rounds. This reduces bias and improves the reliability of qualitative judgments. d. Checklists and Expert Interviews: Checklists ensure systematic consideration of common risks, while expert interviews elicit qualitative insights based on experience, intuition, and domain knowledge.

Categorical Ranking: Risks are assessed using verbal scales for severity and likelihood (e.g., rare, unlikely, possible, likely, almost certain). Each category is defined to ensure consistency. Strengths of Qualitative Risk Analysis: Simple to apply with minimal resources. Valuable when data is scarce or unavailable (Luo et al., 2023). Encourages team participation and expert insight. Facilitates communication with non-technical stakeholders. Limitations: Subjective and prone to bias or inconsistency. Lacks numerical precision, which limits detailed cost-benefit analysis. May oversimplify complex risk interdependencies. 3. Quantitative Risk Analysis: Definition: Quantitative risk analysis involves numerical methods to estimate risk levels, often in terms of probability distributions, expected monetary value (EMV), or statistical likelihoods. It offers a more rigorous and data-driven understanding of risk. Key Features: Requires detailed, high-quality data.Provides measurable outcomes. Supports simulation and scenario analysis. Useful for complex or high-stakes decisions. Methods and Tools: a. Monte Carlo Simulation: A probabilistic modeling technique that uses random sampling to simulate a range of outcomes for uncertain variables. It generates probability distributions for risk impacts, offering insights into worst-case, best-case, and most-likely scenarios. b. Fault Tree Analysis (FTA): A top-down, deductive technique used to analyze the probability of a specific system failure. FTA models are used extensively in engineering and safety-critical systems.

Event Tree Analysis (ETA): A forward-looking, inductive method that explores the consequences of an initiating event and calculates the likelihood of each outcome. d. Bayesian Networks: These are graphical models that use Bayesian inference to update the probability of events based on new evidence (Maranzoni et al., 2023). Useful in dynamic environments where risks evolve over time. e. Expected Monetary Value (EMV): A financial technique used to calculate the weighted average of all possible outcomes, each multiplied by its probability. It helps compare alternative decisions under uncertainty. f. Sensitivity Analysis: This evaluates how changes in one or more variables affect risk outcomes. It helps identify the most influential factors contributing to uncertainty. Strengths of Quantitative Risk Analysis: Objective and reproducible results. Enables detailed financial and operational decision-making. Supports advanced modeling and optimization techniques. Helps identify risk drivers and prioritize controls based on numerical impact. Limitations: Requires significant data and computational resources. Complex methods may be difficult for non-experts to interpret. Models can be misleading if assumptions or data inputs are flawed (Hubballi et al., 2025). 4. Comparative Overview: Qualitative vs. Quantitative Approaches: These approaches are not mutually exclusive.

RISK EVALUATION: PRIORITIZING HAZARDS AND CONTROLS

Risk evaluation is a pivotal phase within the broader framework of risk management, serving as the crucial bridge between risk analysis and the development of risk mitigation strategies. It involves comparing estimated levels of risk against predefined criteria to determine their significance and the necessity for control measures (Pascarella G., et al, 2021). By systematically prioritizing hazards and identifying the most appropriate controls, organizations can allocate resources effectively, enhance operational safety, comply with legal and regulatory requirements, and maintain stakeholder confidence. Whether applied in healthcare, engineering, manufacturing, finance, information technology, or environmental management, risk evaluation provides the foundation for evidence-based decision-making. This detailed exploration unpacks the principles, methods, and real-world applications of risk evaluation, highlighting its significance in the systematic prioritization of hazards and selection of controls to minimize adverse impacts. Understanding Risk Evaluation: Risk evaluation is the process of assessing the outcomes from risk analysis—whether qualitative, quantitative, or both—and determining their acceptability or the need for further action. It integrates technical assessments with organizational context, stakeholder values, regulatory frameworks, and risk appetite. The main goals of risk evaluation are: To determine which risks are tolerable and which require mitigation. To support the selection of appropriate risk control measures. To prioritize risks for resource allocation and action. Risk evaluation is not solely about ranking hazards by severity but involves considering multiple dimensions, such as the probability of occurrence, magnitude of impact, detectability, reversibility, and cost of control (Achumie et al., 2022). 2. Risk Evaluation Criteria: Establishing well-defined evaluation criteria is essential for consistent and rational decision-making. These criteria are often influenced by: Legal requirements: Laws and industry standards often define acceptable risk thresholds (e.g., occupational exposure limits). Organizational policies: Internal risk tolerance levels or codes of conduct may set stricter benchmarks. Societal values: In areas like healthcare or environmental protection, public perceptions and ethics play a key role. Economic considerations: The cost-benefit tradeoff of mitigating a risk versus accepting it. Commonly used evaluation frameworks include: ALARP (As Low As Reasonably Practicable): A widely adopted principle in high-risk industries such as aviation and nuclear power, stating that risk should be reduced unless the cost is grossly disproportionate to the benefit. Risk matrices: Combining likelihood and consequence to categorize risks and inform thresholds for action. Risk acceptance criteria: Predefined numerical or qualitative limits that help determine when a risk is considered acceptable, tolerable with controls, or unacceptable.

Figure 1. Steps of Risk Assessment

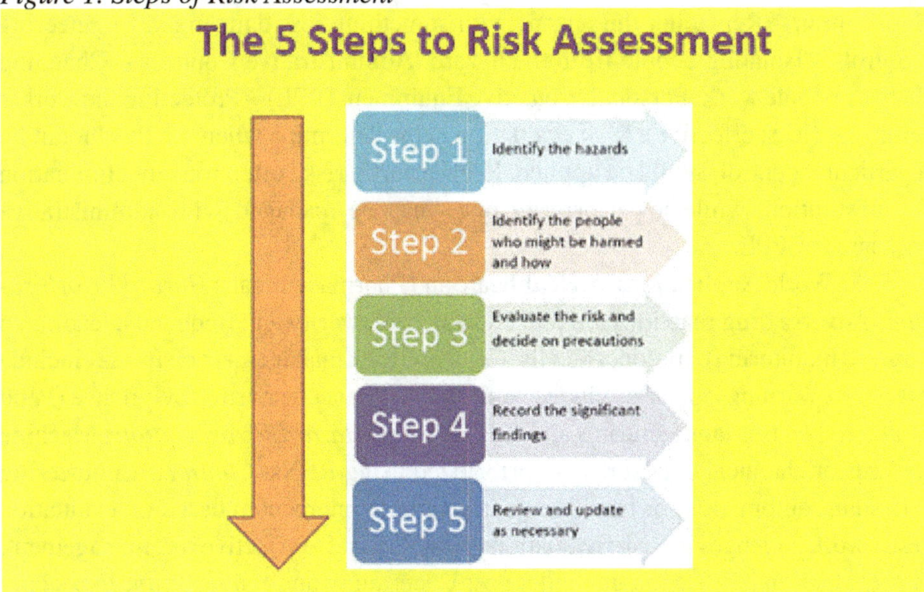

Prioritizing Hazards: Once hazards have been identified and analyzed, risk evaluation ranks them based on their significance. This prioritization helps organizations: Focus efforts on the most critical threats. Avoid over-investing in negligible risks. Ensure compliance with legal and industry mandates. Optimize the use of limited safety or financial resources. Key Approaches to Prioritization: a. Risk Ranking Systems: Systems like Failure Modes and Effects Analysis (FMEA) assign scores for severity, occurrence, and detectability of each risk, and multiply them to calculate a Risk Priority Number (RPN). Higher RPNs indicate greater risk. b. Multi-Criteria Decision Analysis (MCDA): MCDA methods consider multiple factors beyond just probability and severity—such as societal impact, stakeholder concern, or innovation potential—using weighted scoring systems for prioritization. c. Pareto Analysis (80/20 Rule): Based on the principle that a small number of hazards may account for a large proportion of potential harm. Helps identify which risks to address first for maximum effect. d. Bowtie Method: A visual tool that maps the causes and consequences of a central risk event. It aids in identifying critical control points for prioritization (Wahyuni et al., 2025). e. Heat Maps: Visual representations using color gradients to show the relative severity of various risks, enabling intuitive comparison and prioritization. 4. Control Measures: Hierarchy and Selection: Once hazards are prioritized, the next step is selecting and implementing controls to manage the most significant risks. The hierarchy of controls, a framework widely used in occupational health and safety, ranks controls based on effectiveness: Elimination – Physically removing the hazard (most effective). Substitution – Replacing the hazard with something less dangerous. Engineering Controls – Isolating people from the hazard. Administrative Controls – Changing the way people work. Personal Protective Equipment (PPE) – Protecting the worker with gear (least effective). Risk evaluation helps determine where on this hierarchy a particular control should be applied. High-priority risks often warrant elimination or substitution, while lower-priority ones may be managed with administrative measures or PPE.

Real-World Applications: a. Healthcare and Pharmaceuticals: *Hazard Prioritization:* Adverse drug reactions are ranked based on severity and frequency, leading to targeted monitoring or product recalls. *Controls:* Risk management plans may include black box warnings, prescribing restrictions, or enhanced pharmacovigilance (Peng et al., 2022). b. Manufacturing and Engineering: *Hazard Prioritization:* Machine hazards or chemical exposures are prioritized using RPNs. *Controls:* Engineering redesigns, automation, or machine guarding are implemented based on evaluation results. 6. Challenges in Risk Evaluation: While critical to effective risk management, risk evaluation is not without challenges: Uncertainty and Subjectivity: Especially in qualitative assessments, judgments may be biased or based on incomplete data. Dynamic Risk Landscapes: New hazards may emerge, and existing risks may change over time due to evolving conditions. Complex Interdependencies: Some risks interact in non-linear ways, complicating prioritization. Stakeholder Conflicts: Different stakeholders may have varying perceptions of acceptable risk, complicating consensus. Risk evaluation plays a vital role in shaping proactive, evidence-based responses to the multitude of hazards that organizations face (Sakthivel et al., 2025). By enabling the structured prioritization of threats and guiding the selection of proportionate and effective controls, it supports organizational resilience, regulatory compliance, and stakeholder trust. The integration of qualitative and quantitative methods, the application of transparent criteria, and the adoption of digital technologies all contribute to making risk evaluation a powerful tool in today's complex and fast-paced environments. Whether aiming to protect public health, financial stability, technological infrastructure, or environmental integrity, a well-executed risk evaluation framework ensures that resources are focused where they are needed most, thereby minimizing harm and enhancing performance across all sectors.

ELIMINATION AND SUBSTITUTION: REMOVING HAZARDS

In the hierarchy of risk controls, elimination and substitution stand as the most effective strategies for managing workplace hazards and reducing occupational risk. These two methods aim to address hazards at their source rather than merely controlling exposure or minimizing consequences. While often used interchangeably in casual discussions, elimination and substitution represent distinct yet complementary approaches to risk mitigation: elimination involves completely removing a hazard from a process, activity, or environment, whereas substitution entails replacing a hazardous substance, material, or procedure with one that is significantly less dangerous (Syeda et al., 2022). These strategies are fundamental to proactive safety management and are increasingly emphasized in regulations, industry standards, and risk assessment frameworks across sectors such as manufacturing, healthcare, construction, chemical processing, and pharmaceuticals. This detailed discussion explores the rationale, implementation, challenges, and real-world impact of elimination and substitution as top-tier risk control measures. Understanding Elimination and Substitution: a. Elimination: The Ultimate Control: Elimination refers to physically removing a hazard from the workplace so that exposure risk is entirely negated. This might involve removing a toxic chemical from a process, discontinuing a dangerous task, or designing equipment or procedures in a way that the hazard never arises. Since no exposure exists, the risk is reduced to zero. Examples: Designing machinery that operates without sharp edges or rotating parts. Automating a task that previously required manual handling of hazardous materials.

Redesigning a workflow to avoid working at heights. b. Substitution: Replacing with Safer Alternatives: Substitution involves replacing a hazardous agent or process with a safer one. While the hazard may not be completely eliminated, the risk is significantly reduced through the use of less harmful materials or less risky methods. Examples: Replacing lead-based paint with a water-based, non-toxic alternative. Using robots instead of human workers in high-risk zones. Switching from solvent-based cleaners to biodegradable, low-VOC (volatile organic compound) detergents (Carton et al., 2023). Why Elimination and Substitution Are Preferred: The hierarchy of controls, a widely accepted framework in occupational health and safety, ranks hazard controls in order of effectiveness: The reason for placing elimination and substitution at the top of the hierarchy is simple: they address hazards at the source, before exposure occurs, reducing dependence on human behavior or engineering systems. Unlike PPE or administrative measures that may fail due to misuse or non-compliance, eliminated or substituted hazards present a much lower risk of recurrence. Integrating Elimination and Substitution into Risk Assessment: Effective application of elimination and substitution requires their consideration early in the design and planning stages of any process or product. This is often referred to as inherently safer design, a concept promoted by process safety professionals and regulators. During hazard identification and risk assessment: Ask whether the task or material is necessary. Consider alternative methods or materials with a lower risk profile. Conduct a comparative risk analysis to evaluate the net benefit of substitution. Example: In chemical safety, if a volatile and flammable solvent is being used in cleaning operations, teams might evaluate whether the cleaning is necessary at all (elimination) or whether a non-flammable solvent could be used instead (substitution).

Challenges and Barriers to Implementation: While elimination and substitution are ideal, they are not always feasible due to technical, operational, or economic constraints: Cost of Redesign: Redesigning a product or process to eliminate a hazard may require significant investment (Hassan, 2023). Availability of Alternatives: Suitable substitute materials or technologies may not exist or may be unproven. Regulatory Hurdles: Some industries, particularly pharmaceuticals or aerospace, face strict regulatory frameworks that limit change. Product Performance Concerns: Substitutes may not offer the same effectiveness or durability as original materials. Organizations must weigh these constraints against the potential risk reduction, often using structured decision-making frameworks such as cost-benefit analysis or risk-benefit analysis. Elimination and substitution are foundational pillars of effective risk control and should be the first consideration in any risk management strategy. By removing hazards entirely or replacing them with less harmful alternatives, organizations can dramatically reduce the potential for injury, illness, and environmental harm (Grant et al., 2021). Although implementation may present technical or economic challenges, the long-term benefits—including compliance, productivity, reputation, and human well-being—are substantial. Whether in chemical laboratories, hospital operating rooms, construction sites, or industrial plants, the adoption of elimination and substitution reflects a proactive, sustainable approach to safety and risk reduction. As industries evolve and new hazards emerge, the continual pursuit of safer alternatives through innovation and collaboration remains essential to protecting workers, communities, and the environment.

ENGINEERING CONTROLS: DESIGNING SAFER SYSTEMS

Engineering controls are a critical component of the hierarchy of hazard control measures and serve as the first line of defense after elimination and substitution. While elimination and substitution aim to remove or replace hazards entirely, engineering controls focus on physically isolating people from hazards or modifying equipment, processes, or environments to minimize risk (van Gelder et al., 2021). These controls do not rely on human behavior or administrative rules for effectiveness, making them more reliable and sustainable over time. Through thoughtful design, integration of protective systems, and the use of modern technologies, engineering controls can drastically reduce exposure to hazardous substances, machinery, noise, radiation, and other workplace dangers. This comprehensive discussion explores the principles, types, implementation strategies, and real-world applications of engineering controls across various industries. 1. Understanding Engineering Controls: Engineering controls are mechanical or physical systems designed to eliminate or reduce exposure to hazards in the workplace. They function by: Enclosing hazardous processes or equipment, Isolating the source of the hazard from the worker, Ventilating or filtering harmful substances, Automating dangerous tasks to minimize human involvement, or Redesigning workplace layouts to eliminate points of contact with hazards. Unlike administrative controls (e.g., training or shift rotation) or PPE (e.g., gloves, respirators), engineering controls do not depend on individual compliance, making them more consistent in mitigating risks (Vincoli, 2024). 2. Principles of Effective Engineering Control Design: To be effective, engineering controls must be: Technically Feasible: Practical to implement given current technology and infrastructure. Reliable: Operate consistently under normal and extreme conditions. Integrated: Built into the process or system itself rather than added on later. Maintainable: Designed for easy inspection, cleaning, repair, and replacement. Fail-Safe: Incorporate redundancies or alarms in case of malfunction. A successful engineering control design should account for all modes of exposure, including inhalation, dermal contact, ingestion, or physical trauma. 3. Types of Engineering Controls: Engineering controls vary widely depending on the industry, hazard type, and operational context. Some common categories include: a. Ventilation Systems: Local Exhaust Ventilation (LEV): Captures contaminants at or near the source (e.g., fume hoods, downdraft tables). General/Dilution Ventilation: Reduces airborne contaminant concentrations by mixing and diluting them with clean air.

High-Efficiency Particulate Air (HEPA) Filtration: Removes fine particles, biological agents, or hazardous aerosols. b. Machine Guards and Barriers Fixed Guards: Provide a permanent physical barrier to moving parts. Interlocked Guards: Disable machinery when the guard is removed. Two-Hand Controls: Require both hands to be on controls, keeping them away from danger zones (Brauer, 2022). c. Enclosures and Isolation: Soundproof Enclosures: Reduce noise exposure in industrial settings. Biological Safety Cabinets: Protect lab workers from infectious materials. Radiation Shielding: Uses lead or other materials to block harmful emissions. d. Automation and Robotics: Replace manual tasks that expose workers to hazardous conditions (e.g., handling toxic chemicals, working in confined spaces, lifting heavy loads). e. Ergonomic Redesign: Adjustable workstations, anti-fatigue mats, or mechanical assist devices reduce strain and injury risk. f. Fail-Safe Systems and Alarms: Include automatic shut-offs, over-pressure relief valves, or temperature cut-outs to prevent catastrophic failure. 4. Engineering Controls in Key Industries: Needleless systems and safety-engineered sharps prevent needlestick injuries. Negative pressure rooms contain airborne infectious agents like tuberculosis or COVID-19. Lead aprons and shielding in radiology departments protect staff from X-ray exposure. Construction: Guardrails and toe boards on scaffolds prevent falls. Dust suppression systems minimize silica exposure during cutting or grinding. Noise barriers shield workers from high-decibel machinery. Manufacturing: Machine enclosures isolate workers from noise, heat, or mechanical movement. Robotic arms reduce the need for manual handling of dangerous components. Interlocked safety doors prevent access to operating machines. Chemical Industry: Closed transfer systems for liquid chemicals prevent spills and vapor exposure. Explosion-proof enclosures reduce fire and blast risk. Gas detection and suppression systems alert operators to leaks or combustion risk.

Challenges in Implementation: Despite their effectiveness, several challenges may hinder the adoption of engineering controls: High Capital Costs: Initial investments in equipment, design, and installation can be substantial. Retrofit Complexity: Modifying existing systems is often more difficult than integrating controls during initial design. Maintenance Requirements: Controls require regular inspection and upkeep to remain effective. Resistance to Change: Employees or management may resist process redesigns or new technologies. Limited Space or Infrastructure: Some facilities may lack the space or utilities to support new systems. To overcome these challenges, organizations must prioritize risk-based decision-making, explore innovative engineering solutions, and ensure strong leadership support (Onyeke et al., 2022). Engineering controls form a cornerstone of proactive hazard management, offering robust, consistent, and often automated solutions to complex safety challenges. By physically isolating workers from harmful exposures, improving environmental conditions, and enhancing process reliability, engineering controls significantly reduce the likelihood of injury, illness, and equipment damage. Although their implementation may present upfront costs and logistical hurdles, the long-term benefits—improved worker safety, regulatory compliance, reduced liabilities, and operational efficiency—are compelling. As technologies advance and industries evolve, the role of engineering controls will only grow, underscoring the need for continued innovation, investment, and integration in designing safer systems for tomorrow's workplaces.

ADMINISTRATIVE CONTROLS: PROCEDURES AND TRAINING

Administrative controls are essential components in the hierarchy of hazard control strategies and serve as a critical layer of protection when hazards cannot be eliminated, substituted, or fully controlled by engineering means. Unlike physical controls, administrative controls focus on modifying human behavior through the implementation of policies, procedures, training programs, scheduling strategies, supervision, and organizational practices (Mayta-Huiza et al., 2023). Their goal is to reduce the frequency, duration, and intensity of exposure to hazardous conditions by regulating how tasks are performed and by whom. Although they are generally considered less effective than engineering controls—primarily because they rely on human compliance—they play a vital and often complementary role in achieving a safe and healthy working environment. This comprehensive overview explores the principles, components, implementation strategies, limitations, and best practices for administrative controls across diverse industries. Understanding Administrative Controls: Administrative controls are non-physical interventions that aim to influence behavior and decision-making to reduce risks. They include a wide range of strategies such as: These controls are most effective when used in combination with engineering controls and personal protective equipment (PPE), forming a multi-layered safety strategy that enhances resilience against failure in any single control system. The Role of Policies and Procedures: At the core of administrative controls are policies and procedures that standardize safety expectations and ensure consistency across organizational activities. These may include: Standard Operating Procedures (SOPs): Clearly written, step-by-step instructions for performing specific tasks safely. Work Permits and Checklists: Used in high-risk activities like confined space entry, hot work, and electrical maintenance. Hazard Communication Policies: Ensure that workers are aware of chemical hazards through Safety Data Sheets (SDSs) and proper labeling. Incident Reporting Protocols: Encourage timely reporting and response to near-misses and accidents. Documented procedures must be accessible, regularly reviewed, and revised in response to changes in the work environment, technology, or regulations (Murariu et al., 2021). 3. Training and Competency Development: Training is one of the most critical forms of administrative control, as it equips workers with the knowledge and skills to recognize and mitigate hazards. Effective training programs include: Orientation Training: Introduces new employees to workplace hazards, rules, and emergency protocols. Job-Specific Training: Tailored to the unique risks of particular roles or equipment. Refresher Training: Reinforces concepts periodically or after incidents. Emergency Response Training: Prepares staff to react swiftly in fire, chemical spill, or evacuation scenarios. Soft Skills and Human Factors Training: Includes communication, situational awareness, and stress management. Training must be interactive, practical, and tailored to the literacy and language levels of the workforce. Competency should be verified through assessments, observations, and certifications.

Work Scheduling and Exposure Control: One powerful administrative approach to risk mitigation is limiting the duration and timing of exposure. Examples include: Job Rotation: Prevents repetitive strain injuries and reduces exposure to hazardous tasks by rotating workers. Rest Breaks: Reduce fatigue, which is a significant contributor to accidents (Bartolomé et al., 2022). Shift Scheduling: Avoids long or overnight shifts that may impair cognitive performance. Restricted Work Zones: Controls access to high-risk areas through signage, barriers, and security protocols.

These methods manage how long, how often, and how intensely workers interact with hazards, especially in environments where complete hazard elimination is not feasible. 5. Supervision and Safety Culture: Strong supervision plays a vital role in ensuring the effectiveness of administrative controls. Responsibilities include: An organization's safety culture—the shared values, beliefs, and norms around safety—directly influences the effectiveness of administrative controls. A strong culture promotes: Administrative controls are a vital aspect of occupational risk management, providing essential safeguards through policy-making, structured training, behavioral guidance, and organizational discipline. While they do not physically eliminate or isolate hazards, they shape how work is performed in a way that minimizes risk. In many situations—especially when engineering controls are impractical—administrative measures like training, SOPs, and scheduling are the most feasible and cost-effective defense available (Nani et al., 2021). However, because they rely on human adherence, they require constant reinforcement, strong leadership, and a proactive safety culture. When strategically integrated into a comprehensive hazard control plan, administrative controls can significantly enhance workplace safety and support a culture of continuous improvement, awareness, and accountability.

PERSONAL PROTECTIVE EQUIPMENT (PPE): SELECTION AND USE

Personal Protective Equipment (PPE) serves as the last line of defense in hazard control hierarchies, offering a critical safeguard when risks cannot be fully eliminated through engineering or administrative controls. PPE includes a broad spectrum of wearable gear—ranging from gloves, respirators, goggles, and hard hats to specialized suits and hearing protection—designed to protect workers from physical, chemical, biological, radiological, and ergonomic hazards (Brisbine et al., 2022). While PPE does not eliminate the hazard itself, it significantly reduces the likelihood of injury, illness, or fatality by acting as a barrier between the user and harmful exposures. The proper selection, use, maintenance, and training around PPE are vital for its effectiveness. This comprehensive discussion explores the importance of PPE, principles for appropriate selection, standards and classifications, training and compliance, industry-specific applications, and best practices for implementation. The Importance of PPE in Hazard Control: PPE is an indispensable part of occupational health and safety frameworks, especially in environments where residual risks persist despite administrative or engineering interventions. PPE becomes particularly crucial in scenarios involving: By creating a physical barrier, PPE limits the absorption, inhalation, ingestion, or contact of harmful substances. It also protects against mechanical injuries such as cuts, burns, and abrasions. Despite being the least effective control when used in isolation (as per the National Institute for Occupational Safety and Health [NIOSH] hierarchy), it is often essential for personal safety in high-risk environments (Seçkiner et al., 2021). Principles of PPE Selection: PPE selection is a systematic process grounded in hazard identification and risk assessment. The following principles guide proper selection: Risk Assessment: Determine the types and levels of hazards present in the workplace through Job Hazard Analysis (JHA) (Haraga et al., 2021). Fit and Comfort: PPE must properly fit the user to ensure safety without impeding performance or mobility. Material Compatibility: Materials must be resistant to specific hazards—e.g., nitrile gloves for chemicals, Kevlar for cuts. Regulatory Compliance: All equipment must meet established safety standards (e.g., ANSI, OSHA, ISO, EN). Task Specificity: PPE should be tailored to the specific job task—e.g., welding helmets for welding, not general face shields. Durability and Maintenance: Choose PPE that is durable, easy to maintain, and appropriate for repeated use if necessary. User Acceptance: Consideration of worker preferences, gender, size range, and comfort promotes consistent use. Improper selection of PPE can create a false sense of security, increase the risk of injury, or result in non-compliance due to discomfort or impracticality.

Training must be interactive and reinforced regularly, especially when new types of PPE are introduced or when procedures change. Fit-testing is essential for items like respirators, where a proper seal is critical to effectiveness. Additionally, workers must be encouraged to report damaged PPE and to request replacements without hesitation (Rafindadi et al., 2022). 5. Maintenance, Storage, and Replacement: The efficacy of PPE depends not only on its proper selection and use but also on its maintenance and storage. Key aspects include: Routine Inspection: Check for tears, degradation, or contamination before and after each use. Cleaning and Sanitizing: Follow manufacturer and regulatory guidelines to avoid chemical breakdown or contamination spread. Storage Conditions: Keep PPE in clean, dry areas away from heat, sunlight, and chemicals that could degrade materials. Shelf Life Monitoring: Be aware of expiration dates, particularly for items like respiratory filters or chemical suits. Replacement Protocols: Replace PPE based on wear-and-tear, inspection outcomes, or manufacturer-recommended cycles. Failing to maintain PPE appropriately can compromise its protective integrity, potentially exposing workers to the very hazards the equipment is meant to guard against (Samal et al., 2025). Personal Protective Equipment (PPE) is an essential component of occupational safety, providing critical protection when more effective hazard elimination strategies are not feasible. While PPE is the last line of defense, it remains indispensable in industries where workers face a wide array of hazards.

CONTINUOUS MONITORING AND REVIEW OF RISK ASSESSMENT

Continuous monitoring and review form the backbone of a dynamic and responsive risk management system. Risk is not a static concept; it evolves in response to internal operations, technological changes, regulatory shifts, market fluctuations, and external environmental factors. Therefore, organizations must not treat risk assessments as one-time exercises but as ongoing processes that require real-time vigilance, regular reevaluation, and iterative improvement. Continuous monitoring allows organizations to detect emerging threats, assess the effectiveness of existing controls, identify gaps, and implement timely corrective actions. When coupled with periodic reviews, this creates a feedback loop that enhances situational awareness, organizational agility, and long-term resilience (Eastham et al., 2022). The Rationale for Continuous Monitoring in Risk Assessment: The fundamental purpose of continuous monitoring in risk assessment is to ensure that risk profiles remain current and reflective of actual conditions. Even well-structured risk management plans can become obsolete if not revisited regularly. The business environment, workforce dynamics, regulatory landscape, and technology infrastructure are constantly in flux. Without ongoing surveillance, previously acceptable risks may escalate, and new, unforeseen hazards may go undetected. This proactive stance enables risk managers to move beyond reactive problem-solving and embrace predictive, data-driven approaches. Components of a Continuous Risk Monitoring System: An effective continuous monitoring and review system typically includes the following components: a. Key Risk Indicators (KRIs) KRIs are measurable metrics that signal potential risk exposure or control weaknesses. For example, in finance, an increase in customer complaints may indicate reputational risk; in healthcare, a rise in adverse event reports could signal patient safety concerns. The rise of digital transformation has significantly enhanced the capacity for continuous monitoring. Several tools and technologies are now embedded in risk management strategies: a. Integrated Risk Management Platforms (IRMs): These platforms consolidate risk data across departments and geographies, offering centralized dashboards and automated alerts when risk thresholds are breached. b. Artificial Intelligence (AI) and Machine Learning (ML): AI-driven tools can analyze vast datasets to detect anomalies, predict trends, and identify patterns indicative of emerging risks. c. Internet of Things (IoT) Devices: In sectors like manufacturing, construction, and healthcare, IoT devices continuously track variables such as temperature, humidity, pressure, vibration, or personnel movement to identify hazards. d. Data Visualization and Business Intelligence Tools: Graphical dashboards and heat maps allow stakeholders to interpret risk data quickly, identify hotspots, and prioritize mitigation efforts.

Industry Applications of Continuous Risk Monitoring: Hospitals monitor infection control data, patient falls, medication errors, and hospital-acquired conditions in real-time. Continuous monitoring is critical for ensuring patient safety and regulatory compliance (Aziz et al., 2023). Sensors monitor equipment conditions, energy usage, and environmental emissions. Predictive maintenance strategies are employed to prevent breakdowns and reduce downtimeBanks and insurance companies use AI to detect fraudulent transactions, credit risks, or market volatility. Continuous monitoring also supports compliance with Basel III and other financial regulations. Wearable devices track worker movements, fatigue levels, and exposure to hazardous conditions. Environmental monitors ensure air quality and noise levels stay within safe thresholds. e. Information Technology: Cybersecurity systems continuously scan networks for vulnerabilities, intrusions, and unusual user behavior. SIEM (Security Information and Event Management) tools are central to IT risk management. Ultimately, continuous monitoring empowers organizations to navigate uncertainty with confidence and agility. Continuous monitoring and review of risk assessments are no longer optional in today's volatile, uncertain, complex, and ambiguous (VUCA) environment—they are essential for sustainable and responsible operations (Tehrani et al., 2022). By establishing robust systems for real-time data collection, regular analysis, and adaptive response, organizations can stay ahead of emerging threats, reinforce control effectiveness, and ensure compliance with dynamic regulatory standards. This living, breathing approach to risk management transforms it from a compliance-driven checkbox to a strategic advantage, capable of shaping decisions, protecting assets, and enabling innovation. When paired with a culture of accountability and continuous improvement, continuous monitoring becomes a catalyst for resilience, agility, and long-term success.

CONCLUSION

In an increasingly interconnected and complex world, the role of technology in risk assessment and prevention has become not only beneficial but essential. As organizations and industries grapple with a wide range of risks—from occupational hazards and operational failures to cyber threats and environmental crises—technology offers advanced tools and methodologies to identify, evaluate, mitigate, and monitor risks more accurately and efficiently than ever before. The integration of technologies such as artificial intelligence (AI), machine learning (ML), the Internet of Things (IoT), big data analytics, digital twins, cloud computing, geographic information systems (GIS), and blockchain has transformed the traditional risk management landscape.

These innovations not only enhance the precision and scope of assessments but also foster a proactive, real-time, and data-driven approach to hazard prevention. This evolution signals a pivotal shift from reactive to predictive and prescriptive models of risk control, redefining safety, compliance, and resilience across sectors. Artificial Intelligence and Machine Learning for Predictive Risk Analysis: AI and ML are revolutionizing how organizations understand and anticipate risk. Unlike traditional models that rely on static data and historical trends, AI systems continuously learn from vast datasets to identify patterns, anomalies, and correlations that may signal impending risks. Predictive analytics powered by machine learning algorithms can forecast potential equipment failures, occupational accidents, financial losses, or supply chain disruptions, enabling organizations to intervene before incidents occur. For example, in manufacturing, AI models can analyze sensor data to predict machinery breakdowns, reducing downtime and maintenance costs. In healthcare, ML algorithms help identify patients at high risk of complications by analyzing electronic health records and real-time vitals. In cybersecurity, AI detects unusual network behavior and blocks threats in real time. These capabilities not only improve safety outcomes but also optimize resource allocation and operational efficiency. Internet of Things (IoT) and Real-Time Monitoring: The integration of IoT devices into risk management systems has ushered in a new era of continuous monitoring and real-time hazard detection. IoT sensors embedded in equipment, buildings, vehicles, and wearable devices collect data on environmental conditions, system performance, and human behavior. This data is transmitted to central platforms for analysis, alerting safety personnel or control systems when thresholds are exceeded. In the oil and gas industry, for instance, IoT sensors detect gas leaks, pressure changes, and temperature spikes, allowing for immediate intervention. In construction, wearable tech can monitor worker fatigue, exposure to noise or harmful substances, and proximity to dangerous zones. In logistics, smart tracking systems optimize delivery routes and monitor the integrity of perishable goods. This real-time visibility transforms hazard detection from periodic checks to a continuous safety net. Big Data Analytics for Informed Decision-Making: Big data analytics plays a critical role in consolidating and analyzing large volumes of structured and unstructured data from diverse sources—including internal systems, public databases, social media, regulatory reports, and third-party platforms. By uncovering hidden risks, assessing trends, and evaluating the effectiveness of controls, big data supports more informed and strategic decision-making. Organizations use analytics platforms to visualize risk exposure, identify root causes of incidents, prioritize mitigation efforts, and simulate the impact of various scenarios. In finance, analytics identify fraudulent patterns or systemic weaknesses. In healthcare, population-level data helps monitor disease outbreaks and resource allocation. In climate risk management, satellite and meteorological data inform emergency preparedness and response strategies.

This capability ensures that decisions are evidence-based, timely, and aligned with organizational objectives. In manufacturing, digital twins assess how production line changes affect safety and efficiency. This approach helps identify potential failures, optimize system design, and enhance training effectiveness without the risk of real-world experimentation. Cloud Computing and Centralized Risk Platforms: Cloud computing has enabled the centralization of risk data, tools, and collaboration platforms, making risk management more accessible, scalable, and collaborative. Cloud-based risk assessment systems allow geographically dispersed teams to share data, perform analyses, and coordinate responses in real time. This fosters consistency, speed, and transparency in risk mitigation efforts. Cloud solutions are particularly valuable for multinational organizations with operations in multiple regions. They can consolidate safety audits, compliance records, incident logs, and environmental monitoring data into a unified dashboard accessible to all stakeholders. Additionally, Software as a Service (SaaS) platforms offer customizable modules for hazard identification, control implementation, regulatory tracking, and performance analytics, making them cost-effective and easy to update. Technology is not merely supporting risk assessment and hazard prevention—it is fundamentally reshaping how they are conducted. From predictive AI models and real-time IoT sensors to immersive training simulations and blockchain-enabled traceability, today's technological tools empower organizations to manage risks with unprecedented speed, accuracy, and agility.

REFERENCES

Achumie, G. O., Oyegbade, I. K., Igwe, A. N., Ofodile, O. C., & Azubuike, C. (2022). A conceptual model for reducing occupational exposure risks in high-risk manufacturing and petrochemical industries through industrial hygiene practices. *International Journal of Hygiene and Safety*, *1*(1), 26–37. DOI: 10.54660/IJSS-ER.2022.1.1.26-37

Alvand, A., Mirhosseini, S. M., Ehsanifar, M., Zeighami, E., & Mohammadi, A. (2023). Identification and assessment of risk in construction projects using the integrated FMEA-SWARA-WASPAS model under fuzzy environment: A case study of a construction project in Iran. *International Journal of Construction Management*, *23*(1), 1–23. DOI: 10.1080/15623599.2021.1877875

Aziz, L. A. R., & Andriansyah, Y. (2023). Artificial intelligence in modern banking: An exploration of AI-driven approaches for enhanced fraud prevention, risk management, and regulatory compliance. *Reviews of Contemporary Business Analytics*, *5*(2), 112–130. https://core.ac.uk/download/pdf/987654321.pdf

Bartolomé, E. J. N., Carreño, O. F. M., Medina López, E. G., & Muñoz Heredia, C. E. (2022). Administrative management and its impact on the company's productivity, training plan. *Journal of Business and Entrepreneurial Studies*, *6*(3), 1–12. DOI: 10.37956/jbes.v6i3.300

Brauer, R. L. (2022). Safety and health for engineers. Wiley. https://books.google.com/books/about/Safety_and_Health_for_Engineers.html?id=xyz

Brisbine, B. R., Radcliffe, C. R., Jones, M. L. H., Stirling, L., & Coltman, C. E. (2022). Does the fit of personal protective equipment affect functional performance? A systematic review across occupational domains. *PLoS One*, *17*(11), e0278174. Advance online publication. DOI: 10.1371/journal.pone.0278174 PMID: 36449531

Carton, W., Hougaard, I. M., & Markusson, N. (2023). Is carbon removal delaying emission reductions? *Wiley Interdisciplinary Reviews: Climate Change*, *14*(3), e789. Advance online publication. DOI: 10.1002/wcc.789

Crotty, J., & Daniel, E. (2022). Cyber threat: Its origins and consequence and the use of qualitative and quantitative methods in cyber risk assessment. *Applied Computing and Informatics*, *20*(2), 123–140. DOI: 10.1108/ACI-07-2022-0178

Eastham, J. A., Auffenberg, G. B., Barocas, D. A., Chou, R., Crispino, T., Davis, J. W., Eggener, S., Horwitz, E. M., Kane, C. J., Kirkby, E., Lin, D. W., McBride, S. M., Morgans, A. K., Pierorazio, P. M., Rodrigues, G., Wong, W. W., & Boorjian, S. A. (2022). Clinically localized prostate cancer: AUA/ASTRO guideline, part I: Introduction, risk assessment, staging, and risk-based management. *The Journal of Urology*, *208*(1), 10–18. DOI: 10.1097/JU.0000000000002757 PMID: 35536144

EFSA Scientific Committee, More, S., Bampidis, V., Benford, D., Bragard, C., Halldorsson, T., Hernandez-Jerez, A. F., Hougaard Bennekou, S., Koutsoumanis, K., Lambre, C., Machera, K., Nielsen, S. S., Schrenk, D., Turck, D., Younes, M., & et al. (. (2021). Guidance on risk assessment of nanomaterials to be applied in the food and feed chain: Human and animal health. *EFSA Journal*, *19*(8), e06768. Advance online publication. DOI: 10.2903/j.efsa.2021.6768 PMID: 34377190

Fjeld, R. A., DeVol, T. A., & Martinez, N. E. (2023). Quantitative environmental risk analysis for human health. Springer. https://books.google.com/books/about/Quantitative_Environmental_Risk_Analysis.html?id=xyz

Grant, N., Hawkes, A., & Mittal, S. (2021). Confronting mitigation deterrence in low-carbon scenarios. *Environmental Research Letters*, *16*(11), 114012. Advance online publication. DOI: 10.1088/1748-9326/ac2e3f

Haraga, A. R., Cazac, A. M., & Lohan, N. M. (2021). Selection of personal protective equipment—a complex issue of multi-criteria analysis. MATEC Web of Conferences, 343, Article 01012. https://doi.org/DOI: 10.1051/matecconf/202134301012

Hassan, H. R. (2023). A review on different arsenic removal techniques used for decontamination of drinking water. *Environmental Pollutants and Bioavailability*, *35*(1), 45–58. DOI: 10.1080/26395940.2023.2174567

Hubballi, R. B., Attaluri, V., Selvakumar, P., Basava Aradhya, S. G., Dinesh, N., & Seelam, P. K. (2025). Overview of current AI technologies in education. In Kyei-Blankson, L., & Ntuli, E. (Eds.), *Transformative AI practices for personalized learning strategies* (pp. 1–26). IGI Global Scientific Publishing., DOI: 10.4018/979-8-3693-8744-3.ch001

Huy, D. T. N., Thach, N. N., Chuyen, B. M., Nhung, P. T. H., Thang, T. D., & Anh, T. T. (2021). Enhancing risk management culture for sustainable growth of Asia commercial bank-ACB in Vietnam under mixed effects of macro factors. *Entrepreneurship and Sustainability Issues*, *8*(3), 291–307. DOI: 10.9770/jesi.2021.8.3(18)

Kotte, K. R., Attaluri, V., & Selvakumar, P. (2025). Revolutionising retail banking with AI and virtual technologies: AI-driven financial training. In Nair, A. J., Manohar, S., Limbu, Y. B., & Huhmann, B. A. (Eds.), *Intersecting natural language processing and FinTech innovations in service marketing* (pp. 421–440). IGI Global Scientific Publishing., DOI: 10.4018/979-8-3693-9944-6.ch020

Kunnathully, K., Satyanarayana, P., Anute, N., Sharma, M., Sureshkrishna, G., & Manjunath, T. C. (2025). Propaganda, social media and AI. In Vieira, A., Joaquim, A., & Duarte, A. (Eds.), *Digital populism and the use of neo-propaganda and fake news* (pp. 19–38). IGI Global Scientific Publishing., DOI: 10.4018/979-8-3693-9999-6.ch002

Luo, N., Yu, H., You, Z., Li, Y., Zhou, T., Jiao, Y., Han, N., Liu, C., Jiang, Z., & Qiao, S. (2023). Fuzzy logic and neural network-based risk assessment model for import and export enterprises: A review. *Journal of Data Science and Intelligent Systems*, *5*(2), 89–104. DOI: 10.47852/bonviewJDSIS32021078

Maranzoni, A., D'Oria, M., & Rizzo, C. (2023). Quantitative flood hazard assessment methods: A review. *Journal of Flood Risk Management*, *16*(2), e12876. Advance online publication. DOI: 10.1111/jfr3.12855

Mayta-Huiza, D. A., Beru, A. C., Apaza, S. L. V., Chávez-Choque, M. E., & Pineda Yucra, W. (2023). Impacts of supervision, information and communication between management control and administrative management procedures. *International Journal of Membrane Science and Technology*, *10*(3), 1–15. DOI: 10.15379/ijmst.v10i3.1573

Murariu, M., & Bedrule-Grigoru ă, M. V. (2021). Training internal managerial control personnel in the context of digitalization. INTED2021 Proceedings, 1, 4567–4575. https://library.iated.org/view/MURARIU2021TRA

Naim, A. (2022). Role of artificial intelligence in business risk management. *American Journal of Business Management, Economics, and Banking*, *1*, 55–66. https://www.neliti.com/publications/606778

Nani, D. A., & Safitri, V. A. D. (2021). Exploring the relationship between formal management control systems, organisational performance and innovation: The role of leadership characteristics. *Asian Journal of Business and Accounting*, *14*(1), 207–224. DOI: 10.22452/ajba.vol14no1.8

Nobanee, H., Al Hamadi, F. Y., Abdulaziz, F. A., Abukarsh, L. S., Alqahtani, A. F., AlSubaey, S. K., Alqahtani, S. M., & Almansoori, H. A. (2021). A bibliometric analysis of sustainability and risk management. *Sustainability (Basel)*, *13*(6), 3277. Advance online publication. DOI: 10.3390/su13063277

Onyeke, F. O., Odujobi, O., & Adikwu, F. E. (2022). Innovative approaches to enhancing functional safety in distributed control systems (DCS) and safety instrumented systems (SIS) for oil and gas applications. Open Access Research Journal of Engineering Safety, 2(1), 55–70. https://www.researchgate.net/publication/Innovative_Approaches_SIS_DCS_Safety_2022

Panchal, S., & Shrivastava, A. K. (2022). Landslide hazard assessment using analytic hierarchy process (AHP): A case study of National Highway 5 in India. *Ain Shams Engineering Journal, 13*(3), 101672. Advance online publication. DOI: 10.1016/j.asej.2021.10.021

Pascarella, G., Rossi, M., Montella, E., Capasso, A., De Feo, G., Botti, G., Nardone, A., Montuori, P., Triassi, M., D'Auria, S., & Morabito, A. (2021). Risk analysis in healthcare organizations: Methodological framework and critical variables. *Risk Management and Healthcare Policy, 14*, 2897–2911. DOI: 10.2147/RMHP.S309098 PMID: 34267567

Peng, J., Zhang, S., Han, Y., Bate, B., Ke, H., & Chen, Y. (2022). Soil heavy metal pollution of industrial legacies in China and health risk assessment. *The Science of the Total Environment, 816*, 151632. Advance online publication. DOI: 10.1016/j.scitotenv.2021.151632 PMID: 34780826

Rafindadi, A. D., Napiah, M., Othman, I., & Alarifi, H. (2022). Significant factors that influence the use and non-use of personal protective equipment (PPE) on construction sites—Supervisors' perspective. *Ain Shams Engineering Journal, 13*(4), 101234. Advance online publication. DOI: 10.1016/j.asej.2021.10.014

Sakthivel, M., Jaganathan, A., & Mohanraj, M. Mohit, & Manjunath, T. (2025). Brand loyalty through AI and personalisation: AI-generated content for financial marketing. In A. J. Nair, S. Manohar, Y. B. Limbu, & B. A. Huhmann (Eds.), Intersecting natural language processing and FinTech innovations in service marketing (pp. 77–96). IGI Global Scientific Publishing. https://doi.org/DOI: 10.4018/979-8-3693-9944-6.ch006

Samal, A., & Bhargav, J. S. K. M., Selvakumar, P., Sharma, M., & T. C. M. (2025). Measuring brand performance with AI tools. In Z. Hussain, M. Sharipudin, A. Albattat, & A. Khan (Eds.), Strategic brand management in the age of AI and disruption (pp. 279–302). IGI Global Scientific Publishing. https://doi.org/DOI: 10.4018/979-8-3693-9461-8.ch011

Scannell, D., Desens, L., Guadagno, M., Tra, Y., Acker, E., Sheridan, K., Rosner, M., Mathieu, J., & Fulk, M. (2021). COVID-19 vaccine discourse on Twitter: A content analysis of persuasion techniques, sentiment and mis/disinformation. *Journal of Health Communication, 26*(12), 885–897. DOI: 10.1080/10810730.2021.1955050 PMID: 34346288

Seçkiner, S. U., & Ünal, H. (2021). Comparing the alternatives for the most favourable personal protective equipment. *International Journal of Occupational Safety and Ergonomics, 27*(4), 789–798. DOI: 10.1080/10803548.2019.1674044 PMID: 31609158

Syeda, S. R., Khan, E. A., Padungwatanaroj, O., Kuprasertwong, N., & Tula, A. K. (2022). A perspective on hazardous chemical substitution in consumer products. *Current Opinion in Chemical Engineering, 36*, 100748. Advance online publication. DOI: 10.1016/j.coche.2021.100748

Tehrani, F., Teymourian, H., Wuerstle, B., Kavner, J., Patel, R., Furmidge, A., Aghavali, R., Hosseini-Toudeshki, H., Brown, C., Zhang, F., Mahato, K., Li, Z., Barfidokht, A., Yin, L., Warren, P., Huang, N., Patel, Z., Mercier, P. P., & Wang, J. (2022). An integrated wearable microneedle array for the continuous monitoring of multiple biomarkers in interstitial fluid. *Nature Biomedical Engineering, 6*(12), 1345–1354. DOI: 10.1038/s41551-022-00887-1 PMID: 35534575

van Gelder, P., Klaassen, P., Taebi, B., Walhout, B., van Ommen, R., van de Poel, I., Robaey, Z., Asveld, L., Balkenende, R., Hollmann, F., van Kampen, E. J., Khakzad, N., Krebbers, R., de Lange, J., Pieters, W., Terwel, K., Visser, E., van der Werff, T., & Jung, D. (2021). Safe-by-design in engineering: An overview and comparative analysis of engineering disciplines. *International Journal of Environmental Research and Public Health, 18*(9), 4567. Advance online publication. DOI: 10.3390/ijerph18126329 PMID: 34208018

Vincoli, J. W. (2024). Basic guide to system safety. McGraw-Hill Education. https://books.google.com/books/about/Basic_Guide_to_System_Safety.html?id=xyz

Wahab, N. A. A., Aqila, N. A., Isa, N., & Husin, N. I. (2021). A systematic review on hazard identification, risk assessment and risk control in academic laboratory. *Journal of Advanced Research in Applied Sciences and Engineering Technology, 24*(1), 47–62. DOI: 10.37934/araset.24.1.4762

Wahyuni, H. C., Rosid, M. A., Azara, R., & Voak, A. (2025). Blockchain technology design based on food safety and halal risk analysis in the beef supply chain with FMEA-FTA. *Journal of Engineering Research, 36*(2), 1–15. DOI: 10.1016/j.jer.2024.02.002

Chapter 6
Leading Indicators in Occupational Health and Safety and Integration With ISO 45001

Rabia Balcı
http://orcid.org/0000-0002-5959-1955
Occupational Health and Safety Program, Amasya University, Turkey

ABSTRACT

Leading indicators in occupational health and safety (OHS) are proactive measures that provide early warning of potential incidents, complementing traditional lagging indicators such as injury rates. This chapter explains the concept of leading indicators and their importance in fostering a prevention-focused safety culture. It discusses how leading indicators can be developed and integrated into an organization's OHS management system, particularly in alignment with the ISO 45001:2018 standard for OHS management. Through a review of recent literature and case examples, the chapter highlights key categories of leading indicators, benefits of their use, and common challenges in implementation. It also illustrates how integrating these forward-looking metrics within the ISO 45001 framework can drive continuous improvement in safety performance and help organizations in their journey toward 'zero harm'.

1. INTRODUCTION

Occupational health and safety (OHS) is one of the determinants of sustainable industrial performance and human well-being. Despite the existence of substantial

DOI: 10.4018/979-8-3373-6370-7.ch006

regulatory and institutional safeguards established over the past decades to improve OHS outcomes, empirical evidence indicates that occupational risks remain widespread. For instance, according to the International Labour Organization (2023), more than 340 million occupational accidents and 160 million work-related diseases occur annually worldwide, resulting in approximately 2.8 million fatalities. These translate into an economic burden exceeding 4% of global GDP. Within the European Union, the limited long-term reduction in fatal occupational injury rates has led scholars and policy institutions to question the effectiveness of predominantly compliance-based and reactive safety management approaches, particularly those relying heavily on lagging indicators (Zwetsloot et al., 2013; Pawłowska, 2015; European Agency for Safety and Health at Work [EU-OSHA], 2023; Golabchi et al., 2024).

Historically, the measurement of OHS performance has relied primarily on lagging indicators-frequency rates, lost-time injuries, and compensation costs-that quantify past failures (Manuele, 2009; Walaski, 2020). Although valuable for benchmarking, these retrospective metrics fail to capture early warning signals of system degradation. As industries become more digitalised and interdependent, the inadequacy of lagging data in predicting incidents has become increasingly evident (Leveson, 2015; Bargar et al., 2025). In response, researchers and practitioners have called for the systematic incorporation of *leading indicators*-proactive, process-oriented measures that monitor the effectiveness of preventive actions (Sinelnikov et al., 2015; Haas et al., 2017; Neamat, 2019).

1.1 From Reactive Control to Proactive Governance

The intellectual evolution from lagging to leading indicators mirrors the broader paradigmatic shift in safety science from linear causation to systems thinking. Heinrich's classic *accident triangle* framed incidents as predictable outcomes of unsafe acts (Heinrich, 1941). However, later theories offered a different perspective on accident causations. However, later theories offered a different perspective on accident causation. James Reason's Swiss-cheese model and Sidney Dekker's drift into failure model reconceptualised accidents as emergent phenomena rather than isolated events (Reason, 2016; Dekker, 2016). In respect of these views, accidents arise from complex interactions within sociotechnical systems, where multiple latent conditions and organizational factors align over time. Leveson (2015) advanced this understanding through her *Systems-Theoretic Accident Model and Processes (STAMP)*, which positions safety as a control function dependent on feedback, communication, and constraint enforcement rather than as a reactive response to failures. This shift has important implications for safety indicator design, as it directs attention away from outcome-based, lagging indicators toward leading indicators that capture the effectiveness of control actions, information flows, and organizational learning processes within complex sociotechnical systems.

This paradigmatic shift in safety thinking has influenced modern safety-management discourse, reframing measurement as a mechanism for learning rather than punishment (Sheehan et al., 2016; Guo & Yiu, 2016). Empirical work across mining, healthcare, and construction demonstrates that proactive indicators- training participation, risk-assessment quality, and safety-communication frequency- serve as valid predictors of reduced accident rates (Haas et al., 2017; Almost et al., 2018; Xu et al., 2021). Recent systematic reviews indicate that organisations adopting predictive, leading-indicator-based metrics report statistically significant, small-to-moderate, improvements in safety climate measures, alongside enhanced operational resilience (Givehchi et al. 2017; Phinias, 2023; Latham et al., 2024).

Concurrently, the global Vision Zero movement, championed by the International Social Security Association (ISSA) and scholars such as Zwetsloot et al. (2020, 2025), has redefined safety excellence as the prevention of all harm through leadership and continuous improvement. Vision Zero advocates proactive measurement systems that highlight "what goes right" and promote organisational learning. Within this paradigm, leading indicators are not auxiliary statistics but central instruments of governance, reinforcing commitment and accountability across organisational levels.

1.2 Need for Integration with ISO 45001

The publication of ISO 45001:2018 institutionalised this proactive ethos within a structured management-system framework based on the Plan-Do-Check-Act (PDCA) cycle. ISO 45001 requires organisations to establish preventive objectives, evaluate performance, and pursue continuous improvement. Nonetheless, studies indicate that many certified firms continue to emphasise outcome metrics, thereby under-utilising the potential of leading indicators (Mohammadfam et al., 2017; da Silva & Amaral, 2019;). This has remained the same for many years as organisations find it difficult to translate leading-indicator data into clear causal insights for decision-making, which in return weakens managerial confidence in in the preventive value of this data.

Recent literature underscores that embedding proactive measures within each PDCA phase can transform safety management from procedural compliance to strategic foresight (Pera et al., 2023; Castillo et al., 2023). In the Plan phase, indicators such as risk-assessment coverage and resource allocation evaluate preparedness; the Do phase monitors training completion and behavioural observations; Check tracks near-miss reporting and audit follow-up; and Act measures the effectiveness of corrective actions. Such integration converts data into actionable intelligence, enabling dynamic control and organisational learning (Karimi et al., 2021; Zulkarnain et al., 2023).

The 2024 Amendment 1 to ISO 45001, introducing climate-change considerations, further strengthens the case for adaptive and anticipatory measurement systems. Environmental volatility-heat stress, extreme weather, and chemical-exposure shifts-necessitates continuous sensing mechanisms within OHS management (Zwetsloot et al., 2025). Leading indicators such as., embedded in the ISO 45001 framework thus provide an institutionalised means to operationalise resilience and sustainability.

1.3 Purpose and Objectives of the Chapter

Building upon these theoretical and empirical insights, this chapter pursues three complementary objectives:

i. **Conceptual clarification**- to delineate the evolution and typology of lagging and leading indicators in the safety-science literature, synthesising key frameworks (Heinrich, Reason, Leveson, Zwetsloot).
ii. **Framework development**- to construct an integrative model linking leading indicators to the ISO 45001 PDCA cycle.
iii. **Applied analysis**- to illustrate, through empirical evidence from manufacturing and other high-reliability sectors, how proactive metrics enhance learning, compliance, and sustainable performance.

Through these objectives, the chapter aims to bridge the measurement gap between conceptual theory and managerial practice by demonstrating how data-driven preventive metrics can accelerate organisational maturity and resilience (Almost et al., 2019; Xu et al., 2023).

1.4 Significance and Contribution

The chapter's contribution is twofold. Theoretically, it integrates leading-indicator research with management-system theory and resilience engineering, positioning proactive metrics as dynamic feedback controls within sociotechnical systems (Leveson, 2015; Hallowell et al., 2013). Practically, it proposes a replicable ISO 45001-aligned framework that organisations can adopt to achieve continuous improvement, regulatory alignment, and safety-culture maturity (Guo & Yiu, 2016; Pera et al., 2023).

Moreover, aligning leading-indicator systems with policy frameworks such as the EU-OSHA Strategic Framework 2021-2027 and UN Sustainable Development Goal 8 (Decent Work and Economic Growth) explicitly connects occupational safety to broader sustainability and decent-work objectives. As such, industrial organisations operating within the EU manufacturing and energy sectors have integrated leading safety indicators such as proactive hazard reporting, safety training participation rates, and near-miss analysis-into their sustainability and ESG (Environmental, Social, and Governance) reporting procedures, thereby operationalising policy alignment in day-to-day safety management practices (Zwetsloot et al., 2020; Barbosa et al., 2019).

1.5 Structure of the Chapter

The remainder of the chapter is organised as follows. Section 2 reviews the theoretical background and synthesises existing indicator models. Section 3 presents the conceptual framework integrating leading indicators into the ISO 45001 PDCA cycle. Section 4 provides practical case insights from multiple industries. Section 5 discusses theoretical, managerial, and policy implications, while Section 6 concludes with reflections on future research trajectories.

By combining conceptual synthesis with empirical evidence, this chapter advances the understanding of how proactive, predictive, and system-embedded measurement can transform occupational safety from a retrospective compliance obligation into a strategic enabler of organisational resilience and sustainable excellence.

2. THEORETICAL AND CONCEPTUAL FRAMEWORK

2.1 Evolution of Safety Performance Measurement

The study of occupational safety measurement has undergone a gradual but profound transformation over the past century. This evolution is critical because it reflects a fundamental shift in how accidents are understood, governed and prevented, moving from the retrospective counting of failures to the proactive management of systemic risk. Early conceptions of accident causation were shaped by technical rationalism and behavioural control, emphasising errors of individual workers rather than system deficiencies. The most influential of these early models, Heinrich's Domino Theory (1941), proposed that accidents result from a linear sequence of unsafe acts and conditions, implying that removal of a single causal "domino" could prevent an event. This view established the foundation for *lagging indicators* such as frequency and severity rates, which dominated industrial safety management for decades (Bennett & Foster, 2005; Manuele, 2009), largely because they aligned with linear causation models and offered regulatory simplicity, ease of quantification, and compatibility with compliance-oriented oversight and reporting practices.

By the 1970s and 1980s, the field began to shift from mechanistic causality to systemic interactionism. Pioneers such as Reason (2016), Perrow (1984), and Turner & Pidgeon (1997) demonstrated that accidents are not random outcomes but emergent properties of complex sociotechnical systems. Reason's Swiss-cheese model articulated that multiple defensive layers- organisational, managerial, and operational- contain inherent "holes" that align under specific conditions to allow accidents. This model reframed failure as the product of latent organisational weaknesses, thus emphasising management-system quality and cultural precursors rather than mere statistical outcomes.

Later developments in resilience engineering and systems theory further redefined the epistemological basis of safety. Leveson (2015), through her *Systems-Theoretic Accident Model and Processes (STAMP)*, conceptualised safety as a control problem: the objective is to maintain system behaviour within safe constraints by continuous feedback and adaptation. Similarly, Hollnagel's (2017) Safety-II framework proposed that safety is achieved not merely by avoiding failure but by ensuring successful performance under variable conditions. These theoretical evolutions collectively led to the emergence of leading indicators- metrics that evaluate the strength and responsiveness of preventive systems rather than counting post-incident losses (Sinelnikov et al., 2015; Leveson, 2015).

2.2 Conceptualising Lagging and Leading Indicators

The conceptual dichotomy between *lagging* and *leading* indicators reflects a broader epistemological divide between *reactive* and *proactive* safety paradigms. Lagging indicators, such as incident rates and compensation costs, are retrospective; they serve as accountability mechanisms but provide limited diagnostic power (Walaski, 2020; Yorio et al., 2020). In contrast, leading indicators are predictive, behaviourally anchored measures designed to detect system weaknesses before harm occurs (Haas et al., 2017; Givehchi et al., 2017).

Leading indicators can be categorised into several typologies. Guo and Yiu (2016) distinguished *operational* (e.g., inspections completed), *behavioural* (e.g., compliance with safety rules), and *organisational* (e.g., leadership commitment) indicators. Leveson (2015) added a *control-theoretic dimension*, where indicators function as feedback loops for maintaining constraints. Rubio-Romero et al. (2018) extended this by linking leading indicators to *resilience capacities*: the ability to anticipate, monitor, respond, and learn. Phinias (2023) synthesised evidence across sectors, concluding that leading indicators improve safety performance by strengthening communication and accountability networks.

Empirical research provides consistent evidence for the predictive validity of leading indicators across high-risk sectors. In the healthcare context, Joan Almost et al. (2018) demonstrate that management engagement and systematic hazard reporting are significant predictors of subsequent reductions in adverse safety incidents. In the construction sector, studies by Xiaowei Xu et al. (2021, 2023) and Matthew

Hallowell et al. (2013) show that near-miss reporting, pre-task risk assessments, and training participation are strongly associated with lower injury rates. Similarly, evidence from high-risk industries such as chemicals and mining indicates that proactive monitoring of process conditions, structured safety observations, and systematic worker feedback contribute to meaningful reductions in severe events (Zulkarnain et al., 2023; Haas et al., 2017).

2.3 Safety-I and Safety-II: Paradigmatic Shift

The transition from *Safety-I* to *Safety-II* provides a theoretical justification for prioritising leading indicators. Safety-I assumes that safety is the absence of accidents, measured by failure frequencies- thus inherently lagging. Safety-II, in contrast, defines safety as the *presence of adaptive capacity*- the ability of systems to succeed under varying conditions (Hollnagel, 2017). In this framework, leading indicators represent *positive performance measures* that reflect system flexibility, learning, and anticipation.

Dekker (2016) and Hollnagel (2017) both argue that resilience in modern socio-technical systems depends on organisations' capacity to detect and correct weak signals. Measurement systems, therefore, must evolve from punitive, outcome-focused metrics to supportive, process-oriented ones. Latham et al. (2024) reinforced this perspective by showing that in the Australian agriculture sector, leading indicators capturing safety-culture dimensions-communication openness, peer learning, and leadership visibility-were more predictive of sustained improvement than injury rates.

This reconceptualisation aligns with Resilience Engineering, which frames safety as a dynamic process of sustaining operations under stress. Rubio-Romero et al. (2018) demonstrated that composite leading indicators can quantify resilience maturity in municipal waste management, while Zwetsloot et al. (2020) connected resilience-based leading indicators to well-being and sustainability outcomes. Together, these findings situate proactive measurement at the heart of organisational adaptability and high reliability, while also indicating that translating the principles of Safety-II into concrete, decision-relevant indicators remains an unresolved methodological challenge.

2.4 Vision Zero and the Proactive Safety Culture Paradigm

The Vision Zero movement operationalises Safety-II and resilience principles within a strategic policy context. Originating from Swedish traffic safety policy and later adopted by the International Social Security Association (ISSA), Vision Zero asserts that *all injuries are preventable* and that *safety is a moral value*, not merely a management objective (Zwetsloot et al., 2017, 2025). It promotes proactive measurement frameworks that integrate safety, health, and well-being. Zwetsloot et al. (2020) proposed a tiered indicator system-covering leadership, engagement, and continuous learning- as a foundation for proactive governance.

Recent studies have advanced Vision Zero through empirical validation. Bargar et al. (2025) used multi-criteria decision-making methods to evaluate ISSA's proactive indicators, confirming their reliability in assessing organisational well-being. Janaćković (2025) identified key leading indicators for small and medium-sized enterprises, emphasising leadership communication and employee participation as pivotal to sustainable safety performance. Together, these studies point out that Vision Zero is realised through the selection and use of context-specific leading indicators, which enable its principles to inform organisational practice without relying on fixed quantitative targets. This pattern underscores the scalability and cross-sectoral relevance of proactive measurement.

2.5 Integrating Leading Indicators with Management Systems

The evolution of theory has increasingly converged toward *system integration*. ISO 45001:2018 codified a risk-based management approach through the PDCA cycle, providing an explicit structure for embedding leading indicators into each stage (da Silva & Amaral, 2019; Mohammadfam et al., 2017). Research by Pera et al. (2023) and Castillo et al. (2023) elaborated conceptual models mapping leading indicators onto ISO 45001 clauses- context, leadership, planning, support, operation, evaluation, and improvement.

In this model, leading indicators act as *feedback mechanisms* within the PDCA process. During the Plan phase, they measure hazard-identification completeness and resource allocation adequacy; in the Do phase, they assess training participation and behavioural compliance; in Check, they track deviations, audits, and near-miss frequencies; and in Act, they quantify the effectiveness of corrective and preventive measures. When implemented cyclically, these indicators form a *learning loop*, converting data into knowledge and knowledge into resilience (Leveson, 2015; Zwetsloot et al., 2020).

2.6 Synthesis

The theoretical literature reveals a clear trajectory: from reactive, event-based models toward proactive, systemic, and learning-oriented paradigms. Leading indicators serve as both *measurement tools* and *cultural catalysts*, reinforcing leadership accountability, worker engagement, and continuous improvement. They embody the principles of Safety-II, Resilience Engineering, and Vision Zero, providing an empirical means of operationalising high-reliability management.

The integration of these theoretical foundations into the ISO 45001 management-system architecture constitutes the conceptual basis for the framework proposed in this chapter. In doing so, it bridges the gap between abstract safety theory and applied management practice, supporting the transformation of occupational safety from a regulatory requirement into a strategic component of organisational sustainability.

3. FRAMEWORK DEVELOPMENT: INTEGRATING LEADING INDICATORS INTO ISO 45001

3.1 Conceptual Foundation

The integration of leading indicators into occupational health and safety management systems (OHSMS) represents a logical culmination of the theoretical evolution discussed earlier. ISO 45001:2018 established a universal structure for OHSMS through its PDCA cycle, mandating that organisations systematically identify hazards, implement controls, evaluate performance, and drive continual improvement (ISO, 2018). However, empirical studies reveal that this cycle is often implemented in a procedural rather than learning-oriented manner, focusing on documentation and compliance rather than feedback and foresight (da Silva & Amaral, 2019; Mohammadfam et al., 2017).

Leading indicators address this gap by introducing *anticipatory control mechanisms* that transform PDCA into a dynamic information-feedback system. In this model, indicators function as nodes of organisational intelligence that monitor system health in real time, capturing the extent to which preventive actions achieve their intended effects (Leveson, 2015; Pera et al., 2023).

3.2 The PDCA-Leading Indicators Integration Model

The proposed framework, depicted conceptually in *Figure 1*, operationalises leading indicators within each phase of the PDCA cycle. Rather than existing as an auxiliary measurement tool, indicators are embedded as integral components of managerial decision-making.

Figure 1. PDCA-Leading Indicators Integration Model (Placeholder: schematic showing the continuous PDCA cycle with indicator categories linked to each stage-Plan, Do, Check, Act-supported by feedback loops.)

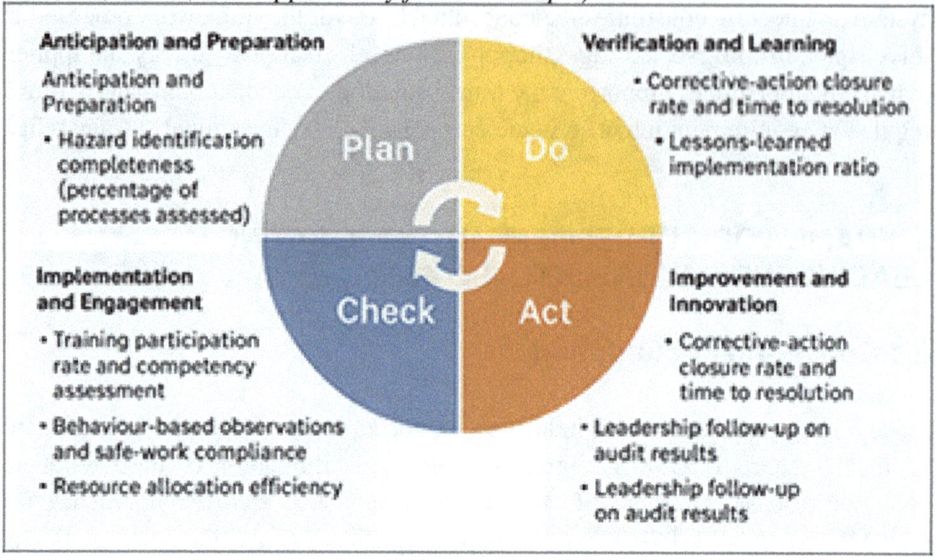

(a) Plan: Anticipation and Preparation

The *Plan* phase establishes the foundation for proactive safety management. Leading indicators here measure the quality of risk anticipation and strategic planning. Examples include:

- ***Hazard identification completeness*** (percentage of processes assessed).
- ***Preventive objective coverage*** (extent to which leading indicators are linked to strategic OHS goals).
- ***Resource allocation efficiency*** (proportion of budget directed to preventive actions).

Karimi et al. (2021) applied Reason's Swiss-cheese model to demonstrate that early-stage deficiencies in planning-such as incomplete risk assessments-serve as latent failures that propagate downstream. Integrating indicators at this stage ensures that planning reflects both organisational context and emerging risks, such as psychosocial and climate-related hazards introduced by ISO 45001:2024/Amd.1 (ISO, 2024).

(b) Do: Implementation and Engagement

The *Do* phase involves the operational execution of controls and programs. Here, leading indicators monitor process integrity and human engagement. Common measures include:

- ***Training participation rate*** and ***competency assessment*** (Guo & Yiu, 2016; Zulkarnain et al., 2023).
- ***Behaviour-based observations*** *and* ***safe-work compliance*** (Hallowell et al., 2013; Haas et al., 2017).
- ***Communication frequency between supervisors and workers***, reflecting the SMART (Specific, Measurable, Achievable, Relevant, Time-bound) communication model (Haas et al., 2021).

These metrics not only verify implementation but also capture behavioural precursors of system drift. According to Dekker (2016), human variability is both a source of error and adaptability; hence, monitoring engagement through proactive indicators enables learning rather than control.

(c) Check: Verification and Learning

In the *Check* phase, data from operational activities are analysed to evaluate process effectiveness and detect weak signals. This involves both *quantitative metrics* (e.g., audit findings, near-miss frequencies) and *qualitative feedback* (e.g., safety-culture surveys). Robson et al. (2017) demonstrated that audit data, when systematically analysed, can be converted into leading indicators predicting OHS performance trends.

Rubio-Romero et al. (2018) introduced the concept of a *Composite Leading Indicator* to assess resilience maturity, integrating dimensions of anticipation, monitoring, response, and learning. In this stage, leading indicators function as the nervous system of the management model, translating dispersed information into actionable insight. Latham et al. (2024) found that in agricultural contexts, feedback loops that translate observation-based data into targeted interventions produce measurable cultural transformation.

(d) Act: Improvement and Innovation

The *Act* phase closes the loop by converting evaluation outcomes into corrective and preventive actions (CAPAs). Here, leading indicators focus on the speed and quality of response, such as:

- *Corrective-action closure rate* and *time to resolution.*
- *Lessons-learned implementation ratio.*
- *Leadership follow-up on audit results.*

As Zwetsloot et al. (2020) emphasised, proactive governance requires that learning from incidents and near misses feeds back into strategic improvement. Within Vision Zero, this corresponds to the "Learning from Success" principle, transforming safety from a static compliance objective into a driver of innovation and well-being (Zwetsloot et al., 2025).

3.3 Indicator Classification and Dimensions

Based on the synthesis of prior frameworks (Guo & Yiu, 2016; Castillo et al., 2023; Pera et al., 2023), the indicators within the proposed PDCA model are grouped into four interrelated dimensions (Table 1). Accordingly, Table 1 summarises the mapping of lagging and leading indicators in the ISO 45001 context, aligning traditional outcome-based measures (e.g.; incident rates, lost-time injuries) with proactive process indicators such as training completion and near-miss reporting.

Table 1. Performance Indicators by PDCA Stage

Category	Focus	Examples of Indicators	Primary PDCA Stage
Operational	Process and control quality	Preventive maintenance rate, equipment inspection frequency	Plan / Do
Behavioural	Human participation and compliance	Observation-based safety index, training completion	Do / Check
Organisational	Leadership and management commitment	Leadership walkaround frequency, budget ratio for prevention	Plan / Act
Learning-oriented	Knowledge acquisition and adaptation	Corrective-action closure time, near-miss reporting ratio	Check / Act

This multi-dimensional classification reflects the systemic nature of safety. Each dimension interacts with the others, forming a continuous improvement network. For example, behavioural indicators depend on organisational leadership, which in turn influences operational compliance. Such interdependencies echo Leveson's (2015) systems-theoretic view that safety is maintained not through isolated barriers but through dynamic control relationships.

3.4 Data Sources, Measurement Logic, and Feedback

The reliability of leading indicators depends on the robustness of data collection and interpretation. Sinelnikov et al. (2015) proposed three validity criteria for effective indicators: sensitivity (ability to detect change), timeliness (response speed), and controllability (actionability). Applying these within ISO 45001 implies integrating data streams from multiple sources-audits, digital sensors, behavioural observations, and management reviews-into a unified analytic dashboard.

Recent technological advancements have facilitated this integration. Costin et al. (2019) introduced an IoT-based framework for generating *active leading indicators* in construction, automatically updating dashboards through real-time data collection. Mousavi et al. (2020) linked Lean Six Sigma maturity to proactive OHS performance, demonstrating that structured feedback mechanisms reduce both variability and latency in safety response. These digital feedback systems operationalise the "Check" and "Act" phases by enabling predictive analytics and visual reporting.

Moreover, embedding indicators into digital management platforms aligns with EU-OSHA's Digitalisation and OSH Strategy (2021-2027), which calls for data-driven decision-making and continuous monitoring of psychosocial and environmental risks. Thus, technology acts as a catalyst transforming the PDCA cycle from periodic evaluation to *continuous adaptation*.

3.5 Integrating Vision Zero and Resilience Engineering Principles

The integration framework presented below incorporates the philosophical foundations of Vision Zero and Resilience Engineering, merging human-centred and system-centred approaches. Vision Zero contributes the ethical imperative-no level of harm is acceptable while Resilience Engineering provides the operational mechanism: anticipate, monitor, respond, and learn (Hollnagel, 2017; Zwetsloot et al., 2020). Within the PDCA cycle, these paradigms interact: Vision Zero's normative goal sets the direction in the Plan phase, while Resilience Engineering informs the adaptive mechanisms that drive implementation, review, and improvement across the Do, Check, and Act phases.

In practical terms, the framework embeds four resilience functions within the PDCA cycle, as summarised in Table 2.

Table 2. Integration of Resilience Engineering Functions with PDCA Phases

Resilience Function	PDCA Alignment	Representative Leading Indicators
Anticipate	Plan	Risk-assessment coverage, emerging-hazard analysis
Monitor	Do	Observation-based reporting, communication frequency
Respond	Check	Near-miss trend analysis, safety-audit findings
Learn	Act	Corrective-action implementation, knowledge-sharing sessions

This integration ensures that the management system not only reacts to deviations but actively enhances its adaptive capacity. Rubio-Romero et al. (2018) empirically validated this alignment in municipal-waste companies, showing that composite resilience indicators significantly correlate with lower incident rates. Similarly, Phinias (2023) and Janaćković (2025) reported that organisations embedding Vision Zero principles into their OHSMS achieved measurable improvements in both performance and worker well-being.

3.6 Implementation Maturity and Continuous Improvement

Successful integration of leading indicators requires a maturity-based implementation approach. This approach, as described by da Silva and Amaral (2019) and Mohammadfam et al. (2017), refers to the structured, stage-wise development of an OHS management system, where organisations progress through clearly defined levels of capability and integration over time. It can be conceptualised as a phased developmental model in four stages:

i. ***Compliance-driven:*** Measurement limited to lagging outcomes.
ii. ***Procedural:*** Leading indicators introduced but not analysed.
iii. ***Integrated:*** Indicators systematically linked to PDCA processes.
iv. ***Optimised:*** Indicators predictive, data-visualised, and decision-driven.

Each maturity level is characterised by differing leadership involvement, data sophistication, and learning culture. Zwetsloot et al. (2025) noted that transitioning from procedural to optimised maturity requires cultural change. In practice, this cultural shift is realized by leaders who regularly communicate the value of proactive safety data, integrate leading indicator results into routine meetings, encourage employee participation in safety initiatives, and establish feedback systems that recognise and reward open reporting and improvement suggestions. Almost et al. (2019) and Haas et al. (2021) also demonstrated that supervisor engagement, transparent communication, and employee empowerment are critical accelerators of this transition. The role of leadership is therefore not limited to resource allocation but extends to modelling proactive behaviour and valuing feedback.

3.7 Practical Implications

Embedding leading indicators in ISO 45001 systems offers tangible managerial benefits. First, it enhances decision quality by providing forward-looking intelligence on safety performance. Second, it strengthens worker participation, a core clause of ISO 45001, by making feedback visible and actionable. Third, it ensures strategic alignment between safety goals and corporate sustainability targets, particularly under ESG reporting frameworks.

Moreover, this integrated model supports climate resilience and digital transformation-emerging dimensions recognised in ISO 45001:2024/Amd.1 and the EU Green Deal. By aligning proactive safety metrics with environmental and technological parameters, organisations can simultaneously improve occupational health and sustainability outcomes.

3.8 Synthesis

The PDCA-Leading Indicators Integration Model transforms ISO 45001 from a compliance checklist into a learning architecture. It operationalises the theoretical principles of Safety-II, Resilience Engineering, and Vision Zero through measurable, actionable, and adaptive indicators. By embedding feedback loops across all PDCA stages, the model converts data into continuous improvement and positions safety as a strategic enabler of organisational resilience.

4. PRACTICAL APPLICATIONS AND CASE INSIGHTS

4.1 Overview

The operationalisation of leading indicators within ISO 45001 management systems has moved from theoretical aspiration to practical reality across several high-risk industries. Evidence from mining, construction, healthcare, and chemical processing demonstrates that embedding proactive metrics in the PDCA cycle enhances risk anticipation, communication, and organisational learning (Haas et al., 2017; Almost et al., 2018; Xu et al., 2023; Zulkarnain et al., 2023). This section synthesises sector-specific experiences to illustrate how the framework proposed in Part 3 functions in practice and how measurable improvements in safety performance, culture, and resilience emerge.

4.2 Automotive Manufacturing

Automotive production combines mechanised assembly lines, robotics, and extensive human-machine interaction-an environment where small process deviations can escalate rapidly. Studies conducted in European and Asian automotive plants applying ISO 45001 found that introducing leading indicators for preventive maintenance, ergonomic assessment, and near-miss analysis reduced incident frequency by 20-30% within two years (Pera et al., 2023; Zwetsloot et al., 2025).

In one multinational automotive plant (case synthesised from EU-OSHA 2023 reports), the PDCA-based indicator model was applied as follows:

- *Plan:* Each department identified critical equipment and defined maintenance reliability indicators.
- *Do:* Supervisors tracked participation in safety huddles and operator feedback frequency.
- *Check:* Near-miss and deviation data were analysed weekly through digital dashboards.
- *Act:* Lessons learned informed procedural revisions and training modules.

This process reflects a broader shift in occupational safety and health policy from lagging, outcome-focused assessments toward more anticipatory and preventive management approaches, as also emphasised at the European policy level (EU-OSHA, 2023). In line with this shift, Hallowell et al. (2013), Mousavi et al. (2019), and Moon et al. (2025) argue that the combined use of leading indicators facilitates feedback and organizational learning processes and is positively associated with safety climate and related cultural dimensions. Notably, Moon et al. (2025), based on a questionnaire survey administered to 630 manufacturing workers across 12 subcontractor firms at a large industrial site, found that accident-free employees reported trust scores (Trust Leading Indicator, TLI) approximately 6% higher than those who had experienced accidents, and that higher trust scores were strongly associated with improved safety climate and leadership perceptions.

4.3 Construction Industry

Construction remains one of the most hazardous sectors, characterised by transient worksites, subcontracting, and dynamic risk profiles. The integration of leading indicators has been studied extensively in this context. Xu et al. (2021, 2023) identified a core set of proactive measures-pre-task risk assessments, safety-meeting frequency, and inspection closure rates-that reliably predicted project safety outcomes.

In a major infrastructure project analysed by Lingard et al. (2017), replacing monthly lagging reports with weekly leading-indicator dashboards (e.g., percentage of pre-task meetings held, hazard observations logged) allowed early intervention, reducing serious incidents by 25%. The PDCA logic underpinned continuous improvement:

- *Plan:* Hazard analyses integrated into work-planning documents.
- *Do:* Supervisors measured behavioural compliance using observation checklists.
- *Check:* Safety performance data reviewed during coordination meetings.
- *Act:* Corrective actions documented and implemented before new work phases.

Phinias (2023) confirmed that such data-driven monitoring builds mutual accountability among contractors, leading to sustained compliance even after project completion. Integration of Vision Zero principles further strengthened worker engagement-employees were trained to view safety as a shared value, not an imposed obligation.

4.4 Healthcare Sector

Healthcare institutions present complex safety challenges, including exposure to biological hazards, ergonomic strain, and psychosocial risks. Traditional lagging metrics, such as injury rates, fail to capture latent process weaknesses. Almost et al. (2018, 2019) pioneered the application of leading indicators in hospitals, using participation in risk assessments, communication frequency, and near-miss reporting as predictors of safety performance.

In their longitudinal study across 32 Canadian healthcare facilities, the introduction of proactive metrics resulted in:

- a 22% increase in hazard reports,
- a 35% improvement in corrective-action closure rates, and
- a 15% decline in recordable injuries over 18 months.

This was achieved by embedding measurement within ISO 45001-aligned PDCA cycles. *Plan* emphasised identification of ergonomic and psychosocial risks; *Do* involved training and team communication exercises; *Check* consisted of regular safety-culture surveys; and *Act* entailed leadership walk-rounds to reinforce feedback.

De Cieri et al. (2015) and Sheehan et al. (2016) further demonstrated that when management leadership is assessed through leading indicators-such as supervisor coaching frequency and resource allocation-positive spill-overs occur across patient-safety and service-quality domains, linking occupational and clinical outcomes.

4.5 Process and Chemical Industries

The process and chemical sectors operate under stringent risk-control requirements, making them ideal testbeds for leading-indicator integration. Zulkarnain et al. (2023) developed a framework for chemical-process industries (CPI) centred on proactive safety-performance metrics such as safety-instrumented system testing rates and management-of-change completion times. Implemented within the ISO 45001 PDCA architecture, these indicators provide early warning signals of process-safety weaknesses and support continuous improvement in abnormal event prevention (ibid.).

Similarly, Cheraghi and Khodadadi-karimvand (2023) applied leading indicators within process-safety management to track procedural compliance and training coverage. Their findings revealed that predictive metrics facilitated earlier identification of operational deviations, avoiding potential major incidents.

These examples show how the PDCA model aligns with Leveson's (2015) systems-control theory-each indicator serves as a feedback loop maintaining system behaviour within safe operational limits. Integration of environmental parameters, as mandated by ISO 45001:2024/Amd. 1, allowed chemical plants to anticipate climate-related hazards such as heat stress and volatile chemical reactions during high-temperature events.

As a result, unlike traditional, compliance-based control, which usually involves verifying adherence to prescriptive rules after the event, this systems-control approach allows operational practices to be adjusted continuously and in real time in response to emerging risks. Consequently, safety management becomes proactive rather than reactive.

4.6 Cross-Sectoral Lessons Learned

Synthesis of the above cases reveals several recurring enablers of success:

- ***Leadership Commitment:*** Across all sectors, managerial visibility and responsiveness are decisive. Zwetsloot et al. (2020) and Bargar et al. (2025) noted that proactive leadership indicators-such as walk-round frequency and feedback turnaround time-correlate strongly with system maturity.

- ***Employee Participation:*** In every case, organisations that empowered workers to contribute to indicator data collection achieved faster improvement. Frick (2011) highlighted that participatory OHSMS produce richer information streams and stronger learning cycles.
- ***Digitalisation:*** Automation of data gathering and dashboard visualisation enhances timeliness and accuracy. The IoT-based systems described by Costin et al. (2019) reduced reporting lag from weeks to hours.
- ***Cultural Integration:*** Sustainable success depends on embedding measurement within everyday behaviour, not treating it as external audit work. This cultural embedding reflects the *Resilience Engineering* principle of learning from normal work (Hollnagel, 2017).

Quantitative evaluations from multiple studies corroborate these findings, as summarised in Table 3.

Table 3. Cross-Sectoral Implementation of Leading Indicators and Their Impact

Sector	Key Leading Indicators Implemented	Observed Impact	Reference
Automotive	Preventive-maintenance rate, near-miss analysis	20-30% reduction in incidents	Pera et al. (2023)
Construction	Pre-task assessments, observation frequency	25% decline in major injuries	Xu et al. (2023)
Healthcare	Communication and reporting indicators	15% drop in recordables	Almost et al. (2018)
Chemical	Process-control and training indicators	40% fewer abnormal events	Zulkarnain et al. (2023)

These results empirically validate the theoretical claim that leading indicators are not merely diagnostic tools but drivers of system learning and adaptive capacity.

4.7 Barriers and Mitigation Strategies

Despite proven benefits, implementation challenges persist. da Silva and Amaral (2019) identified organisational inertia and insufficient analytical capacity as major barriers. In smaller enterprises, lack of expertise and resource constraints hinder indicator selection and data interpretation (Janaćković, 2025). Phinias (2023) observed resistance among construction supervisors who perceived the reporting task as an additional administrative burden.

Mitigation strategies derived from the reviewed cases and include:

- *Simplification of Metrics (SMART Approach)*: Instead of using complex data sets, the focus is on a limited number of high value-added indicators that are appropriate to the risk profile of the organisation. This strategy eases the analytical workload and ensures the most efficient use of limited corporate resources.
- *Competence Development (Data Literacy)*: Not only technical training, but also the ability of managers and employees to read, analyse and take action from data should be improved. This closes skill gaps by increasing confidence in interpreting indicators.
- *Automation of Data Collection Processes (Digital Innovation)*: Manual data entry should be minimised by using mobile applications and Internet of Things (IoT)-based digital tools. This method improves the speed of access to data by directly eliminating administrative workload complaints.
- *Transparent Feedback and Participation*: The data obtained should be shared regularly and transparently with all stakeholders. By demonstrating the tangible benefits of the system to operational security, both organisational inertia should be broken and the commitment of employees to the process should be maintained.

These strategies align with the EU-OSHA Strategic Framework (2021-2027), which emphasises competence development, worker participation, and digital innovation in occupational health and safety management.

4.8 Synthesis

Practical evidence confirms that integrating leading indicators into ISO 45001 transforms occupational-safety management from retrospective evaluation to real-time learning and foresight. Across industries, proactive metrics enhance leadership accountability, worker participation, and systemic resilience. The PDCA-Leading Indicators Integration Model thereby proves adaptable, scalable, and effective in achieving Vision Zero's ambition of *zero harm and continuous improvement.*

The lessons drawn here form the empirical grounding for the next section's analytical discussion on the cultural, managerial, and policy implications of proactive OHS measurement.

5. DISCUSSION

5.1 Overview

The findings across theoretical and empirical sections reveal that the integration of leading indicators into ISO 45001 represents a paradigm shift from *reactive compliance* to *proactive learning*. This transition resonates with global safety philosophies such as Vision Zero and the EU-OSHA 2021-2027 Strategic Framework, both of which emphasise prevention, participation, and continuous improvement as the cornerstones of sustainable occupational safety and health (OSH). By embedding predictive metrics in each phase of the PDCA cycle, organisations evolve from measuring failure to managing foresight-a hallmark of what Hollnagel (2017) calls *Safety-II systems thinking.*

5.2 Leadership and Governance Implications

Leadership commitment emerges as the central determinant of indicator effectiveness. The evidence from automotive, construction, and healthcare sectors (Haas et al., 2017; Almost et al., 2018; Xu et al., 2023) consistently shows that *managerial visibility*-through walk-rounds, open communication, and rapid response-correlates with higher safety maturity levels. Within the ISO 45001 framework, leadership is not a static top-down control but an *enabling process* that fosters trust, dialogue, and distributed accountability.

This aligns with Vision Zero's first golden rule: "Take leadership-demonstrate commitment." Zwetsloot (2025) describe leadership as the "engine of proactive culture," in which data from leading indicators guide strategic decision-making. Such data-driven governance moves beyond lagging statistics and situates safety performance within broader ESG reorting frameworks. Consequently, leading indicators function simultaneously as management intelligence tools and ethical compasses-quantifying the value of human well-being within organisational sustainability.

5.3 Safety Culture Transformation

Traditional safety cultures often rely on retrospective accountability-counting incidents rather than learning from their absence. The widespread introduction of leading indicators has redefined this paradigm. Studies by Phinias (2023) and Latham et al. (2024) show that when teams track proactive metrics such as hazard observations or communication frequency, psychological ownership of safety increases markedly.

The transformation process reflects three cultural dimensions:

i. ***Cognitive:*** Understanding what constitutes proactive safety behaviour.
ii. ***Normative:*** Embedding leading-indicator use as a shared organisational value.
iii. ***Behavioural:*** Sustaining participation through transparent feedback.

Under ISO 45001, these dimensions manifest in clause 5 (*Leadership and Worker Participation*) and clause 9 (*Performance Evaluation*). Embedding leading indicators into these clauses transforms the system from a managerial checklist into a participatory learning architecture.

Furthermore, the integration of Vision Zero principles encourages positive framing-learning from successful operations rather than exclusively analysing deviations. This *appreciative inquiry* approach, central to Safety-II, strengthens morale and resilience, especially in sectors with high emotional and cognitive demands such as healthcare.

5.4 Learning Systems and Organisational Resilience

Leading indicators act as the nervous system of a resilient organisation. By providing continuous sensory input on human, technical, and environmental conditions, they enable early detection of weak signals preceding failure (Leveson, 2015; Rubio-Romero et al., 2018). The practical cases from Part 4 illustrate that when such indicators are integrated into PDCA reviews, learning becomes systematic rather than incidental.

Resilience Engineering offers the theoretical lens for this phenomenon. Hollnagel's four resilience functions-*anticipate, monitor, respond, learn*-map directly onto the PDCA phases. The integration framework proposed in Part 3 operationalises these functions through measurable indicators, converting resilience from an abstract aspiration into a quantifiable management outcome.

Empirical validation from chemical and construction industries confirms this relationship: organisations that maintain regular learning reviews based on proactive metrics exhibit stronger adaptive capacity to external shocks, including climate-related disruptions. This connection underscores ISO 45001:2024/Amd. 1's new emphasis on environmental and climate factors within the OHS context.

5.5 Digital Transformation and Data-Driven Safety

Digitalisation is both an enabler and a challenge in the modern safety landscape. Smart sensors, IoT networks, and data-visualisation tools allow real-time monitoring of leading indicators, automating much of the *Check* and *Act* stages. Costin et al. (2019) demonstrated that IoT-based dashboards shortened response times and provided predictive analytics that pre-empted accidents in construction settings.

However, digital transformation requires robust data-governance frameworks. Without quality assurance and context-sensitive interpretation, the abundance of information risks overwhelming decision-makers. The EU-OSHA (2021-2027) framework therefore calls for a balance between technological innovation and human-centred design. Proactive indicators must measure not only physical safety but also psychosocial well-being, ergonomics, and digital workload-domains increasingly relevant in hybrid and AI-supported workplaces.

Integrating digital metrics such as automated data collection, artificial intelligence, and advanced analytics into ISO 45001 systems is creating a new frontier of smart occupational health and safety management (Maroń et al., 2022; ILO & ISSA, 2022). Via these "smart OHSMSs," predictive analytics feed directly into leadership dashboards, enabling organisations to monitor workplace conditions, anticipate risks, and support timely, evidence-based decisions.

5.6 Integration with Sustainability and ESG Frameworks

As corporations align their strategies with the UN Sustainable Development Goals (SDGs), leading indicators offer a bridge between occupational safety and sustainability performance. ISO 45001's preventive logic directly supports SDG 8 (*Decent Work and Economic Growth*) and SDG 3 (*Good Health and Well-Being*). Proactive safety metrics demonstrate tangible progress toward these goals by quantifying non-financial performance.

For example, reporting the *percentage of corrective actions closed within target time* or *training completion rates* can serve as ESG metrics under the "Social" dimension. These data are auditable, comparable, and can be standardised across industries. Consequently, OHS professionals evolve from compliance officers to strategic partners contributing to corporate sustainability and investor transparency.

5.7 Policy Alignment and Global Convergence

At the policy level, the integration of leading indicators aligns with transnational initiatives promoting preventive governance. The EU-OSHA Strategic Framework 2021-2027 explicitly calls for "better anticipation of new and emerging risks," emphasising digitalisation, demographic change, and green transition as critical priorities. The PDCA-Leading Indicator model provides a micro-level operationalisation of these macro-objectives.

Similarly, the Vision Zero Global Campaign, supported by the International Social Security Association (ISSA), positions proactive indicators as the principal mechanism for continuous improvement and well-being at work (Zwetsloot et al., 2020; Bargar et al., 2025). The convergence of ISO 45001, Vision Zero, and EU-OSHA strategies thus forms a cohesive global governance architecture where data, leadership, and participation reinforce one another.

This convergence also supports harmonisation with other ISO standards-9001 (Quality), 14001 (Environment), and 50001 (Energy)-enabling integrated management systems that address safety as part of holistic organisational excellence.

5.8 Barriers, Risks, and Ethical Considerations

Despite substantial progress, challenges persist in implementing leading indicators equitably and effectively. Key barriers include:

- ***Data Overload:*** Excessive proliferation of metrics without a clear strategic purpose, leading to diluted managerial focus and reduced interpretability.
- ***Cultural Resistance:*** Worker perceptions of leading-indicator tracking as a form of surveillance rather than a tool for improvement.
- ***Resource Constraints:*** Limited analytical capacity, expertise, and financial resources, particularly among small and medium-sized enterprises (SMEs).
- ***Ethical Concerns:*** The use of algorithmic decision-support tools that may introduce bias, opacity, or risks to data privacy.

To mitigate these risks, organisations should adopt a minimal-viable-indicator approach, selecting a balanced set of high-impact metrics explicitly linked to genuine improvement objectives. Moreover, transparency in data use, participatory development of indicators, and ethical oversight of digital systems are essential to maintaining trust and legitimacy, consistent with established governance frameworks such as the EU AI Act-which emphasises transparency, human oversight, and risk-based accountability-and ISO 26000, which frames occupational health, stakeholder participation, and responsible organisational behaviour as core elements of social responsibility (ISO, 2010; European Commission, 2024).

5.9 Toward a Proactive Safety Paradigm

The synthesis of theoretical, empirical, and policy insights points toward a new paradigm: proactive safety governance. This paradigm is characterised by:

i. ***Predictive measurement:*** anticipating risk rather than reacting to harm.
ii. ***Participatory learning:*** involving all organisational levels in indicator generation and interpretation.

iii. ***Digital-resilient systems:*** integrating human and technological intelligence.
iv. ***Ethical leadership:*** aligning safety decisions with human-centred values.

Such a paradigm reframes occupational health and safety as a dynamic system of socio-technical adaptation. Within this system, leading indicators are not auxiliary metrics but *strategic sensors*-measuring the organisation's capacity to foresee, learn, and evolve.

5.10 Synthesis

In conclusion, the integration of leading indicators into ISO 45001 and Vision Zero frameworks represents a decisive advancement in global OHS management. It transforms compliance-oriented management systems into adaptive, evidence-based learning networks. Through leadership commitment, cultural transformation, digital innovation, and policy alignment, the PDCA-Leading Indicator model provides a replicable pathway toward sustainable excellence in safety, health, and well-being at work.

This discussion establishes the conceptual and managerial foundation for the final chapter-Conclusion and Future Research Directions-which will synthesise implications, limitations, and future trajectories for research and practice.

6. CONCLUSION AND FUTURE RESEARCH

6.1 Summary of Core Insights

The integration of leading indicators into occupational health and safety management systems represents a decisive evolution from retrospective to anticipatory governance. As demonstrated throughout this chapter, the PDCA-Leading Indicators Integration Model operationalises the preventive logic of ISO 45001 by embedding foresight and feedback into every organisational process. Drawing on insights from multiple industries such as automotive, construction, healthcare, and chemical manufacturing-the evidence confirms that proactive measurement improves both safety outcomes and organisational learning capacity (Haas et al., 2017; Almost et al., 2018; Xu et al., 2023; Zulkarnain et al., 2023).

At the theoretical level, the framework fuses Safety-II thinking, Resilience Engineering, and Vision Zero principles. At the managerial level, it aligns with ISO 45001's PDCA cycle, transforming static compliance systems into adaptive learning architectures. At the policy level, it directly supports the objectives of the EU-OSHA 2021-2027 Strategic Framework, which calls for proactive anticipation of emerging risks in an era of digitalisation, demographic change, and the green transition. The detailed explanation, implementation stages, and practical implications of this integration model are discussed in Sections 3.5 through 3.8.

6.2 Managerial and Policy Implications

For practitioners and policymakers, this synthesis underscores several actionable insights:

i. ***Strategic Integration:*** Leading indicators must be treated as *strategic performance tools*, not peripheral metrics. Linking them to ESG reporting and corporate sustainability objectives embeds occupational safety into long-term value creation.
ii. ***Leadership and Participation:*** Effective implementation depends on leadership engagement and worker participation. Vision Zero's "Take leadership-demonstrate commitment" and ISO 45001's clause 5 ("Leadership and worker participation") converge to define participatory governance as the foundation of proactive safety culture (Zwetsloot et al., 2025).
iii. ***Digitalisation and Data Governance:*** Digital tools enable continuous monitoring, but must be accompanied by ethical data management. Transparent algorithms, participatory indicator design, and protection of worker privacy are essential for trust.
iv. ***Regulatory Convergence:*** The alignment between ISO 45001:2024/Amd. 1, EU-OSHA strategies, and Vision Zero constitutes a global framework for sustainable OHS governance-one that integrates climate adaptation, psychosocial health, and digital transformation.

These implications illustrate a broader shift: occupational safety and health is no longer a compliance obligation but a strategic competency contributing to resilience, innovation, and sustainable competitiveness.

6.3 Methodological and Theoretical Gaps

Despite major advances, several gaps remain in research and practice:

- ***Measurement Validity:*** Few studies validate the predictive power of leading indicators across diverse contexts. Longitudinal and cross-industry datasets are needed to establish causal relationships.

- ***Psychosocial Dimensions:*** Current indicators focus primarily on physical and procedural risks; psychosocial and cognitive metrics-stress, burnout, digital fatigue-remain underdeveloped.
- ***AI-Enabled Predictive Analytics:*** The rise of machine learning offers opportunities for real-time risk prediction, yet algorithmic transparency and explainability must be assured.
- ***Integration with Sustainability Metrics:*** Future research should explore the linkage between OHS leading indicators and SDG performance metrics, especially under corporate ESG frameworks.
- ***Small-Enterprise Contexts:*** Most empirical work addresses large organisations. Tailoring simplified, resource-efficient indicator systems for SMEs remains an urgent priority.

Addressing these gaps requires interdisciplinary collaboration between engineers, social scientists, and data analysts, supported by international benchmarking initiatives such as those of the ISSA and EU-OSHA.

6.4 Future Research Directions

To advance the state of knowledge, future studies should:

i. ***Develop Dynamic Indicator Models:*** Construct adaptive algorithms that integrate multi-source data (behavioural, environmental, organisational) for predictive analytics within the PDCA cycle.
ii. ***Quantify the "Return on Prevention":*** Employ econometric and system-dynamics models to assess how proactive safety metrics translate into financial and social value.
iii. ***Investigate Cross-Standard Synergies:*** Explore how ISO 45001 indicators can interface with ISO 9001, 14001, and 50001 to create integrated sustainability dashboards.
iv. ***Assess Climate-Related Risks:*** Incorporate meteorological and environmental data into safety indicators, operationalising ISO 45001:2024/Amd. 1's climate-risk provisions.
v. ***Enhance Participatory Methods:*** Study the influence of participatory indicator co-design on motivation, trust, and reporting accuracy among workers.

These priorities, particularly those focused on participatory methods and shared learning, directly support the cultural transformation outlined in Section 5.3, where the integration of leading indicators fosters psychological ownership, open feedback, and a proactive safety mindset.

6.5 Final Thoughts

The transition *from lagging to leading indicators* symbolises more than a methodological innovation; it reflects a philosophical shift in how organisations perceive safety and responsibility. By transforming data into foresight, and foresight into action, leading indicators make visible the latent structures of resilience within complex socio-technical systems.

Integrating these indicators into ISO 45001 and Vision Zero frameworks redefines occupational safety as a living system-adaptive, participatory, and ethically grounded. The challenge ahead is not merely to measure safety more effectively, but to govern it intelligently, ensuring that technology, leadership, and human values evolve together.

Through continued research and international cooperation, the PDCA-Leading Indicators Integration Model can serve as a blueprint for organisations seeking not only to prevent harm but to promote health, innovation, and sustainable excellence in the workplaces of the future.

AI USE DECLARATION

Artificial intelligence tools (such as ChatGPT) were used solely for minor grammar correction, language refinement, and stylistic improvement of the text. No AI-generated content, data interpretation, or substantive writing assistance was used in the development of the ideas, analysis, or conclusions presented in this chapter.

REFERENCES

Almost, J., Caicco Tett, L., Vandenkerkhof, E., Paré, G., Strahlendorf, P., Noonan, J., Hayes, T., Van Hulle, H., Holden, J., Silva e Silva, V., & Rochon, A. (2019). Leading indicators in occupational health and safety management systems in healthcare: A quasi-experimental longitudinal study. *Journal of Occupational and Environmental Medicine*, *61*(12), e486–e496. DOI: 10.1097/JOM.0000000000001738 PMID: 31626068

Almost, J. M., VanDenKerkhof, E. G., Strahlendorf, P., Caicco Tett, L., Noonan, J., Hayes, T., Van hulle, H., Adam, R., Holden, J., Kent-Hillis, T., McDonald, M., Paré, G. C., Lachhar, K., & Silva e Silva, V. (2018). A study of leading indicators for occupational health and safety management systems in healthcare. *BMC Health Services Research*, *18*(1), 296. DOI: 10.1186/s12913-018-3103-0 PMID: 29685147

Barbosa, C., Azevedo, R., & Rodrigues, M. A. (2019). Occupational safety and health performance indicators in SMEs: A literature review. *Work (Reading, Mass.)*, *64*(2), 217–227. DOI: 10.3233/WOR-192988 PMID: 31524191

Bargar, M., Jahangiri, M., Alimohammadlou, M., Karimpour, S., & Kamalinia, M. (2025). Evaluation of ISSA proactive leading indicators for safety, health and well-being: Application of multi-criteria decision-making methods based on hesitant fuzzy. *PLoS One*, *20*(9), e0322575. DOI: 10.1371/journal.pone.0322575 PMID: 40892944

Bennett, J., & Foster, P. (2005). Predicting progress: The use of leading indicators in occupational safety and health. *Policy and Practice in Health and Safety*, *3*(2), 77–90. DOI: 10.1080/14774003.2005.11667663

Castillo, C., Shahriari, M., Casarejos, F., & Arezes, P. (2023). Prioritization of leading operational indicators in occupational safety and health. *International Journal of Occupational Safety and Ergonomics*, *29*(2), 806–814. DOI: 10.1080/10803548.2022.2082693 PMID: 35622396

Cheraghi, S., & Khodadadi-karimvand, M. (2023). Developing and Implementation of Leading and Lagging Indicators in Process Safety Management. *Researcher*, *3*(01), 36–45.

Costin, A., Wehle, A., & Adibfar, A. (2019). Leading indicators-A conceptual IoT-based framework to produce active leading indicators for construction safety. *Safety (Basel, Switzerland)*, *5*(4), 86. DOI: 10.3390/safety5040086 da Silva, S. L. C., & Amaral, F. G. (2019). Critical factors of success and barriers to the implementation of occupational health and safety management systems: A systematic review of literature. *Safety Science, 117*, 123–132. DOI: 10.1016/j.ssci.2019.03.026

Dekker, S. (2016). *Drift into failure: From hunting broken components to understanding complex systems*. CRC press., DOI: 10.1201/9781315257396

European Agency for Safety and Health at Work. (2023). *Improving occupational safety and health through leading indicators: Case studies and practical applications*. Publications Office of the European Union. https://osha.europa.eu

European Agency for Safety and Health at Work. (2023). *Priorities for occupational safety and health research in Europe 2023–2026*. Publications Office of the European Union. https://www.europarl.europa.eu/cmsdata/286347/EU-OSHA%20CAAR%202023.pdf

European Commission. (2021, June 28). *EU Strategic Framework on Health and Safety at Work 2021-2027: Occupational safety and health in a changing world of work* (COM (2021) 323 final). Brussels: European Commission.

European Commission. (2024). *Regulation of the European Parliament and of the Council laying down harmonised rules on artificial intelligence (Artificial Intelligence Act)*. https://eur-lex.europa.eu

Frick, K. (2011). Worker influence on voluntary OHS management systems–A review of its ends and means. *Safety Science, 49*(7), 974–987. DOI: 10.1016/j.ssci.2011.04.007

Givehchi, S., Hemmativaghef, E., & Hoveidi, H. (2017). Association between safety leading indicators and safety climate levels. *Journal of Safety Research, 62*, 23–32. DOI: 10.1016/j.jsr.2017.05.003 PMID: 28882271

Givehchi, S., Hemmativaghef, E., & Hoveidi, H. (2017). Association between safety leading indicators and safety climate levels. *Journal of Safety Research, 62*, 23–32. DOI: 10.1016/j.jsr.2017.05.003 PMID: 28882271

Golabchi, H., Abellanosa, A. D., Lefsrud, L., Pereira, E., & Mohamed, Y. (2024). A comprehensive systematic review of safety leading indicators in construction. *Safety Science, 172*, 106433. DOI: 10.1016/j.ssci.2024.106433

Guo, B. H., & Yiu, T. W. (2016). Developing leading indicators to monitor the safety conditions of construction projects. *Journal of Management Engineering, 32*(1), 04015016. DOI: 10.1061/(ASCE)ME.1943-5479.0000376

Haas, E. J., Connor, B. P., Vendetti, J., & Heiser, R. (2017). A case study exploring field-level risk assessments as a leading safety indicator. *Transactions of Society for Mining, Metallurgy, and Exploration, Inc., 342*(1), 22–28. DOI: 10.19150/trans.8104 PMID: 29503519

Haas, E. J., Manley, T., & Wickizer, J. (2021). Applying Smart Principles to Improve Supervisor Communication as a Leading Indicator of OSH. https://stacks.cdc.gov/view/cdc/215408

Hallowell, M. R., Hinze, J. W., Baud, K. C., & Wehle, A. (2013). Proactive construction safety control: Measuring, monitoring, and responding to safety leading indicators. *Journal of Construction Engineering and Management, 139*(10), 04013010. DOI: 10.1061/(ASCE)CO.1943-7862.0000730

Heinrich, H. W. (1941). *Industrial accident prevention: A scientific approach* (2nd ed.). McGraw-Hill Book Company.

Hollnagel, E. (2017). *Safety-II in practice: developing the resilience potentials*. Routledge., DOI: 10.4324/9781315201023

International Labour Organization & International Social Security Association [ILO-ISSA]. (2024). *Digitalisation and smart safety management systems: Transforming occupational safety and health*. https://www.ilo.org/global/topics/safety-and-health-at-work/resources-library/publications/WCMS_848328/lang--en/index.htm

International Organization for Standardization. (2010). *ISO 26000:2010-Guidance on social responsibility*. https://www.iso.org/standard/42546.html

International Organization for Standardization [ILO]. (2024). *ISO 45001:2024/Amd. 1 Occupational health and safety management systems-Amendment 1: Incorporating climate change considerations.* Geneva, Switzerland: Author.

International Organization for Standardization [ISO]. (2018). *ISO 45001:2018 Occupational health and safety management systems- Requirements with guidance for use.* Geneva, Switzerland: Author.

Janaćković, G. L. (2025). Identification of key leading indicators for improving occupational health and safety systems in small and medium-sized cold storage facilities. *South African Journal of Industrial Engineering, 36*(1), 1–13. https://hdl.handle.net/10520/ejc-indeng_v36_n1_a2. DOI: 10.7166/36-1-3149

Karimi, A., Abbasi, M., Zokaei, M., & Falahati, M. (2021). Development of leading indicators for the assessment of occupational health performance using Reason's Swiss cheese model. *Journal of Education and Health Promotion, 10*(1), 158. DOI: 10.4103/jehp.jehp_1326_20 PMID: 34222533

Latham, A., Cotton, J., & Brumby, S. (2024). A rapid review of leading indicators to measure Australian farm safety culture. *Heliyon, 10*(12), e32736. DOI: 10.1016/j.heliyon.2024.e32736 PMID: 38975133

Leveson, N. (2015). A systems approach to risk management through leading safety indicators. *Reliability Engineering & System Safety, 136*, 17–34. DOI: 10.1016/j.ress.2014.10.008

Lingard, H., Hallowell, M., Salas, R., & Pirzadeh, P. (2017). Leading or lagging? Temporal analysis of safety indicators on a large infrastructure construction project. *Safety Science, 91*, 206–220. DOI: 10.1016/j.ssci.2016.08.020

Manuele, F. A. (2009). Leading & Lagging Indicators: Do they add value to the practice of safety? *Professional Safety, 54*(12), 28–33. https://www.jstor.org/stable/48687557

Maroń, A., Tomaszewska, E., & Chaberek, M. (2022). Smart Safety Management Systems (SSMS): A new paradigm in occupational health and safety. *Safety Science, 148*, 105638. DOI: 10.1016/j.ssci.2022.105638

Mohammadfam, I., Kamalinia, M., Momeni, M., Golmohammadi, R., Hamidi, Y., & Soltanian, A. (2017). Evaluation of the quality of occupational health and safety management systems based on key performance indicators in certified organizations. *Safety and Health at Work, 8*(2), 156–161. DOI: 10.1016/j.shaw.2016.09.001 PMID: 28593071

Moon, K., Kim, K. W., & Lee, J. D. (2025). Exploratory study for the adaptability of trust leading indicator and proactive leading indicator based on prevention culture. *Safety and Health at Work, 16*(2), 193–199. DOI: 10.1016/j.shaw.2025.02.002 PMID: 40575697

Mousavi, S. S., Khani Jazani, R., Cudney, E. A., & Trucco, P. (2020). Quantifying the relationship between lean maturity and occupational health and safety: Antecedents and leading indicators. *International Journal of Lean Six Sigma, 11*(1), 150–170. DOI: 10.1108/IJLSS-04-2018-0043

Neamat, S. D. S. (2019). A comparative study of safety leading and lagging indicators measuring project safety performance *indicators, 1*, 29. https://dx.doi.org/ DOI: 10.25046/aj040639

Pawłowska, Z. (2015). Using lagging and leading indicators for the evaluation of occupational safety and health performance in industry. *International Journal of Occupational Safety and Ergonomics, 21*(3), 284–290. DOI: 10.1080/10803548.2015.1081769 PMID: 26647949

Pera, F., Murino, T., Madonna, M., Di Nardo, M., & Bizzarro, R. (2023). A Conceptual Framework to Defining Leading Indicators to Measure Safety Management System Performance. *Chemical Engineering Transactions*, *99*, 439–444. DOI: 10.3303/CET2399074

Phinias, R. N. (2023). Benefits and challenges relating to the implementation of health and safety leading indicators in the construction industry: A systematic review. *Safety Science*, *163*, 106131. DOI: 10.1016/j.ssci.2023.106131

Reason, J. (2016). *Managing the risks of organizational accidents*. Routledge., DOI: 10.4324/9781315543543

Robson, L. S., Ibrahim, S., Hogg-Johnson, S., Steenstra, I. A., Van Eerd, D., & Amick, B. C.III. (2017). Developing leading indicators from OHS management audit data: Determining the measurement properties of audit data from the field. *Journal of Safety Research*, *61*, 93–103. DOI: 10.1016/j.jsr.2017.02.008 PMID: 28454876

Rubio-Romero, J. C., del Carmen Pardo-Ferreira, M., De la Varga-Salto, J., & Galindo-Reyes, F. (2018). Composite leading indicator to assess the resilience engineering in occupational health & safety in municipal solid waste management companies. *Safety Science*, *108*, 161–172. DOI: 10.1016/j.ssci.2018.04.014

Sheehan, C., Donohue, R., Shea, T., Cooper, B., & De Cieri, H. (2016). Leading and lagging indicators of occupational health and safety: The moderating role of safety leadership. *Accident; Analysis and Prevention*, *92*, 130–138. DOI: 10.1016/j.aap.2016.03.018 PMID: 27060754

Sinelnikov, S., Inouye, J., & Kerper, S. (2015). Using leading indicators to measure occupational health and safety performance. *Safety Science*, *72*, 240–248. DOI: 10.1016/j.ssci.2014.09.010

Turner, B. A., & Pidgeon, N. F. (1997). *Man-made disasters* (2nd ed.). Butterworth-Heinemann., https://worldcat.org/title/36023955

Walaski, P. (2020). The role of leading & lagging indicators in OSH performance management. *Professional Safety*, *65*(08), 29–35. https://onepetro.org/PS/article-abstract/65/08/29/448759/The-Role-of-Leading-amp-Lagging-Indicators-in-OSH

Xu, J., Cheung, C., Manu, P., & Ejohwomu, O. (2021). Safety leading indicators in construction: A systematic review. *Safety Science*, *139*, 105250. DOI: 10.1016/j.ssci.2021.105250

Xu, J., Cheung, C., Manu, P., Ejohwomu, O., & Too, J. (2023). Implementing safety leading indicators in construction: Toward a proactive approach to safety management. *Safety Science*, *157*, 105929. DOI: 10.1016/j.ssci.2022.105929

Yorio, P. L., Haas, E. J., Bell, J. L., Moore, S. M., & Greenawald, L. A. (2020). Lagging or leading? Exploring the temporal relationship among lagging indicators in mining establishments 2006–2017. *Journal of Safety Research*, *74*, 179–185. DOI: 10.1016/j.jsr.2020.06.018 PMID: 32951782

Zulkarnain, N. A., Anuar, N. I. P., Aziz, H. A., & Nawi, W. N. R. M. (2023, September). Determination of leading indicator for proactive safe performance in chemical process industries (CPI). In. AIP Conference Proceedings: Vol. 2792. *No. 1* (p. 040007). AIP Publishing LLC., DOI: 10.1063/5.0148562

Zwetsloot, G., Leka, S., Kines, P., & Jain, A. (2020). Vision zero: Developing proactive leading indicators for safety, health and wellbeing at work. *Safety Science*, *130*, 104890. DOI: 10.1016/j.ssci.2020.104890

Zwetsloot, G. I. (2025). *Vision Zero at Work: Strategies for Sustainable Excellence in Safety, Health and Wellbeing*. CRC Press. DOI: 10.1201/9781003621225

Zwetsloot, G. I., Aaltonen, M., Wybo, J. L., Saari, J., Kines, P., & De Beeck, R. O. (2013). The case for research into the zero accident vision. *Safety Science*, *58*, 41–48. DOI: 10.1016/j.ssci.2013.01.026

KEY TERMS AND DEFINITIONS

ISO 45001: An international standard for occupational health and safety management systems providing a structured, risk-based framework for managing hazards, engaging workers, and achieving continual improvement. It integrates the PDCA cycle and supports proactive performance measurement through leading indicators.

Lagging Indicator: A reactive metric that reflects past performance or adverse outcomes, such as injury rates or lost-time incidents. Lagging indicators are useful for benchmarking but provide limited predictive insight compared to leading indicators.

Leading Indicator: A proactive, preventive metric used to assess the effectiveness of OHS processes before incidents occur. Leading indicators measure activities such as training participation, near-miss reporting, and leadership engagement to anticipate and mitigate risk.

PDCA Cycle (Plan-Do-Check-Act): A continuous improvement framework forming the core of ISO 45001. • Plan involves identifying hazards and setting objectives; • Do implements preventive controls; • Check monitors performance and compliance; • Act ensures corrective actions and learning feed back into system improvement.

Proactive Safety Governance: An integrated management approach that combines leadership commitment, data-driven decision-making, and participatory learning to predict and prevent risks. It shifts OHS management from compliance to strategic foresight.

Resilience Engineering: A systems-theoretic approach to safety that focuses on an organisation's ability to anticipate, monitor, respond, and learn. It views safety as an emergent property of complex socio-technical systems and provides the conceptual foundation for Safety-II.

Safety-I: A traditional safety paradigm defining safety as the absence of accidents or harm. Safety-I focuses on controlling deviations, analysing failures, and reducing incident frequency.

Safety-II: A modern paradigm defining safety as the ability to succeed under varying conditions. It emphasises adaptability, learning from normal operations, and sustaining successful performance rather than merely preventing failure.

Vision Zero: A global prevention strategy promoted by the International Social Security Association (ISSA), asserting that all accidents and work-related harm are preventable. Vision Zero integrates safety, health, and well-being, guided by seven "Golden Rules," including leadership commitment, risk anticipation, and continuous learning.

Chapter 7
Navigating the Intersection of Climate Change and Occupational Health and Safety

Özlem Yurtsever
http://orcid.org/0000-0002-5312-3771
Marmara University, Turkey

ABSTRACT

This chapter asserts the climate crisis presents an urgent challenge to global OHS systems, which fail to address climate risks and structural inequities. The objective is to offer actionable strategies and identify research priorities for vulnerable populations. The chapter details how climate change harms workers through direct hazards like extreme heat and extreme weather, and indirect psychosocial stressors. Vulnerabilities are mapped across climate-sensitive sectors: agriculture, construction, and informal labor. Addressing this requires a shift toward a climate-adaptive safety paradigm, integrating ILO/WHO frameworks with ESG reporting. Effective safeguarding demands combining regulatory reform, workplace adaptation, and technological innovation, including wearable sensors and AI-driven analytics, for real-time risk assessment. Ultimately, the chapter advocates for sustained intersectoral collaboration and long-term, equity-centered strategies to ensure worker protection.

1. INTRODUCTION

The climate crisis ranks among the most urgent challenges of our century. Its impacts span public health, economic stability, and working conditions. Rising tem-

DOI: 10.4018/979-8-3373-6370-7.ch007

peratures, shifting rainfall patterns, and the growing frequency of extreme weather are already altering how and where people work across many industries (ILO, 2024; Marinaccio et al., 2025). Workers in various sectors face heightened threats like extreme heat and a rise in climate-sensitive diseases. Addressing these challenges calls for timely and evidence-based action. Three areas are particularly urgent: revising occupational health and safety (OHS) frameworks to include climate-related risks, applying technological innovations to strengthen protection systems, and developing adaptive strategies that build resilience and support long-term adjustment.

Even moderate warming of 1.5°C, a threshold likely to be breached by 2030, is expected to have cascading effects that extend beyond the environment to worker health and safety. Rising sea levels, food insecurity, and pressures on public health systems amplify occupational vulnerabilities, particularly in agriculture, fisheries, and coastal industries. Developing small island states epitomizes this uncertainty. Marginalized groups, such as Sub-Saharan African agricultural workers, face compound crises where extreme heat and drought undermine both worker health and local food systems (FAO, 2019; IPCC, 2018). Yet, regional-level evidence on how climate change affects occupational health remains limited, especially concerning the differences between 1.5°C and 2°C warming across climate-sensitive outcomes and their implications for labor migration (Ebi et al., 2018).

Climate-sensitive health outcomes such as infectious diseases, deteriorating air quality, and rising mental health challenges carry direct implications for occupational health. Diarrheal diseases, for example, increase risks for workers in water-dependent industries, while heat-related stress contributes to burnout in emergency services and construction sectors. Migration triggered by climate pressures heightens worker vulnerability, as displaced laborers often fall outside formal OHS protections.

As climate risks intensify, policymaker responses are gaining momentum. This growing awareness is beginning to translate into policy action, especially as concrete data from urban and national bodies become available. Workers in climate-sensitive sectors face disproportionate risks, a point elaborated in Section 2.1. These environmental shifts intensify physical and psychosocial risks, threatening worker well-being and economic resilience across sectors. Beyond direct health effects, climate change imposes significant economic burdens through productivity losses, increased healthcare expenditures, and deepening poverty, revealing the macroeconomic imperative of investing in OHS. In the United States, public hearings were held by the Occupational Safety and Health Administration (OSHA) in July 2025 for a proposed rule to protect workers from heat-related injury in both indoor and outdoor occupational settings, an indication that heat is becoming increasingly considered an important occupational hazard (OSHA, 2025). Similarly, urban analyses, such as a 2024 New York City report, highlight how extreme temperatures and air quality degradation disproportionately affect outdoor workers, particularly vulnerable populations like migrants and informal laborers (NYC Mayor's Office of Immigrant Affairs, 2025). These impacts are not distributed evenly; instead, they exacerbate existing inequities by placing a heavier burden on vulnerable groups (Amoako, 2019; Van Selm et al., 2025). This highlights the need for targeted interventions to protect at-risk groups. While these examples are drawn from high-income settings, similar patterns of heat- and pollution-related occupational risks are evident worldwide, particularly in rapidly urbanizing regions of Asia, Africa, and Latin America, where formal protections are often weaker.

These national and urban-level efforts are part of a broader global push to address climate-driven occupational risks, a trend also reflected in projections from international bodies such as the International Labor Organization (ILO). The ILO projects that rising global temperatures could result in labor losses amounting to tens of millions of full-time jobs by 2030, with low- and middle-income countries expected to shoulder the greatest burden (ILO, 2024). Addressing this looming challenge necessitates the urgent incorporation of climate resilience into occupational health and safety frameworks, particularly in sectors such as agriculture, construction, and manufacturing, where the intersection of economic security and worker health is most pronounced.

Recent developments point to a broadening of the OHS agenda, as conventional climate adaptation strategies increasingly intersect with emerging issues such as psychosocial well-being and the implications of new energy technologies (Solberg, 2025). Protective measures are consistently undermined by a persistent lack of coordination across sectors. For example, more than 4,000 occupational heat-related injuries occur each year in Italy alone, representing not only a health indicator but also a heavy socioeconomic cost of inaction. (Gariazzo et al., 2023; Grandi et al., 2023; Schulte et al., 2023). The climate change exposures land most heavily on small and medium-sized enterprises (SMEs) and on informal employees, who are often excluded from formal protection mechanisms and thus exposed to biased exposure (Amoadu et al., 2023; Schulte et al., 2016).

Bridging this protection gap requires concerted effort across numerous levels: ranging from international standards and national law-making to community-based interventions. Therefore, this chapter begins by laying out the institutional foundations, starting with international and national OHS frameworks and recent ILO initiatives that explicitly connect climate change with worker safety. The analysis then moves to the business arena, where Environmental, Social, and Governance (ESG) reporting is increasingly becoming one of the important tools of accountability and resilience. At the organizational level, firms are integrating traditional engineering and administrative controls with emerging technologies. This hybrid approach not only enhances existing protections but also offers novel mechanisms for risk prediction before harm actually occurs. For instance, wearable sensors and (Artificial

Intelligence) AI-driven analytics facilitate real-time tracking of heat stress, and Internet of Things (IoT)-based networked safety networks extend these capabilities.

While technology offers powerful tools, these solutions by themselves are insufficient to address the full scope of the problem. Building true resilience in the face of climate pressures demands a holistic approach, integrating environmental, social, psychological, and organizational elements into how a workplace functions and adapts (Buller et al., 2018; Ferrari et al., 2025). The need to do so is exemplified by the World Meteorological Organization's State of the Global Climate report, published in March 2025 (WMO, 2025). The report notes that extreme heat, wildfires and floods, which most affect workers in sectors such as agriculture, construction and emergency response, are intensifying throughout the year.

To address these pressures, this chapter follows three lines of inquiry: first, tracing how climate change undermines occupational health, second, mapping sectoral vulnerabilities, and third, evaluating responses that vary from regulatory reform to workplace-level adaptation. Meeting these challenges requires insights from multiple disciplines, including climate science, epidemiology, occupational medicine, and policy analysis.

Overall, the objective is practical and theoretical. Practically, the chapter tries to provide organizations with information that they can utilize in the near future. Conceptually, though, it establishes research priorities for the future, particularly for vulnerable groups where the evidence is currently inadequate. The central argument is that current OHS frameworks are not adequate for the climate era, not only because they are reactive, but also because they fail to address structural inequities and fragmented governance. There are two changes that need to be made by a better response: a proactive safety paradigm and a model of governance based on occupational justice and intersectoral coordination.

2. CLIMATE CHANGE PATHWAYS AFFECTING OCCUPATIONAL HEALTH AND SAFETY

Climate change affects health and safety at work in both direct and indirect ways. On one side are the immediate physical burdens, like higher temperatures and more frequent extreme weather events, that put workers in danger right away. On the other hand, there are indirect effects that come from changes in the environment and larger social and economic problems that can change how people work and how vulnerable they are. These processes exacerbate existing occupational risks and introduce new hazards that current occupational health and safety systems are frequently ill-equipped to manage (Adam-Poupart, Labrèche, et al., 2013; ILO, 2024). These causal pathways are shown in Figure 1.

Figure 1. Pathways from Climate Change to Occupational Health and Safety Risks

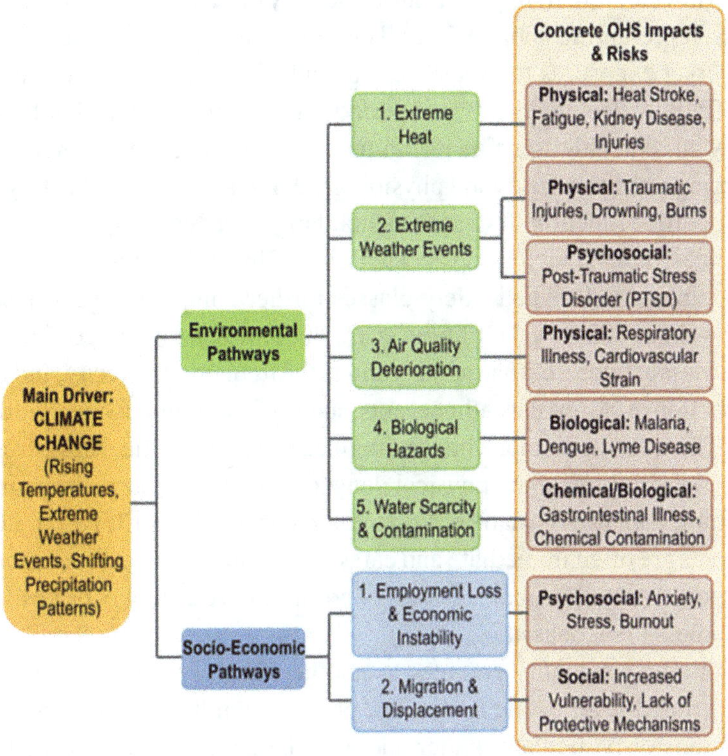

At the most immediate level, workers routinely encounter extreme heat conditions, which compromise thermoregulation, impair cognitive and physical performance, and significantly elevate the risk of heat-related illnesses, including heat exhaustion, heat stroke, and dehydration. These physiological stresses are often exacerbated by factors such as the metabolic load of physical labor, inadequate rest breaks, and insufficient access to cooling infrastructure. Such conditions are particularly prevalent in many low- and middle-income countries (A. D. Flouris et al., 2018; Ireland et al., 2023).

The global burden of heat stress on the workforce is substantial. Approximately 2.41 billion workers (71% globally) are exposed to excessive heat. In three major regions—Africa, Arab States, and Asia and the Pacific— the prevalence of heat exposure exceeded the global average, with 92.9% in Africa, 83.6% in the Arab States, and 74.7% in Asia and the Pacific. Each year, excessive heat results in an estimated 22.85 million occupational injuries and 18,970 work-related deaths (A. Flouris et al., 2024).

OHS related chronic health burden is equally alarming. In 2020, 26.2 million people worldwide were living with chronic kidney disease attributable to workplace heat stress, representing about 3% of all chronic kidney disease cases. Importantly, 90% of worker exposures to excessive heat and 80% of heat-related injuries occur beyond officially declared heatwave periods. This demonstrates that heat stress is not merely an acute, high-profile phenomenon but a pervasive hazard embedded in daily working conditions. Beyond physiological impacts, extreme heat significantly impairs cognitive function. It causes mental fatigue, reduced vigilance, and slower reaction times, thereby increasing accident risks. Physical symptoms such as sweaty palms, dizziness, and fogged safety glasses further compound workplace hazards (A. Flouris et al., 2024; A. D. Flouris et al., 2018; Ireland et al., 2023).

Beyond temperature, climate change alters the frequency, severity, and geographical distribution of extreme weather events, such as floods, hurricanes, wildfires, and droughts. These events disrupt work environments, damage critical infrastructure, and expose workers to immediate physical dangers, including injury, drowning, burns, and structural collapses (Adam-Poupart, Labrèche, et al., 2013; IOSH, 2025). The OHS impacts are often immediate and catastrophic but can also manifest as delayed health effects due to chemical releases, exposure to toxic smoke, or post-traumatic stress disorders (Tsakonas et al., 2025).

Deteriorating air quality emerges as another critical pathway. Climate change amplifies ground-level ozone formation, prolongs wildfire seasons, and increases concentrations of particulate matter and aeroallergens. Outdoor workers are particularly vulnerable, with elevated risks of respiratory and cardiovascular strain as well as reduced work capacity under poor air quality conditions (Chandra, 2024; Kjellstrom et al., 2017). Climatic alterations such as premature spring onset, modified precipitation regimes, elevated temperatures, and rising atmospheric CO_2 levels further intensify pollen seasons (EPA, 2025a). Agricultural and other outdoor workers consequently face heightened exposure to aeroallergens, aggravating conditions such as allergic rhinitis and asthma, among others.

Indoor environments are not exempt. Heightened humidity from altered weather patterns promotes mold proliferation and microbial growth, while compromised ventilation and construction-related contaminants (e.g., formaldehyde from particle boards, fiberglass dispersal) exacerbate respiratory risks. The Canadian Centre for Occupational Health and Safety underscores source control, identifying, isolating, and eliminating contaminants, as a cornerstone mitigation strategy (CCOHS, 2025; EPA, 2025b; Raulf & Annesi-Maesano, 2025).

These atmospheric changes intersect with ecological transformations to expand vector habitats, elevating occupational exposure to malaria, dengue, and Lyme disease, particularly among forestry and landscape workers in newly vulnerable regions (Amoadu et al., 2023; Manni, 2023). Parallel hydrological disruptions manifest as both water scarcity and contamination crises: drought conditions impair hydration access, while extreme precipitation events introduce pathogenic and chemical pollutants into water supplies, increasing risks of gastrointestinal and dermal occupational illnesses (Marinaccio et al., 2025).

Critically, these pathways also interact in compounding ways, as thermal extremes potentiate air pollutant formation, flood events mobilize hazardous materials, and drought-driven crop failures intensify psychosocial stressors. This interconnectivity of hazards renders traditional single hazard approaches inadequate. It underscores the need to examine sector-specific vulnerabilities where these complex interactions are most pronounced (Schulte et al., 2016).

2.1. Sector-Specific Vulnerabilities

While climate change affects all workers, certain sectors and worker populations face disproportionate vulnerabilities due to their work environments and socioeconomic circumstances. Climate impacts vary significantly across occupations, with risk severity determined by work type, exposure duration, geographic location, and existing workforce protections (Amoadu et al., 2023; ILO, 2024). Outdoor-based sectors, most notably agriculture, forestry, fisheries, and construction, remain among the most directly threatened, though climate-related hazards are increasingly recognized in indoor workplaces. Healthcare settings and informal economies are also increasingly affected.

Agriculture and forestry workers confront particularly severe climate impacts. Southeast Asian studies document that peak Wet-Bulb Globe Temperatures (WBGT), a composite index for heat stress, regularly surpass international occupational limits during hot seasons, reducing worker productivity by over 20% (Kjellstrom et al., 2017). Changing ecosystems compound these risks. For example, flooded rice fields and deforested zones create ideal conditions for disease vectors, substantially increasing malaria and dengue transmission among outdoor workers (Ghimire, 2024). Forestry crews face escalating wildfire threats, with Australian and Canadian fire suppression teams experiencing longer fire seasons. These bring combined hazards of extreme heat, heavy smoke inhalation, and elevated accident risks during emergency operations (Gibb et al., 2024).

Maritime industries show parallel vulnerability patterns, where ocean warming, rising sea levels, and intensified storm surges create multifaceted occupational threats. Norwegian aquaculture studies reveal that unstable vessel conditions from turbulent seas significantly increase drowning risks and musculoskeletal injury rates among fish farm workers (Thorvaldsen et al., 2025). Tropical regions present additional heat stress challenges, with onshore processing and net maintenance activities frequently occurring at WBGT levels above 32°C, thus exceeding safe exposure thresholds and compromising worker safety (Lin et al., 2022). These marine-specific hazards demonstrate how climate change simultaneously jeopardizes both physical safety and long-term occupational health in fisheries.

The construction sector is especially sensitive to climate change. Construction work is often carried out on heat-retaining materials like asphalt and steel, further intensifying heat-related dangers. A case–crossover study in Italy showed that extreme heat days significantly increase the risk of occupational injuries, especially falls from height and equipment-related accidents (Gariazzo et al., 2023). Beyond heat, construction work is disrupted by extreme precipitation, flooding, and high winds, which can damage scaffolding, compromise structures, and raise electrical hazards (Moda et al., 2024).

Manufacturing and industrial sectors, though often indoors, are not immune to climate-related risks. In many low- and middle-income countries, production facilities lack adequate cooling, often producing indoor heat levels as high as—or even higher than—outdoor conditions (Rînjea et al., 2022). Such environments increase error rates, reduce concentration, and heighten accident risks, while climate-related power disruptions further compound hazards through unsafe shutdowns or equipment failures (ILO, 2023).

Healthcare and emergency response workers face a unique convergence of occupational hazards under climate change. During extreme weather disasters, hurricanes, floods, wildfires, these workers are simultaneously exposed to physical danger, psychological strain, and infectious disease risks (Tsakonas et al., 2025). For example, during the 2021 British Columbia heat dome, emergency medical personnel faced both extreme environmental heat and overwhelming caseloads, leading to burnout and heat-related illnesses among staff. This dual exposure underscores the compounded pressures that climate change imposes on already overextended health systems. In low-resource settings, the lack of personal protective equipment (PPE) and heat-adapted uniforms exacerbates vulnerability (Ghimire, 2024).

The informal and precarious work sectors, including street vendors, waste pickers, and day laborers, are among the most exposed and least protected. Often lacking shaded rest areas, potable water, or climate-appropriate PPE, these workers face high rates of heat exhaustion, dehydration, and fainting episodes, with severe economic consequences as such events directly reduce daily earnings (Dodman et al., 2023; Kadungure et al., 2023). Beyond individual health risks, structural vulnerabilities compound these exposures. Informal settlements, home to over a billion people worldwide, are characterized by insecure tenure, poor housing, inadequate sanitation, and limited water access, leaving workers doubly exposed to extreme heat, floods, storms, and air pollution (Hussainzad & Gou, 2024; Sepadi, 2025). This underscores the structural inequities faced by informal workers, reinforcing the need for systemic interventions.

In addition to these sector-specific vulnerabilities, climate change also exacerbates existing health and safety issues among certain workers and may create new, unanticipated risks, particularly for groups that are already marginalized or underserved. In the United States, for example, socioeconomic status and citizenship greatly influence who is most at risk. Almost eight in ten non-elderly workers without a high school diploma, and about six in ten with only a high school degree are employed in at-risk occupations. By contrast, fewer than half of those with some college degree, and only one in four of those with a bachelor's degree or higher face similar risks. Immigration status compounds these disparities: roughly six in ten noncitizen immigrant workers hold at-risk jobs, compared to just 41% of U.S.-born citizens. Racial and ethnic differences are especially striking—about 77% of Hispanic noncitizen workers, two-thirds of Black noncitizen workers, and 80% of Native Hawaiian and Other Pacific Islander (NHOPI) noncitizen workers are employed in vulnerable occupations (Ndugga et al., 2023). Moreover, workers in climate-exposed jobs are disproportionately uninsured, further limiting access to care when health problems arise.

Table 1 summarizes the primary climate-related hazards, their occupational health and safety implications, and illustrative examples from the literature, offering a sector-by-sector overview of vulnerabilities across diverse regions and economies.

Table 1. Sector-Specific Climate Vulnerabilities and Associated Health Impacts

Sector	Main Climate-Related Hazard	Primary OHS Impact	Example Source
Agriculture & Forestry	Extreme heat, drought, vector expansion	Heat stress, vector-borne diseases, wildfire smoke inhalation	(Kjellstrom et al., 2017)
Fisheries & Aquaculture	Storm surges, sea-level rise, extreme heat	Drowning risk, musculoskeletal injuries, heat illness	(Lin et al., 2022)
Construction	Heat on reflective surfaces, heavy rain, windstorms	Falls, equipment accidents, schedule disruption	(Gariazzo et al., 2023)
Manufacturing	Indoor heat build-up, HVAC load, power disruptions	Cognitive impairment, accidents, chemical exposure	(Rînjea et al., 2022)
Healthcare & Emergency Response	Disasters, high caseloads, infectious outbreaks	Heat strain, burnout, disease exposure	(Tsakonas et al., 2025)
Informal Work	Lack of infrastructure, prolonged sun exposure	Dehydration, heat stroke, economic loss	(Kadungure et al., 2023)

These sector-specific findings make clear that climate change is not simply an added workplace concern but a force that intensifies existing vulnerabilities. It makes structural problems worse, like when manufacturing doesn't have enough cooling systems or when informal economies don't have any formal protections. It also puts the most pressure on already overburdened workers. This reality underscores the inadequacy of current safety frameworks and highlights the urgent need for risk management strategies that are both more targeted and equitable.

2.2. Occupational Hazards Intensified by Climate Change

The ILO defines climate change as a powerful force that is changing occupational risk patterns across the globe. This shift occurs across four main dimensions (ILO, 2024):

- Physical hazards - including extreme heat and weather events
- Chemical exposures - from altered substance behaviors
- Biological risks - through changing disease patterns
- Psychosocial stress factors - driven by climate instability

Building on Section 2, extreme heat is the most serious physical threat, and it is widely known to be both acute and well-documented. Studies consistently demonstrate a correlation between increasing temperatures and a variety of heat-related health issues. Evidence indicates that these risks encompass acute conditions, such as heat stroke and dehydration, as well as gradual, long-term consequences, including renal impairment. The burden falls most heavily on workers in agriculture, construction, and manufacturing. (Gibb et al., 2024; Ireland et al., 2023). Long-term data from Italy, for example, indicate a 23% rise in construction-related injuries during summer heat waves, underscoring how thermal stress threatens both health and safety on the job (Gariazzo et al., 2023). On a global scale, the ILO estimates that by 2030 heat stress could reduce total working hours by around 2%, the equivalent of 80 million full-time jobs, with the most severe impacts expected in low- and middle-income countries (ILO, 2024).

Extreme weather events make these thermal risks even worse in several ways. They create both acute trauma hazards and chronic air quality impacts. For example, during floods, emergency personnel are immediately at risk of drowning, and outdoor workers are exposed to cumulative cardiopulmonary damage from wildfire smoke (Cosme, 2025; Tsakonas et al., 2025).This layered physical risk profile demands protection strategies that address both acute safety threats and long-term health impacts.

In a similar fashion, climate change intensifies chemical hazards across industrial and agricultural workplaces. Flooding can trigger the release of toxic substances from compromised storage sites, while extreme heat accelerates the volatilization of chemicals, heightening inhalation risks for industrial workers (ILO, 2023). In agriculture, rising temperatures and shifting rainfall patterns influence the degradation of pesticides, with the potential to increase exposure among field workers (Zhang & Kim, 2025). The established safety standards are complicated by these dynamics. They often require more frequent monitoring and adjustments to exposure thresholds. The relationship between climate conditions and chemical exposure risks is particularly intricate. Higher ambient temperatures not only promote the environmental spread of chemicals but also increase absorption in the human body through elevated respiration, perspiration, and skin permeability.

Biological risks are also evolving. Warmer climates and altered precipitation regimes have expanded the range of vector-borne diseases, including malaria, dengue, and Lyme disease, placing forestry, agriculture, and outdoor workers at greater risk (Adam-Poupart, Labrèche, et al., 2013). Zoonotic spillover risks are heightened in animal husbandry and wildlife management sectors as ecosystem disruptions increase the frequency of human-animal interactions (Ghimire, 2024). For example, in Sub-Saharan Africa, malaria incidence among agricultural workers has risen by 10% due to extended mosquito seasons, a pattern echoed in other tropical regions where prolonged warm and wet conditions extend vector activity (WHO, 2024). In addition, waterborne pathogens can proliferate in poorly managed workplaces lacking adequate sanitation, particularly during prolonged flooding events.

Perhaps less visible but equally critical are psychosocial hazards. Climate-related occupational stressors manifest along several dimensions. First, job insecurity due to climate-induced disruptions in industries such as fisheries, agriculture, and tourism increases economic anxiety and livelihood instability (Galappaththi et al., 2022). Second, working under extreme environmental conditions—whether prolonged heat or disaster response, has been linked to higher rates of anxiety, depression, and burnout, which impair cognitive performance and elevate accident risks (Brooks & Greenberg, 2023; Sirakaya, 2024). For example, healthcare staff during recent heatwaves reported elevated burnout, underscoring the importance of adaptive staffing and integrated mental health support (Wibowo et al., 2025). Third, beyond economic and workplace stressors, climate disasters themselves act as traumatic events, sometimes leading to long-term conditions such as Post-Traumatic Stress Disorder (PTSD) (Roe et al., 2024). These effects extend beyond individuals to families and workplaces, especially where displacement and migration occur. Drawing on lessons from the COVID-19 pandemic, adaptations such as flexible work arrangements and enhanced mental health support could offer viable strategies for mitigating these psychosocial disruptions.

Table 2 synthesizes key findings from the literature on how climate change is amplifying these hazard categories, providing examples of affected sectors, health outcomes, and proposed mitigation measures. This holistic view underscores the interconnected nature of climate-related occupational risks and the need for integrated risk management strategies that address both direct and indirect pathways of harm.

Table 2. Climate Change–Intensified Occupational Hazards, Associated Sectors, and Documented Impacts

Hazard Type	Climate-Change Driver	Affected Sectors	Documented Impacts	Example Source
Physical hazards (heat, UV, extreme weather)	Rising temperatures, increased extreme events	Agriculture, construction, manufacturing, emergency response	Heat stress, injuries, reduced productivity, mortality	(Gibb et al., 2024)
Chemical hazards	Flooding, heat-induced volatilization	Industrial manufacturing, agriculture	Toxic exposure, respiratory illness, chemical burns	(ILO, 2023)
Biological hazards	Vector range expansion, water contamination	Forestry, fishery, healthcare, animal husbandry	Infectious diseases (malaria, dengue, leptospirosis), zoonoses	(Adam-Poupart, Labrèche, et al., 2013)
Psychosocial hazards	Job insecurity, disaster trauma	Fisheries, tourism, agriculture, emergency response	Anxiety, depression, burnout, cognitive impairment	(Brooks & Greenberg, 2023)

Building on this analysis, Table 3 presents projected increases in occupational hazards under 1.5 °C and 2 °C warming scenarios, emphasizing the urgency of proactive risk management to mitigate future impacts (IPCC, 2018; WHO, 2023).

Table 3. Comparison of OHS Risks Under 1.5°C and 2°C Warming Scenarios

Warming Scenario	Risk Type	Affected Sectors	Estimated Impacts
1.5°C	Heat Stress	Agriculture, Construction	1.5% global working hours lost; 15 million jobs impacted
1.5°C	Vector-Borne Diseases	Forestry, Fisheries	5% increase in malaria/dengue cases
2°C	Heat Stress	Agriculture, Construction	2.1% global working hours lost; 80 million jobs impacted
2°C	Extreme Weather	Emergency Response, Transport	20% increase in injury rates

Although most projections emphasize 1.5 °C and 2 °C warming scenarios, exceeding 3 °C by 2100 would dramatically intensify occupational health risks. Such extreme conditions are projected to generate unprecedented levels of heat stress and widen the spread of vector-borne diseases, placing enormous strain on OHS systems (Calvin et al., 2023). These scenarios highlight the urgency of moving beyond short-term temperature thresholds toward strategies of long-term adaptation. As a risk multiplier, climate change not only intensifies known hazards but also creates new risks that test the limits of current OHS systems. Outdoor and informal workers are particularly affected, serving as a reminder that reactive approaches are no longer adequate. The way forward must be proactive and climate-adaptive, with an emphasis on identifying risks early enough to prevent escalation. Current safeguards are insufficient, especially for vulnerable groups, which highlights the need for solutions that can be applied fairly to marginalized workers and scaled across high-risk industries.

Progress on three interrelated fronts is necessary to move forward in this direction. First, safety regulations need to evolve to match the pace of emerging threats. Second, workplace practices must adapt to minimize immediate exposures. Third, technology should be deployed to anticipate hazards before they occur. When combined, these tactics provide a workable framework for enhancing occupational health and safety resilience.

3. MITIGATION AND ADAPTATION STRATEGIES FOR CLIMATE-RELATED OHS RISKS

The evolving nature of workplace risks under climate change calls for adaptation and mitigation to be pursued together. Hazards ranging from heat stress to psychosocial strain demand multi-layered responses that operate across governance, regulatory systems, and day-to-day workplace practices (Ferrari et al., 2023; ILO, 2024; Marinaccio et al., 2025). However, OHS systems are still not well adapted to this new reality in many parts of the world. Evidence consistently shows that climate effects are strongly integrated into workers' health, productivity, and economic performance (A. Flouris et al., 2024; Gibb et al., 2024). Coordinated progress in three crucial areas, which are enforcing corporate responsibility, implementing successful workplace interventions, and strengthening policy, is necessary to close these serious protection gaps.

Mitigation and adaptation in occupational health and safety are not substitutes but complementary strategies. Mitigation focuses on addressing hazards at their source, while adaptation involves redesigning work patterns and environments to cope with risks that cannot be eliminate (Amoadu et al., 2023; IOSH, 2025). How they are achieved depends extensively on context. Factors like available resources, workforce demographics, and industry-specific exposures all determine what can be done. Because of this, there are solutions ranging from sophisticated AI-driven forecasting systems to grassroots safety networks for informal workers. To be successful, these responses must be scalable. International labor standards and global climate commitments are a crucial starting point, but they only become truly effective when they are translated into local policies and actions that fit the specific context of a community. (Grandi et al., 2023; Schulte et al., 2023). In order to develop comprehensive and contextually appropriate solutions, it is necessary to create a synergy between technology-facilitated strategies and community-based initiatives (Koch et al., 2022; Ojha, 2022). Their resilience depends on addressing structural gaps that technical solutions cannot close. Some of these vulnerabilities are weak labor protections, inadequate training provisions, and insufficient access to personal protective equipment. Infusing flexibility in OHS systems, through continuous learning, ongoing monitoring, and adaptable standards, prepares them to protect workers in the short term. At the same time, it lays the foundations for long-term resilience (Holt, 2025; Schulte et al., 2016).

The remainder of this chapter investigates the most important pillars of implementation. It addresses the role of regulatory frameworks, describes effective risk management strategies at the workplace level, and examines the importance of targeted training schemes. These elements together form a realistic roadmap to building climate resilient OHS systems that will protect workers today and for generations to come.

3.1. Policy and Regulatory Frameworks

Policy and regulation provide the foundation for developing climate resilient OHS systems. They establish not only the legal duties of employers and governments but also the broader vision that guides both adaptation and mitigation efforts. At the international level, organizations such as the ILO, (World Health Organization), and IPCC (Intergovernmental Panel on Climate Change) have begun to frame climate change explicitly as a major source of occupational risk, urging countries to integrate environmental realities into workplace protections (Grandi et al., 2023; ILO, 2024).

The ILO, for instance, has historically placed OHS at the core of its concerns since its founding in 1919. Over time, it has developed a range of international OHS directives. Key ILO conventions, such as the Occupational Safety and Health Convention (No. 155), the Working Environment (Air Pollution, Noise and Vibration) Convention (No. 148), and the Chemicals Convention (No. 170), establish the right to a safe and healthy working environment as a fundamental principle. Recently, the ILO highlighted that climate change creates a combination of serious health hazards for 70% of the world's workers and estimated that improved safety measures to prevent heat-related injuries could save $361 billion globally (ILO, 2024). Meanwhile, the IPCC has identified direct human health effects linked to climate change, such as heat stress and injuries or deaths from floods, wildfires, and storms. Similarly, the WHO has emphasized knowledge gaps in climate and health research, particularly concerning occupational health, air quality, and infectious diseases (Ebi et al., 2018).

National governments have translated these global directives into sector-specific regulations, although the degree of implementation varies widely depending on political will, economic resources, and the maturity of existing OHS infrastructures (Jia & Jia, 2024; Schulte et al., 2023). A widening gap is evident in the global regulatory response to climate risks. High-income countries have generally moved further ahead, introducing measures such as legally binding heat exposure limits, structured work–rest cycles, and guaranteed access to shaded or cooled spaces for outdoor workers. By contrast, many low- and middle-income countries struggle to establish, let alone enforce, comparable safeguards. In settings where informal work dominates, safety regulations are often broad in wording or weakly applied, leaving large gaps where climate-related hazards continue to grow unchecked (Dodman et al., 2023; Kadungure et al., 2023).

At organization level, one important trend has been the incorporation of OHS indicators into (Environment Social Governance) ESG reporting. For multinationals in particular, this has become central to accountability (D'Silva, 2025; Holt, 2025). The reasoning is straightforward: sustainability cannot be reduced to environmental terms alone. It must also encompass fair labor conditions, transparent governance, and effective worker protections. Environment, Health and Safety (EHS) systems provide the operational basis for implementing these broad principles, but their strategic value is yet not clearly understood in most corporations.

In practice, the alignment between ESG and EHS is becoming more apparent. Both agendas are simultaneously advanced by climate risk assessments, workplace safety procedures, mental health programs, and sustainable waste management. Data generated within EHS frameworks now flows directly into ESG disclosures, shaping corporate strategy and investment decisions. Meanwhile, regulators are tightening requirements, placing firms under growing pressure to embed climate resilience into OHS structures. The Corporate Sustainability Reporting Directive (CSRD) of the European Union and the new climate disclosure regulations of the U.S. Securities and Exchange Commission are two examples of instruments that are changing compliance environments and elevating the standard for corporate accountability (Parsakia & Tabar, 2024).

Alongside mandatory regulations, voluntary certification frameworks, most notably ISO 45001 for occupational safety management, are beginning to incorporate climate-related risks into their standards (A. Flouris et al., 2024; Levi et al., 2018). Collectively, these systems encourage organizations to adopt continuous, forward-looking processes for identifying and addressing climate-sensitive hazards.

Nevertheless, translating formal standards into genuine protection remains uneven. Agriculture, construction, and fishing are particularly exposed, as regulatory oversight is often limited and large portions of the workforce consist of informal or migrant laborers (Shan, 2022; Zhang & Kim, 2025). These sectors rely heavily on temporary and off-the-books workers who frequently fall outside formal safety systems. Closing these gaps requires more than regulatory reform; it demands stronger enforcement bodies, adequate resources, and cross-sector accountability.

For OHS systems to stay relevant, they must operate as dynamic frameworks. This means constantly taking in new climate evidence, identifying emerging risks, and adjusting protections to the needs of specific industries and regions. Their long-term strength depends on risk assessment and management practices that are explicitly informed by climate conditions and that put worker health at the forefront. This principle lays the groundwork for the discussion that follows.

3.2. Risk Assessment and Management Strategies

Addressing climate-related OHS risks requires more than traditional safety audits. It calls for a comprehensive strategy that integrates climate science with workplace practice. The focus of occupational safety must shift from a reactive stance to a model of continuous adaptation, where hazard detection, risk evaluation, and preventive actions are updated in step with changing environmental conditions (Holt, 2025; Parsakia & Tabar, 2024).

Central to this model is accurate hazard identification. Traditional risk assessments were built on the assumption of relatively stable environmental conditions. Climate change has rendered this assumption increasingly untenable. With recurring heatwaves, shifting humidity, and declining air quality, static, one-off evaluations no longer suffice. Instead, workplaces now require ongoing monitoring that can capture risks as they evolve in real time.

To make protective measures effective under these circumstances, risk assessment needs to be understood as a dynamic and iterative process rather than a single compliance exercise. Among the advanced tools, the WBGT index has gained prominence by integrating multiple environmental factors into a single heat stress metric. By combining temperature, humidity, wind speed, and radiant heat, WBGT provides a much more realistic picture of the physiological strain on workers than any single measure alone. Evidence from construction, agriculture, and mining consistently shows that site-specific WBGT readings offer a more precise indication of worker strain than regional weather data, which can mask localized hazards (Kim & Yoo, 2023; Sharma et al., 2022).

To effectively manage workplace risks, employers should conduct proactive and climate-informed hazard assessments. This involves evaluating cumulative heat stress from environmental conditions (temperature, humidity, sunlight, air speed) and from workplace sources (e.g., hot tar ovens, furnaces), especially across consecutive days.

However, climate related OHS risk assessment must go beyond environmental metrics. Task-specific factors, including work intensity, shift duration, and the thermal properties of PPE, can also play a decisive role in shaping risk. In sectors such as firefighting and chemical manufacturing, impermeable PPE not only reduces heat dissipation but can exacerbate cardiovascular strain, especially in high-humidity conditions (Chandra, 2024). Individual factors, such as age, acclimatization level, hydration status, underlying health conditions, and medication use, further shape vulnerability profiles (Alemu Gelaye et al., 2021).

A critical but often overlooked factor is cumulative exposure over time. Epidemiological evidence shows that approximately 80–90% of heat-related occupational illnesses occur outside declared heatwave periods, underscoring the importance of continuous exposure monitoring even under "moderate" weather conditions (Manni, 2023). Wearable devices capable of real-time core temperature, heart rate, and skin conductance measurements are becoming invaluable tools in this regard, especially when combined with AI-driven predictive models that forecast risk spikes based on microclimate, workload, and health data (Adamopoulos, Valamontes, et al., 2025).

Effective mitigation requires coordinated action across three tiers:

1. Engineering controls, e.g., installing reflective barriers, improving ventilation, using misting systems in outdoor worksites, retrofitting vehicles with cooling systems, and designing shaded rest stations (Karthick et al., 2023).
2. Administrative controls, such as implementing work-rest cycles, rescheduling strenuous tasks to cooler hours, job rotation, and mandating hydration breaks (Silva et al., 2024).
3. Personal protective strategies, ensuring PPE is climate-adapted, breathable, and lightweight where possible, and integrating cooling vests or moisture-wicking garments in extreme heat settings (Buller et al., 2018). A critical component of personal protective strategies is hydration management. Ensuring easy access to potable, cool water (<15°C) and, during prolonged sweating, electrolyte-containing sports drinks is essential for maintaining hydration. Hydration stations should be strategically placed to minimize productivity loss while maximizing intake compliance (Van Selm et al., 2025). Alcohol, high-sugar, and high-caffeine beverages should be actively discouraged due to their diuretic effects.

Climate change also elevates non-thermal occupational hazards (Adamopoulos, Syrou, et al., 2025; Ngajilo et al., 2023):

- Air quality deterioration from wildfire smoke and urban smog requires integration of portable particulate monitors and respiratory protection protocols.
- Vector-borne diseases, such as malaria and dengue, are expanding geographically, placing outdoor workers at greater risk. Preventive measures include repellents, protective clothing, and environmental control of breeding sites

- Water scarcity and contamination, especially after floods or in drought-affected areas, require safe storage and purification measures (e.g., chlorination, boiling). Complementary worker hygiene training is also critical.

Satellite-based monitoring provides real-time detection of wildfires, storms, and UV index fluctuations, offering early warnings that can be directly translated into workplace safety measures. At the local level, IoT-enabled sensors can trigger automated workplace hazard notifications (Koch et al., 2022). The integration of these systems into mobile applications allows direct communication with workers, ensuring rapid dissemination of protective instructions.

Effective training initiatives remain fundamental for climate-resilient workplaces. Comprehensive programs should address five core elements: identifying climate-aggravated hazards, recognizing heat-related illness symptoms, executing emergency protocols, implementing graduated acclimatization periods, and understanding how personal health factors influence risk susceptibility (Adam-Poupart, Labreche, et al., 2013). Pre-season update training programs are vital for preparing workers ahead of high-risk climatic periods. Keeping protective knowledge up to date is essential. Just as important is linking training to daily practice so that workers can turn guidance into immediate action. A practical illustration is the "buddy system". The buddy system, where workers in pairs monitor each other for early warning signs, has demonstrated measurable success in the forestry, construction, and agriculture sectors. This method facilitates timely medical intervention, reducing the number of serious health incidents (Yovi et al., 2023).

Organizational culture plays an equally important role. Workplaces that foster open dialogue and weave adaptive routines into their daily operations consistently report fewer accidents (Ghatak & Lalitha, 2015). The integration of OHS indicators into ESG reporting helps drive this cultural shift, reframing climate safety from a temporary concern into a core business responsibility. This perspective incentivizes sustained investment in worker protection over reactive, short-term fixes.

Long-term protection depends on integrating resilience into organizations' core operations, even though technical measures are still crucial. Resilience is not just one reaction to harm; it is an ongoing ability to adapt as risks evolve. To build that capacity, climate change needs to be taken into account at every level of decision-making, from corporate governance to daily routines on the front lines (Dodman et al., 2023; Holt, 2025). This kind of comprehensive integration is becoming a defining feature of modern occupational health as the effects of climate change become more evident.

4. RESILIENCE AND ADAPTIVE CAPACITY IN OHS

Resilience has become a major focus of occupational health as climate change accelerates. The goal is not only to reduce immediate harm but to equip workers and organizations to withstand mounting environmental pressures. Resilience is not the same as mitigation, which tries to cut down on greenhouse gas emissions at their source. Instead, it is about how workplaces and communities can prepare for, deal with, and recover from climate shocks without losing important functions (Holt, 2025; Schulte et al., 2016). Meeting this challenge requires replacing static, reactive safety models with frameworks that adapt to shifting conditions. Effective planning must be climate-informed, shaping regulatory standards, guiding infrastructure decisions, directing workforce training, and steering the adoption of new technologies.

Building resilience calls for coordinated action on several fronts. A lot of companies are adding community involvement and new technologies to their management systems. The strongest protections result when climatic considerations are incorporated into operations on a day-to-day level, like designing facilities with future risks in mind, adapting training to new hazards, and revising emergency protocols constantly as estimates evolve. Such practices do more than safeguard employees, they also strengthen organizational stability and competitiveness in a context where disruption is increasingly common.

In practical terms, resilient OHS systems must be prepared to respond to both sudden climate shocks and the slower accumulation of long-term stress. Acute events, such as heatwaves, floods, and wildfires call for rapid response tools, including early warning systems, flexible work schedules, and clear evacuation procedures. Longer-term pressures, such as rising average temperatures, prolonged droughts, and the spread of vector-borne diseases, require steady adaptation through redesigned workflows, climate-resilient building materials, and improved workplace ventilation and shading (Abokhashabah et al., 2021; Adam-Poupart, Labreche, et al., 2013).

At a macro level, building resilience is contingent on strong government leadership to foster a climate-aware health system. Historically, health considerations have been poorly integrated into broader climate action plans, creating a critical vulnerability. While tools such as the WHO's "vulnerability and adaptation assessment" are vital, their true value emerges when they are used not just as risk inventories but as catalysts for systemic foresight, compelling sectors to integrate climate intelligence into all levels of planning. This transforms the health system's posture from being merely reactive to becoming genuinely anticipatory.

This systemic foresight, however, cannot be built on infrastructure and surveillance tools alone. A people-centered approach is essential to its legitimacy and efficacy. A system's ability to withstand shocks is closely tied to inclusivity and equity. Interventions address more than just material needs when they are co-designed with historically marginalized and local groups. They build trust and social cohesion, which are essential foundations for collective action. This confirms that the most durable forms of resilience are socially constructed as much as they are technical.

In occupational safety, this social dimension holds particular significance in sectors with limited formal protection. Industries characterized by informal labor and decentralized workplaces, such as agriculture, forestry, and fisheries, exemplify this dynamic. In these settings, community-driven initiatives, yield measurable improvements in worker safety (Kadungure et al., 2023; Moda et al., 2024). In contexts where institutional oversight is limited, grassroots solidarity and mutual support play a vital role in addressing critical protection gaps.

At the same time, technological tools can facilitate a shift in risk management, enabling organizations to move beyond reactive compliance toward a proactive, resilient approach. Wearable sensors, for instance, are pivotal in this transition, not merely by monitoring vital signs but by changing the focus to leading physiological indicators (Buller et al., 2018; Ojha, 2022). When combined with satellite-based weather tracking and geographic information systems, these tools allow for highly localized risk assessments and enable safety managers to design protections that are tailored to specific sites. In practice, mining, construction, and agricultural workers are already receiving automated hazard alerts through IoT-enabled mobile devices, where environmental data are processed instantly to trigger timely protective actions (Sharma et al., 2022). This creates a systemic capability for continuous, evidence-based adaptation. It moves beyond rigid, predefined safety protocols.

In practice, resilience unfolds not as a static state but as a process of continuous, dynamic learning. This principle is demonstrated across diverse sectors. Canada's fishing industry, for example, embodies this by frequently updating safety protocols in response to unpredictable weather, treating resilience as an ongoing adaptive process (Shan, 2022). Similarly, farmers in Southeast Asia show adaptive learning by combining agricultural innovation (heat-tolerant rice varieties) with ergonomic adjustments to ease physical strain (Yovi et al., 2023). In Australia, construction companies piloting AI-driven systems treat resilience as an iterative experiment in risk optimization, constantly adjusting to real-time data (Fatima et al., 2023).

Taken together, these examples show how sector-specific adaptation can turn climate pressure into occasions for safer and more sustainable practices. Yet their effectiveness hinges on key conditions, most importantly, active worker involvement and policies flexible enough to adjust as risks evolve. The evidence indicates that the most robust resilience emerges from a synergy between technological systems and localized social capital. For instance, agricultural cooperatives that coordinate harvesting to avoid peak heat hours illustrate the protective power of collective decision-making. Similarly, construction projects that use wearable monitoring devices demonstrate how real-time data can shift safety management from reactive responses toward anticipatory protection.

Ultimately, managing climate-related occupational risks requires more than a series of local experiments. It depends on cross-sector knowledge sharing, the uptake of appropriate technologies, and the integration of community-driven approaches. Adaptation measures vary significantly across sectors due to differences in hazard exposure, technological capacity, and socio-economic conditions. Table 4 summarizes implemented climate adaptation measures across key sectors, from agriculture and forestry to fisheries, construction, healthcare, and the informal economy, along with the technologies employed and indicators of success.

Table 4. Sector-Specific Adaptation Strategies for Climate-Resilient Occupational Health and Safety

Sector	Adaptation Measures	Technology/Tools Used	Key Outcomes & Success Indicators	Example Sources
Agriculture & Forestry	Shifted work hours to cooler periods; introduction of heat-resilient crop varieties; provision of shade structures	Mobile weather alert apps; low-cost WBGT meters; evaporative cooling vests	Reduced incidence of heat-related illnesses; improved productivity during peak heat periods	(Ngajilo et al., 2023; Yovi et al., 2023)
Construction	Rotating crews; mandatory hydration breaks; shaded rest areas; high-visibility PPE with cooling elements	Wearable physiological sensors; AI-based workload optimization software	Lowered worker core temperatures; reduced lost-time injury rates; improved compliance with OHS policies	(Moda et al., 2024; Sharma et al., 2022)
Fisheries & Aquaculture	Enhanced storm and wave early warning systems; redesigned vessel safety equipment; climate-specific safety drills	Satellite-based marine forecasting; GPS-linked safety beacons	Decreased accident rates during extreme weather; faster rescue response times	(Lin et al., 2022; Shan, 2022)
Healthcare & Emergency Response	Revised PPE protocols for extreme heat; flexible staffing models during climate emergencies	Smart textiles for thermal comfort; IoT-enabled air quality monitoring in ambulances and clinics	Improved heat tolerance during emergency operations; reduced staff fatigue and turnover	(Adam-Poupart, Labreche, et al., 2013; Holt, 2025)
Informal Sector	Community-based training; micro-financing for protective equipment; local safety committees	Solar-powered cooling shelters; community health hotlines	Increased awareness of climate risks; greater uptake of protective practices	(Amoako, 2019; Kadungure et al., 2023)

As shown in Table 4, effective adaptation strategies are often achieved through a combination of engineering solutions, administrative measures, targeted training programs, and technological innovations. Emerging technologies such as wearable heat-stress sensors, AI-driven workload management, and real-time environmental monitoring significantly enhance workplace adaptation strategies by enabling proactive risk identification and mitigation.

The diversity of sectoral responses to climate risks highlights the importance of tailoring adaptation measures to local realities while drawing from global best practices. These sectoral responses provide valuable insights for future OHS policies. As illustrated in Figure 2, adaptation strategies at the sectoral level often generate lessons that feed into broader policy frameworks and identify research gaps, reinforcing the cyclical relationship between practice, governance, and scholarship. In the next section, these findings are synthesized to identify key conclusions and outline future research priorities that bridge theoretical frameworks with real-world implementation.

Figure 2. Adaptation–Policy–Research Flow

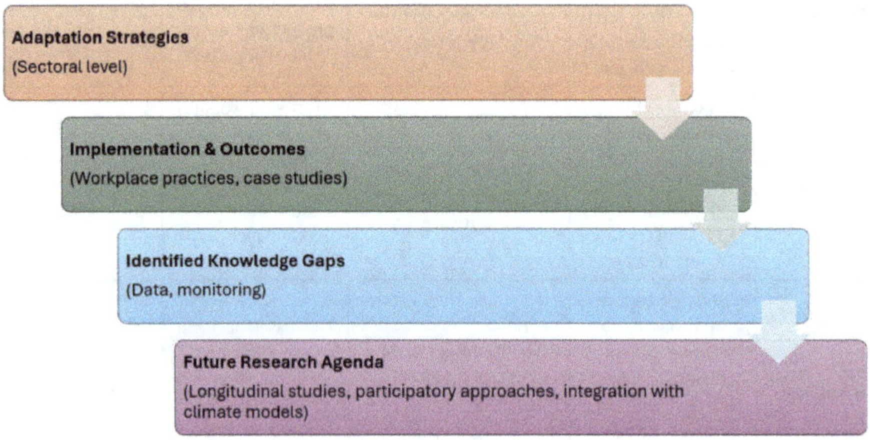

The evidence presented across previous sections demonstrates that climate change is not only a public health challenge but also an urgent OHS concern that requires proactive, integrated, and adaptive responses. The diverse strategies implemented across sectors reveal that resilience is possible when policy frameworks, workplace innovations, and community engagement align. Yet, persistent gaps in implementation, monitoring, and interdisciplinary collaboration, as well as limited funding and uneven enforcement capacity, continue to limit the scalability of these successes. This makes it essential to synthesize key lessons learned and define the future research and policy priorities that will guide the next generation of climate-resilient and future-ready OHS systems.

5. CONCLUSION AND FUTURE RESEARCH DIRECTIONS

Climate change is no longer a distant concern reserved for policy debate; it has become an everyday occupational reality, intensifying risks across all sectors of work. The central argument of this chapter is that traditional OHS frameworks remain inadequate because they were not designed to anticipate climate-driven risks. Addressing this gap calls for a twofold transformation: one that strengthens the ability to predict emerging hazards while also embedding principles of equity and justice. Safeguarding workers in this new context will depend on strategies that combine adaptation, technological innovation, and the meaningful participation of employees.

Building such resilience requires more than incremental adjustments. It depends on bringing together public health, labor governance, and climate science in a genuinely integrated framework, where workplace safety and climate adaptation are treated as inseparable challenges. Based on this synthesis, several key conclusions emerge:

- Climate risks are systemic occupational risks. The chapter demonstrated that climate-driven hazards are deeply integrated with existing workplace systems, not isolated, peripheral events.
- Vulnerability is structurally determined and inequitable. The evidence confirmed that climate risks are not distributed evenly, but instead exacerbate existing inequalities, disproportionately burdening marginalized and informal workers.
- Resilience requires continuous adaptation rather than static defenses. The most effective OHS systems are those that foster continuous learning, where rather than replacing, technology supports, worker participation.
- A strong economic case for proactive investment is crucial. The analysis highlighted that overcoming implementation barriers, especially for SMEs, depends on framing climate resilient OHS not as a cost but as a necessary investment in economic and social stability.

These conclusions provide five key recommendations for moving from analysis to action. The recommendations, detailed below and summarized in Table 5, outline a multi-level strategy for building climate resilient OHS systems:

i. Strengthen Policy and Governance Frameworks: Governments and international organizations such as the ILO, WHO, and IPCC need to explicitly integrate climate-related risks into OHS legislation. This should include mandatory climate risk assessments, incentives for climate-resilient workplace design, and the inclusion of OHS as a core pillar in national climate adaptation plans. At the corporate level, OHS metrics should be part of ESG reporting to ensure accountability and transparency.
ii. Implement Comprehensive Workplace Interventions: Employers must move beyond short-term, reactive measures and adopt climate-informed risk assessments. Effective interventions should combine engineering solutions (such as ventilation and cooling systems), administrative practices (like modified work–rest cycles and acclimatization protocols), and climate-appropriate PPE. Standard OHS programs also need to embed early warning systems, adaptive scheduling, and accessible mental health support.
iii. Leverage Technology and Innovation: Digital tools should play a greater role in OHS management. Real-time monitoring systems, predictive hazard models, and personalized early warning platforms can all strengthen organizational resilience while giving workers immediate protection against climate-related risks.
iv. Focus on Vulnerable Populations and Foster Collaboration: Special protections are needed for groups most at risk, including outdoor workers, informal laborers, and marginalized communities. Protecting these populations requires collaboration across health, labor, environmental, and economic sectors to ensure equitable and fair resource distribution.
v. Address Critical Research Gaps: Research efforts should focus on interdisciplinary, participatory projects that bring together climate science, occupational epidemiology, and economics. There is a pressing need for longitudinal studies, sectoral comparisons, and more data collection in low- and middle-income countries to develop context-specific, evidence-based adaptation strategies and to build the economic case for sustained investment.

Table 5. Summary of Recommendations for Climate-Resilient OHS

Recommendation	Responsible Actors	Expected Impact
Strengthening Policy and Governance Frameworks	Governments, International bodies (ILO, WHO, IPCC)	Coherent legislation, climate-OHS integration, ESG accountability
Implement Comprehensive Workplace Interventions	Employers, OHS managers	Reduced climate-exacerbated risks, improved worker safety & mental health
Leverage Technology and Innovation	Employers, Tech developers, Researchers	Real-time hazard prediction, expanded resilience capacity
Focus on Vulnerable Populations & Foster Collaboration	Governments, NGOs, Cross-sectoral bodies	Equity, protection of informal/outdoor workers, fair resource allocation
Address Critical Research Gaps	Research funders, Academia, International consortia	Stronger evidence base, participatory solutions, business case for investment

As illustrated in the integrative framework in Figure 3, these recommendations are not isolated actions but interconnected components of a functioning system. Resilience cannot be secured through piecemeal measures; it relies on the coordinated interaction of policy, workplace practice, technology, and science, all grounded in a principle of participation.

Figure 3. Integrative Framework for Climate-Resilient OHS

In conclusion, climate change represents one of the most profound challenges facing occupational health systems. Yet it also offers a pivotal opportunity: by embedding climate resilience into policy, corporate governance, and workplace culture, societies can safeguard workers while advancing long-term economic stability. The time for gradual, piecemeal adjustments has passed. What is needed now is a fundamental shift toward an anticipatory safety model grounded in decisive, coordinated, and equitable action, ensuring no worker is left behind as the impacts of climate change intensify.

REFERENCES

Abokhashabah, T., Jamoussi, B., Summan, A. S., Abdelfattah, E., & Ahmad, I. (2021). Effects of Heat Stress Exposure and Climate Change on Health and Safety of Outdoor Workers. *Current World Environment, 16*(3), 836–847. DOI: 10.12944/CWE.16.3.15

Adam-Poupart, A., Labreche, F., Smargiassi, A., Duguay, P., & Busque, M.-A., GAGN^|^Eacute; C., RINTAM^|^Auml;KI, H., Kjellstrom, T., & Zayed, J. (. (2013). Climate Change and Occupational Health and Safety in a Temperate Climate: Potential Impacts and Research Priorities in Quebec, Canada. *Industrial Health, 51*(1), 68–78. DOI: 10.2486/indhealth.2012-0100 PMID: 23411758

Adam-Poupart, A., Labrèche, F., Smargiassi, A., Duguay, P., Busque, M.-A., Gagné, C., & Zayed, J. (2013). *Impacts of climate change on occupational health and safety*. Institut de recherche Robert-Sauvé en santé et en sécurité du travail.

Adamopoulos, I., Syrou, N., Mpourazanis, G., Constantinidis, T. C., & Dounias, G. (2025). The Association of the Global Climate Crisis with Environmental Risks and the Impact of Heat Stress on Occupational Safety, Health, and Hygiene. *One Health, 2024*, 2. DOI: 10.3390/msf2025033002

Adamopoulos, I., Valamontes, A., Tsirkas, P., & Dounias, G. (2025). Predicting Workplace Hazard, Stress and Burnout Among Public Health Inspectors: An AI-Driven Analysis in the Context of Climate Change. *European Journal of Investigation in Health, Psychology and Education, 15*(5), 65. DOI: 10.3390/ejihpe15050065 PMID: 40422294

Alemu Gelaye, K., Debalke, G., Awoke Ayele, T., Fekadu Wolde, H., Sisay, M. M., Teshome, D. F., Akalu, T. Y., & Daba Wami, S. (2021). Occupational Health Problems among Seasonal and Migrant Farmworkers in Ethiopia: A Cross-Sectional Study. *Risk Management and Healthcare Policy, 14*, 4447–4456. DOI: 10.2147/RMHP.S323503 PMID: 34744466

Amoadu, M., Ansah, E. W., Sarfo, J. O., & Hormenu, T. (2023). Impact of climate change and heat stress on workers' health and productivity: A scoping review. *The Journal of Climate Change and Health, 12*, 100249. DOI: 10.1016/j.joclim.2023.100249

Amoako, J. (2019). *Women's occupational health and safety in the informal economy: Women's occupational health and safety in the informal economy: Maternal market traders in Accra, Ghana Maternal market traders in Accra, Ghana* [Master of Arts Thesis, University of Northern Iowa]. https://scholarworks.uni.edu/etd/951

Brooks, S. K., & Greenberg, N. (2023). Climate change effects on mental health: Are there workplace implications? *Occupational Medicine*, *73*(3), 133–137. DOI: 10.1093/occmed/kqac100 PMID: 36170162

Buller, M. J., Welles, A. P., & Friedl, K. E. (2018). Wearable physiological monitoring for human thermal-work strain optimization. *Journal of Applied Physiology*, *124*(2), 432–441. DOI: 10.1152/japplphysiol.00353.2017 PMID: 28798200

Calvin, K., Dasgupta, D., Krinner, G., Mukherji, A., Thorne, P. W., Trisos, C., Romero, J., Aldunce, P., Barrett, K., Blanco, G., Cheung, W. W. L., Connors, S., Denton, F., Diongue-Niang, A., Dodman, D., Garschagen, M., Geden, O., Hayward, B., Jones, C., . . . Péan, C. (2023). *IPCC, 2023: Climate Change 2023: Synthesis Report. Contribution of Working Groups I, II and III to the Sixth Assessment Report of the Intergovernmental Panel on Climate Change [Core Writing Team, H. Lee and J. Romero (eds.)]. IPCC, Geneva, Switzerland.* (First). Intergovernmental Panel on Climate Change (IPCC). DOI: 10.59327/IPCC/AR6-9789291691647

CCOHS. (2025, June 24). *CCOHS: Indoor Air Quality - General.* Canadian Centre for Occupational Health and Safety. https://www.ccohs.ca/oshanswers/chemicals/iaq/iaq_intro.html

Chandra, N. (2024). Heat and health of occupational workers: A short summary of literature. *Journal of Occupational Health*, *66*(1), uiae018. Advance online publication. DOI: 10.1093/joccuh/uiae018 PMID: 38604180

Cosme, C. (2025). Climate change and occupational health: Escalating risks and prevention strategies. *Field Actions Science Reports. The Journal of Field Actions*, *27*(Special Issue), 36–40.

D'Silva, M. N. (2025). Towards Net-Zero: Analyzing Sustainability Reports of Global Airlines. *International Journal of Environmental Sciences*, *11*(5s), 781–813.

Dodman, D., Sverdlik, A., Agarwal, S., Kadungure, A., Kothiwal, K., Machemedze, R., & Verma, S. (2023). Climate change and informal workers: Towards an agenda for research and practice. *Urban Climate*, *48*, 101401. DOI: 10.1016/j.uclim.2022.101401

Ebi, K., Campbell-Lendrum, D., & Wyns, A. (2018). *The 1. 5 health report.* World Health Organization. https://cdn.who.int/media/docs/default-source/climate-change/who-the-1-5-healthreport.pdf?sfvrsn=61b2098_3&download=true

EPA. (2025a, August). *Climate Change and the Health of Workers* [Overviews and Factsheets]. https://www.epa.gov/climateimpacts/climate-change-and-health-workers

EPA. (2025b, August). *Indoor Air Quality and Changing Outdoor Environments* [Overviews and Factsheets]. https://www.epa.gov/indoor-air-quality-iaq/indoor-air-quality-and-changing-outdoor-environments

FAO. (2019). *Regional Overview of Food Security and Nutrition in Europe and Central Asia 2019. Structural Transformations of Agriculture for Improved Food Security*. Nutrition and Environment. https://openknowledge.fao.org/server/api/core/bitstreams/41dc9831-6937-4025-afed-2df4e6618730/content

Fatima, S. H., Rothmore, P., Giles, L. C., & Bi, P. (2023). Impacts of hot climatic conditions on work, health, and safety in Australia: A case study of policies in practice in the construction industry. *Safety Science, 165*, 106197. DOI: 10.1016/j.ssci.2023.106197

Ferrari, G. N., Leal, G. C. L., Ossani, P. C., & Galdamez, E. V. C. (2025). Investigation of the usage of machine learning to explore the impacts of climate change on occupational health: A systematic review and research agenda. *Frontiers in Public Health, 13*, 1578558. DOI: 10.3389/fpubh.2025.1578558 PMID: 40589827

Ferrari, G. N., Leal, G. C. L., Thom De Souza, R. C., & Galdamez, E. V. C. (2023). Impact of climate change on occupational health and safety: A review of methodological approaches. *Work (Reading, Mass.), 74*(2), 485–499. DOI: 10.3233/WOR-211303 PMID: 36314181

Flouris, A., Azzi, M., Graczyk, H., Nafradi, B., & Scott, N. (2024). *Heat at work: Implications for safety and health : a global review of the science, policy and practice* [1st ed.]. ILO.

Flouris, A. D., Dinas, P. C., Ioannou, L. G., Nybo, L., Havenith, G., Kenny, G. P., & Kjellstrom, T. (2018). Workers' health and productivity under occupational heat strain: A systematic review and meta-analysis. *The Lancet. Planetary Health, 2*(12), e521–e531. DOI: 10.1016/S2542-5196(18)30237-7 PMID: 30526938

Galappaththi, E. K., Susarla, V. B., Loutet, S. J. T., Ichien, S. T., Hyman, A. A., & Ford, J. D. (2022). Climate change adaptation in fisheries. *Fish and Fisheries, 23*(1), 4–21. DOI: 10.1111/faf.12595

Gariazzo, C., Taiano, L., Bonafede, M., Leva, A., Morabito, M., de' Donato, F., & Marinaccio, A. (2023). Association between extreme temperature exposure and occupational injuries among construction workers in Italy: An analysis of risk factors. *Environment International, 171*, 107677. DOI: 10.1016/j.envint.2022.107677 PMID: 36495676

Ghatak, A., & Lalitha, N. (2015). *Occupational Health Among Workers in the Informal Sector in India* (GIDR Working Paper Series No. 10092; p. 37). Gujarat Institute of Development Research. https://gidr.ac.in/pdf/WP%20228.pdf

Ghimire, M. (2024). Climate Change Impacts of Health Workers Without Safety Measures on Working Places. *Journal of Economics and Management*, *4*(1), 1–8. DOI: 10.3126/jem.v4i1.72884

Gibb, K., Beckman, S., Vergara, X. P., Heinzerling, A., & Harrison, R. (2024). Extreme Heat and Occupational Health Risks. *Annual Review of Public Health*, *45*(1), 315–335. DOI: 10.1146/annurev-publhealth-060222-034715 PMID: 38166501

Grandi, C., Lancia, A., & D'Ovidio, M. C. (2023). Climate Change: An Issue That Should Be Part of Workers' Information and Training Duties Envisaged by EU Directives on Occupational Health and Safety. *Atmosphere (Basel)*, *14*(7), 1183. DOI: 10.3390/atmos14071183

Holt, K. L. (2025). Occupational Health in the Age of Climate Change: A Review of Emerging Challenges and Strategic Interventions. *Contemporary Journal of Applied Sciences*, *3*(5), 265–282.

Hussainzad, E. A., & Gou, Z. (2024). Climate Risk and Vulnerability Assessment in Informal Settlements of the Global South: A Critical Review. *Land (Basel)*, *13*(9), 1357. DOI: 10.3390/land13091357

ILO. (2023). *Chemicals and climate change in the world of work: Impacts for occupational safety and health* ([1 ed.]). International Labour Organization. https://www.ilo.org/sites/default/files/wcmsp5/groups/public/%40ed_dialogue/%40lab_admin/documents/publication/wcms_887111.pdf

ILO. (2024). *Ensuring safety and health at work in a changing climate* (p. 132). International Labour Organization. https://www.ilo.org/sites/default/files/2024-07/ILO_SafeDay24_Report_r11.pdf

IOSH. (2025). *The heat is on: Protecting worker health and safety from the impacts of climate change*. IOSH. https://iosh.com/about/what-we-do/white-papers/the-heat-is-on-full-report

IPCC. (2018). *Global Warming of 1.5°C: IPCC Special Report on Impacts of Global Warming of 1.5°C above Pre-industrial Levels in Context of Strengthening Response to Climate Change, Sustainable Development, and Efforts to Eradicate Poverty* (1st ed.). Cambridge University Press., DOI: 10.1017/9781009157940

Ireland, A., Johnston, D., & Knott, R. (2023). Heat and worker health. *Journal of Health Economics*, *91*, 102800. DOI: 10.1016/j.jhealeco.2023.102800 PMID: 37562282

Jia, J. A., & Jia, N. I. (2024). Impacts of Climate Change on Occupational Health and Safety: A Comprehensive Review. *Asian Journal of Advanced Research and Reports, 18*(10), 136–154. DOI: 10.9734/ajarr/2024/v18i10761

Kadungure, A., Machemedze, R., Malaya, W., Banda, N., Loewenson, R., & Sverdlik, A. (2023). Towards climate change resilience and informal workers' health in Zimbabwe: An action-research case study. *Cities & Health, 7*(6), 1055–1071. DOI: 10.1080/23748834.2023.2261755

Karthick, S., Kermanshachi, S., Pamidimukkala, A., & Namian, M. (2023). A review of construction workforce health challenges and strategies in extreme weather conditions. *International Journal of Occupational Safety and Ergonomics, 29*(2), 773–784. DOI: 10.1080/10803548.2022.2082138 PMID: 35622383

Kim, A., & Yoo, G. (2023). Field-based Heatwave Risk Assessment of Outdoor Workers Measured by Wearable Sensors. *Asia-Pacific Journal of Atmospheric Sciences, 59*(4), 447–458. DOI: 10.1007/s13143-023-00326-4

Kjellstrom, T., Lemke, B., & Otto, M. (2017). Climate conditions, workplace heat and occupational health in South-East Asia in the context of climate change. *WHO South-East Asia Journal of Public Health, 6*(2), 15. DOI: 10.4103/2224-3151.213786 PMID: 28857058

Koch, M., Matzke, I., Huhn, S., Gunga, H.-C., Maggioni, M. A., Munga, S., Obor, D., Sié, A., Boudo, V., Bunker, A., Dambach, P., Bärnighausen, T., & Barteit, S. (2022). Wearables for Measuring Health Effects of Climate Change–Induced Weather Extremes: Scoping Review. *JMIR mHealth and uHealth, 10*(9), e39532. DOI: 10.2196/39532 PMID: 36083624

Levi, M., Kjellstrom, T., & Baldasseroni, A. (2018). Impact of climate change on occupational health and productivity: A systematic literature review focusing on workplace heat. *La Medicina del Lavoro, 109*(3), 163–179. DOI: 10.23749/mdl.v109i3.6851 PMID: 29943748

Lin, M.-S., Lin, Y.-C., Huang, T.-J., & Chen, M.-Y. (2022). Health Inequality among Fishery Workers during Climate Change: A National Population-Based and Retrospective Longitudinal Cohort Study. *International Journal of Environmental Research and Public Health, 19*(16), 10281. DOI: 10.3390/ijerph191610281 PMID: 36011913

Manni, G. (2023). *Heat Stress and Im/migrant FarmworkerHealth in the US and Mexico: A Scoping Review* (MECC Case Studies No. Nr.0002). CliMoHub. https://www.climohub.org/wp-content/uploads/2024/10/MECC-Case-Studies-_-Gregory-Manni.pdf

Marinaccio, A., Gariazzo, C., Taiano, L., Bonafede, M., Martini, D., D'Amario, S., de'Donato, F., Morabito, M., Ariani, F., Binazzi, A., Bogi, A., Bonafede, M., Buccelli, R., Crisci, A., de'Donato, F., De Francesco, F., Del Ferraro, S., Galzerano, P., Gariazzo, C., & Viaggi, D. (2025). Climate change and occupational health and safety. Risk of injuries, productivity loss and the co-benefits perspective. *Environmental Research*, *269*, 120844. DOI: 10.1016/j.envres.2025.120844 PMID: 39832550

Moda, H. M., Zailani, M. B., Rangarajan, R., Hickey, P., Abubakar, M., Maina, J., & Makarfi, Y. I. (2024). Safety awareness and adaptation strategies of Nigerian construction workers in extreme heat conditions. *PLOS Climate*, *3*(4), e0000380. DOI: 10.1371/journal.pclm.0000380

Ndugga, N., Pillai, D., & Published, S. A. (2023, June 26). Climate-Related Health Risks Among Workers: Who is at Increased Risk? *KFF*. https://www.kff.org/racial-equity-and-health-policy/issue-brief/climate-related-health-risks-among-workers-who-is-at-increased-risk/

Ngajilo, D., Adams, S., Kincl, L., Guernsey, J., & Jeebhay, M. F. (2023). Occupational Health and Safety in Tanzanian Aquaculture – Emerging Issues. *Journal of Agromedicine*, *28*(2), 321–333. DOI: 10.1080/1059924X.2022.2058139 PMID: 35337252

NYC Mayor's Office of Immigrant Affairs. (2025). *2024 Annual Report On New York City's Immigrant Population and Initiatives of the Office*. https://www.nyc.gov/assets/immigrants/downloads/pdf/MOIA-2024-Annual-Report_4.4.25.pdf

Ojha, A. (2022). *Heat Strain Risk Assessment of Field Workers: Integrating Wearable Sensing and Artificial Intelligence* [Master Thesis, The Pennsylvania State University]. https://etda.libraries.psu.edu/catalog/27626azo5242

OSHA. (2025). *Heat Injury and Illness Prevention in Outdoor and Indoor Work Settings Rulemaking | Occupational Safety and Health Administration*. Occupational Safety and Health Administration. https://www.osha.gov/heat-exposure/rulemaking

Parsakia, K., & Tabar, S. H. S. A. (2024). The Future of Occupational Health: Anticipating Risks in the Evolving Workplace. *Journal of Foresight and Health Governance*, *1*(2), 30–45.

Raulf, M., & Annesi-Maesano, I. (2025). Occupational allergy and climate change. *Current Opinion in Allergy and Clinical Immunology*, *25*(2), 83–87. DOI: 10.1097/ACI.0000000000001060 PMID: 39869536

Rînjea, C., Chivu, O. R., Darabont, D.-C., Feier, A. I., Borda, C., Gheorghe, M., & Nitoi, D. F. (2022). Influence of the Thermal Environment on Occupational Health and Safety in Automotive Industry: A Case Study. *International Journal of Environmental Research and Public Health, 19*(14), 8572. DOI: 10.3390/ijerph19148572 PMID: 35886424

Roe, E. A., Khov, E., Kim, H., Demelo, B., Zavitz, S., & Cao-Lei, L. (2024). The impact of climate change-related disasters on mental health and epigenetics: A narrative review. *Frontiers in Epigenetics and Epigenomics, 2*, 1464276. DOI: 10.3389/freae.2024.1464276

Schulte, P. A., Bhattacharya, A., Butler, C. R., Chun, H. K., Jacklitsch, B., Jacobs, T., Kiefer, M., Lincoln, J., Pendergrass, S., Shire, J., Watson, J., & Wagner, G. R. (2016). Advancing the framework for considering the effects of climate change on worker safety and health. *Journal of Occupational and Environmental Hygiene, 13*(11), 847–865. DOI: 10.1080/15459624.2016.1179388 PMID: 27115294

Schulte, P. A., Jacklitsch, B. L., Bhattacharya, A., Chun, H., Edwards, N., Elliott, K. C., Flynn, M. A., Guerin, R., Hodson, L., Lincoln, J. M., MacMahon, K. L., Pendergrass, S., Siven, J., & Vietas, J. (2023). Updated assessment of occupational safety and health hazards of climate change. *Journal of Occupational and Environmental Hygiene, 20*(5–6), 183–206. DOI: 10.1080/15459624.2023.2205468 PMID: 37104117

Sepadi, M. M. (2025). Impact of Climate Change on Informal Street Vendors: A Systematic Review to Help South Africa and Other Nations (2015–2024). *Atmosphere (Basel), 16*(2), 179. DOI: 10.3390/atmos16020179

Shan, D. (2022). Enforcement of fishing Occupational Health and Safety (OHS) standards: Challenges in Atlantic Canada. *Marine Policy, 145*, 105282. DOI: 10.1016/j.marpol.2022.105282

Sharma, M., Suri, N. M., & Kant, S. (2022). Analyzing occupational heat stress using sensor-based monitoring: A wearable approach with environmental ergonomics perspective. *International Journal of Environmental Science and Technology, 19*(11), 11421–11434. DOI: 10.1007/s13762-021-03862-6 PMID: 35106066

Silva, J. P., Macassa, G., Barros, H., & Ribeiro, A. I. (2024). Local Climate Change Adaptation under the Lenses of Public Health: A Case Study from Porto, Portugal. *Portuguese Journal of Public Health*, •••, 1–15. DOI: 10.1159/000540747 PMID: 40607201

Sirakaya, Assoc. Prof. Y. (2024). Climate Crisis and Human Psychology: Profound Impacts on Economic Concerns and Psychological Health. *International Journal of Social Science and Human Research, 07*(11), 8610–8628. DOI: 10.47191/ijsshr/v7-i11-67

Solberg, E. D. (2025, February). *Revealed: The top four trends for health and safety in 2025* [IOSH]. Health and Safety International. https://www.healthandsafetyinternational.com/article/1904790?utm_source=website&utm_medium=social

Thorvaldsen, T., Kongsvik, T., Holmen, I. M., Størkersen, K., & Holen, S. (2025). Revisiting occupational health and safety in Norwegian fish farming – Results from an employee survey. *Aquaculture (Amsterdam, Netherlands)*, *603*, 742441. DOI: 10.1016/j.aquaculture.2025.742441

Tsakonas, K., Badyal, S., Takaro, T., & Buse, C. (2025). Occupational health impacts of climate-related emergency events on health system workers in British Columbia. *International Journal of Disaster Risk Reduction*, *118*, 105227. DOI: 10.1016/j.ijdrr.2025.105227

Van Selm, L., Williams, S., de'Donato, F., Briones-Vozmediano, E., Stratil, J., Sroczynski, G., Tonne, C., De Sario, M., & Requena-Méndez, A. (2025). Occupational Heat Stress Among Migrant and Ethnic Minority Outdoor Workers: A Scoping Review. *Current Environmental Health Reports*, *12*(1), 16. DOI: 10.1007/s40572-025-00481-y PMID: 40123011

WHO. (2023, October 12). *Climate change*. World Health Organization Newsroom. https://www.who.int/news-room/fact-sheets/detail/climate-change-and-health

WHO. (2024, December 11). *Fact Sheet About Malaria*. World Health Organization. https://www.who.int/news-room/fact-sheets/detail/malaria

Wibowo, R., Satow, M., Quartucci, C., Weinmann, T., Koller, D., Daanen, H. A. M., Nowak, D., Bose-O'Reilly, S., & Rakete, S. (2025). Impact of heat stress and protective clothing on healthcare workers: Health, performance, and well-being in hospital settings. *Annals of Work Exposures and Health*, *69*(6), 665–675. DOI: 10.1093/annweh/wxaf026 PMID: 40448948

WMO. (2025). *State of the Global Climate 2024* (Vol. 1–1368). World Meteorological Organization.

Yovi, E. Y., Nastiti, A., & Kuncahyo, B. (2023). Heat-Related Knowledge, Risk Perception, and Precautionary Behavior among Indonesian Forestry Workers and Farmers: Implications for Occupational Health Promotion in the Face of Climate Change Impacts. *Forests*, *14*(7), 1455. DOI: 10.3390/f14071455

Zhang, M., & Kim, R. (2025). Occupational health in agriculture: A re-emerging frontier in worker protection. *Global Health Journal (Amsterdam, Netherlands)*, *9*(2), 65–71. DOI: 10.1016/j.glohj.2025.06.001

Chapter 8
Learning From COVID-19: Building a Robust Pandemic Preparedness Management System

Rashmi Singh
http://orcid.org/0000-0002-9620-0343
Sanskriti University, India

Abhishek Singh Bhadouria
Rajiv Academy for Technology and Management, India

Mohd Zahid
Rajiv Academy for Technology and Management, India

Junaid Hassan
Sanskriti University, India

ABSTRACT

This study aims to examine the importance of a preparedness management system in organizations to ensure a safe and healthy work environment for employees working on-site, particularly during health crises such as COVID-19. The study emphasizes the need for integrating COVID-19 based health and safety policies into organizational practices. Primary data were collected from 120 employees across different organizations using a structured questionnaire and factor analysis was done using SPSS software. The findings reveal that effective preparedness planning, clear health protocols, employee awareness and management support play a significant role in enhancing workplace safety and employee confidence. The study concludes that the

DOI: 10.4018/979-8-3373-6370-7.ch008

adoption of a systematic preparedness management framework and health-focused policies is essential for improving organizational resilience, safeguarding employee well-being and ensuring continuity of operations during emergency situations.

1. INTRODUCTION

The current estimated population of India is about 1.4 billion (World Population Review, 2021), and the country is facing the covid-19 epidemic from the year 2020. Covid-19's first case in India was confirmed in Kerala on 27 January 2020 (Andrews et al., 2020). Since then, breaking the chain of coronavirus has become the main concern for the government officials and other higher-level authorities of India, and following the safety, protocols have become the duty of the individuals residing in India. Even after adopting proper safety measures, a pandemic that started in 2020 has tremendously affected many, and the condition got even more terrifying in the second wave of the year 2021. According to Worldometer, the number of deaths has increased to 346,784, and the number of registered active cases increased to 28,809,339 till 6 June 2021 (Worldometer, 2021) and are still rising. Now, the question arises that what went wrong in India when the situation was handled appropriately in the first wave of covid-19? Moreover, the answer is the mismanagement and lack of proper planning to deal with the covid-19's second wave, resulting in an increased number of active cases and deaths.

Other countries are slowly reviving their economy from the losses held due to the covid-19 pandemic, and India is still finding its way and is struggling to reduce the virus spread. Covid-19 cases in India that started to show up in March 2020 continued to increase and have taken a terrible phase until December 2020. The number of reported cases has shown a little slowdown from January to March 2021, but without any appropriate preparedness management system, the cases again started to increase in April 2021(Wikipedia, 2021).

India has started its covid-19 vaccination campaign in January 2021. As of 5 June 2021, India has managed 231,322,417 doses overall (first and second dose) of the presently permitted vaccine (Wikipedia, 2021). In the first phase of covid -19, the threat of risk was more on people of age group 45 or 45+; therefore, the government of India started the vaccination process for this age group. Later, the lack of a preparedness management system leads the country into the second phase of covid -19, where the threat has shifted to the younger generation of 18+. Therefore, looking into the seriousness of the matter, the government permitted the vaccination of this age group in May 2021. Whereas India being the second-largest populated country globally, is facing a shortage of vaccination doses. Hence, if the situation continues, there will be no way out, and the third wave will continue to have a much more terrifying situation as it is predicted to spread the infection among the children of the age group below 18 and result in even more deaths. Therefore, focusing on the criticality of the situation, our paper aims to generate the idea of providing covid-19 license and covid-19 passport to the individuals residing in India based on their current health situation.

There are various categories of people residing in India. These people belong to different age groups and having different levels of income. Based on this criterion, the Indian population is separated into higher-level, Middle-class, and lower-middle-class. They are further bifurcated based on the different occupational groups: service class, Business class, and daily earners. Furthermore, viewing the situation of covid-19, they can be again bifurcated based on the people doing jobs in which their presence is essential, such as doctors, Police, Cleaners, and Reporters. At the same time, some jobs can be handled remotely, such as Teachers, Contract agents, Online business operators, etc., and lastly, some jobs are not much essential and could be suspended for some time to reduce the spread of the virus, such as Gyms, Saloons, Amusement parks, etc. Therefore, concentrating on these categories, our paper aims to drive a Suitable covid-19 preparedness management system for the organizations, which could help them to easily differentiate the employees who are covid-19 negative and be called back for work on-site and the employees who are covid-19 positive and will require time to recover and should stay home. Hence, the above-presented concept will help bring out the real sense of the new normal and help reduce the coronavirus cases in organizations and India overall.

2. IMPACT OF FIRST COVID WAVE AND STRATEGIC MANAGEMENT IN THE CORPORATE SECTOR

Since the end of 2019, the world faces an unexpected pandemic called covid-19 (Jalali et al., 2020). The pandemic, which started in China, did not take much time to spread worldwide, for which the world was unprepared. The covid-19 disease has shocked the world, especially business, as now they must take their ventures into the "unknowns and unknowns." The covid-19 pandemic has also majorly affected the global GDP growth between 2.3% to 4.8% (World Economic, 2020; ADB, 2020). Furthermore, it also decreased Foreign Direct Investment by 5% to 15% (UNCTAD, 2020). The covid-19 disease has affected the countries, people, economy, and organizations worldwide because they had no clue how things will turn up for them, how it will affect their lives, and how long the disease will continue. Many of the major businesses get affected due to the pandemic, especially the hospitality, construction, personal service, and transportation businesses, and the government restrictions to social distancing and lockdown have also made the employees stay at home and work remotely. Businesses got digitalized overnight, and meetings started virtually to avoid the virus spread and maintain the health and safety of employees (Shirey, 2021). Since the starting of the pandemic stock changes, indexes in the UK (FTSE 100) dropped about 19.3%, in the USA (Dow Jones), it dropped about 13.3%, in Japan, it dropped about 5.2% (Jones et al., 2020), and in India (BSE Sensex) it dropped about 6.6% (Acharya & N, 2021).

The pandemic has severely affected the different sectors of the economy. However, the most harmful effects are visible on various businesses as they are facing varieties of issues like shortage of raw material, Supply chain disruptions, decrease in demand, cancellation of orders which in turn, leads to a huge loss for them and thus, putting them not only into the financial pressure but also pressurizing them in terms of managing their staff and adjusting according to the requirements of the customers (Kuckertz & Brändle, 2021). Compared to the already well-known and established businesses, MSMEs and the businesses in the initial development phase experienced added challenges (Bartik et al., 2020; Prasad et al., 2015; Shafi et al., 2020). They now have to build, operate, and compete against the already established organizations and be involved in the crisis management they were not prepared (Salamzadeh and Dana 2021). According to the US Census, Small Business Pulse Survey, about 50% of the business has largely been negatively impacted by covid-19, and only 15-20% of the businesses have a sufficient amount of cash in hand to meet their expenses (Bohn, Mejia, & Lafortune, 2020). As per a survey conducted on 6,000 small businesses by Bartik et al. 2020, more than 43% of small businesses temporarily stopped.

"The covid-19 impact will not going to end in one day- the business will continue to have long-term implications in the future" (Chowdhury et al., 2020). During the first wave, businesses were facing problems related to cash shortage as their sale volume has decreased tremendously. Similarly, businesses also faced problems meeting their regular expenses, such as employees' wages and salary, interest on the loan, and paying rent due to less cash inflow to the company (Chowdhury et al., 2020). The pandemic has also left many companies under financial pressure and will likely cause bankruptcy (Donthu & Gustafsson, 2020). In 2020 business activities in the USA dropped by 22% during the covid-19 first wave (Fairlie 2020). It dropped about 10.1% to 14.8% in Canada in 2020 (Beland et al., 2020). In India, it dropped about 15% in 2019 to 5.3% in 2020 (GEM, 2020), Whereas in African-American business owner activities dropped about 41%, and in Latinx by 32% (Fairlie, 2020). In a small survey conducted by Indusguru, about half of the companies see more than 5% of their employees getting the covid-19 infection. Hence, the study has pointed that out of 2.54 crore covid-19 active cases, India caseloads account for 1.9% of the total population. If we compare this, companies' caseload average of 5% is about three times more than a national average (Business Today, 2021).

Businesses like SMEs have faced an even more pathetic situation during covid-19 first wave (Fairlie, 2020). They had to deal with the large unemployment rate and reduced growth due to the new project's conception deterioration. Unlike large businesses, these types of small businesses cannot recover quickly (Miocevic, 2021). Studies highlight that those small businesses are still emerging from the losses due to crisis and the lack of preparedness management has led them towards an even more devastating effect of the second wave of coronavirus.

During the pandemic, the immediate concern for BOD was the health and safety of its employees. Therefore, they tried to create an environment that demonstrates trust, care, openness, fairness, and timely communication to all its stakeholders. Other than this, they also tried to focus on the long-term implications of the business while tackling its short-term problems, such as meeting the expenses and maintaining the liquidity. Moreover, the BOD has taken this pandemic as an opportunity to recognize the flows of their business and come up with a solution for it. The businesses adopted an integrated approach and have shifted their emphasis from shareholders to stakeholders. Similarly, businesses also re-designed their pay structure, corporate governance mechanism, partnership, and value chains into sustainable and fairer ones to reduce the gap between the shareholders and stakeholders. In a survey held on 77% of the European practitioners, sustainable corporate governance was the major factor that helped the business emerge through this pandemic (Eklund, 2021).

"Strategies such as on-time information sharing, flexible orientation, Supply chain triple-A components implementations- alignment, adaptability and agility, employees support and recovery plans are some of the strategies being adopted during the first wave of covid-19" (Paul & Chowdhury, 2020). In addition to this, various strategic management tools like "risk management, data analysis, MBO, problem-solving, ICT and HR management techniques" has been used to handle the situation well (Luisetto & Latiyshev, 2020). An evolutionary model BMI (Business model innovation) has also gained momentum among researchers and development executives to help the industries revive from the revolting effects of coronavirus by generating sustainable competitive advantage to the firms in the rapidly changing environment (Breier et al., 2021).

In May 2020, the cases started to decline globally, and the lockdown ended. The economies and the business slowly started to open up by adopting "new normal." The businesses started to re-frame their business structure, business authority, and employee compensation system as per the new normal standards. For this, corporates started by developing sustainability and values in their re-designed structures and strategies (Eklund, 2021). Similarly, businesses have adopted various other strategies to deal with the crisis and remain flexible in the face of uncertainty.

Furthermore, the way the recovery will take place is still raising the concerns of India's different business and business authorities. The virus could be a renaissance, and the high level of preparedness for this renaissance is necessary for the recovery of businesses. This requires detailed planning of the recovery of businesses now, rather than just sitting and waiting for the epidemic to pass (McKinsey, 2020).

3. LACK OF COVID PREPAREDNESS AND INFORMATIVE ASSESSMENT

The various countries have adopted an extensive range of safety and preparedness measures since coronavirus was first registered in China. Sanitization, social distancing, avoiding foreign and non-essential traveling, lockdown, isolation, and washing hands have been used to stop the spread of the virus (Ali, 2020). However, the virus could not end and is still taking the lives of many people all over the globe as the spread of the virus could not be eliminated very fast. Although vaccines are available, the factors like vaccination hesitancy among people and children and the shortage of vaccines in countries like India, having the world's second-largest population, are two major factors that make the vaccination program unsuccessful and thus increase the virus spread.

After the first wave of covid-19, when the cases in India started to reduce, and the government has suspended the restrictions on lockdown and traveling, the people become very casual and started traveling, stopped following the mandatory restrictions of wearing the mask, avoiding non-essential traveling, and avoiding large gatherings.

Figure 1. Responses to non-essential travel

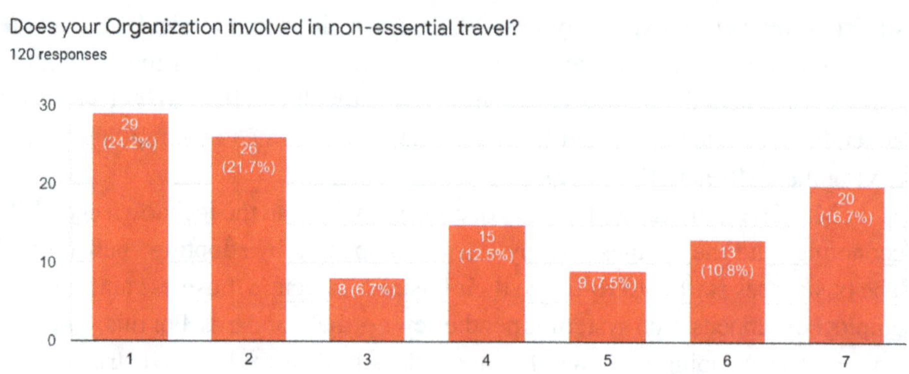

Figure-1 depicts that as per our study, 16.7% of the respondents highly agreed, 10.8% agreed, and 7.5% were somewhat agreed positively that after the first wave, their company has involved in non-essential travel, resulting in the covid-19 second wave whereas 24.2% highly disagreed, 21.7% disagreed, 6.7% somewhat agreed and 12.5% were neutral.

Moreover, the above-presented scenario has raised a serious question: Are we ready to reopen corporates and other economic activities in India and other countries with the same conditions as India?

Failure to follow the covid ethics and incapability to prepare themselves by taking learnings from the first wave led the countries into the second phase of coronavirus, which was even more terrifying and resulted in the deaths of even more people worldwide. Top industrialist ET emphasized taking learning from the experience of the previous year (2020). Ramesh Chauhan, chairperson of "Bisleri International," has also highlighted the irony of holding marriages, rallies, and Kumb Melas during the global pandemic (Vijayraghavan and Philip, 2021). One of the articles published by Economic Times has presented the mismatch announced by the government and followed on the ground. For instance, Organizations are pushing the vaccination of their employees, but the vaccination of daily workers and the lower-level staff is still a question mark.

Table 1. Lack of covid preparedness management and information assessment in India

Reason	Author	Lack of preparedness management and information assessment
The expert's warning went unheard.	**Sonali Acharjee**	• Not keeping track of virus transformations leads to the second wave. • No corresponding nationwide efforts have been made to categorize covid samples to distinguish the dominant variant and find life-threatening virus transformation. • Do not keep track of variants affecting the countries like the UK, South Africa, and Brazil, which were later detected in India. • Restrictions on importing resources and lack of fund allocation to the labs like INSACOG for sequencing. • No appointment of officers by states for the assortment and transference of samples to labs and lack of availability of cold storage to preserve the samples resulted in not meeting the targets of sequencing samples. • People are not following the covid-19 protocols.
Missing the alarm bells		• B.1.617 strain was already discussed in the meeting held with 21 members of the National Task Force of covid in April, and the safety guidance was sent to the government of India by the task force, which they underestimated, and this led to the spread of transformed variant resulting in more active cases.
Lessons from the UK		• The UK has formed the UK research association-COG-UK and allocated four times more funds than India, 329 crores. It helped the UK know where the virus is coming from in the country, finding out the variant connected to more intense disease, and helped determine the reasons for spreading the virus in certain areas. India is also building its capabilities through CCMB, but most labs are not getting the samples from the state to find out the variant and its spread.
Not Supplementing the medical workforce	**Kaushik Deka**	• The life-saving devices like ventilators and anesthetists were unable to operate because of less workforce. • The lack of proper medical infrastructure and shortage of medical facilities like ICUS, life-saving drugs, oxygen, etc. • Out of 157 medical colleges, only 27 and out of 22 AIIMS, only six are operational.
The vanishing hospital beds	**Kaushik Deka**	• During the first wave of covid-19, the patients struggled to find beds; even then, when the second wave started, many states like Delhi had witnessed the shortage of beds, and patients had to share the beds.
Going out of track	**Rahul Noronha, Kiran D. Tare, Romita Datta, and Amitabh Srivastava**	• The authorities of India have used the protocol T3 (test, treat and track) to fight against covid-19. However, no state has seen to be following these protocols and other covid ethics seriously, which resulted in the major spread of the virus.
No proper room that works	**Sandeep Unnithan**	• The patients calling the covid centers for the availability of beds were directed to the hospitals with no beds. • The rise in the reports of corruption and nepotism for black marketing of oxygen cylinders and life-saving drugs.

continued on following page

Table 1. Continued

Reason	Author	Lack of preparedness management and information assessment
Lack of preparedness for testing	**Sonali Acharjee**	• Lack of RT-PCR machines creates a huge difference in collected data and the data being processed, leading to a delay in the treatment of patients. • The delayed reports have made the patient's condition worse as they cannot seek proper medical care.
The vaccine complacency	**Amarnath K. Menon, P.B. Jayakumar**	• Although having the world's highest production of vaccination, the government of India has underestimated the demand assuming that corona is over after the first wave. • They do not involve private firms in vaccine production, which led to the shortage of vaccines in India.
The absent state	**Kaushik Deka**	• India's administration (State and Union government) did not take experts' advice seriously of avoiding large gatherings, and they seemed to be planning for Kumb Mela, which has the largest gathering in India. • There have been elections, social gatherings, non-essential traveling, and breaking other covid protocols in India that the Union Home Minister must stop by focusing on disaster management. • Massive election rallies have been taken place in the different states of India, and the election commission did not take any action on this and breaking the protocols of covid-19.

Source: India Today, 2021

Despite many health warnings by WHO in 2011, the world was unprepared to respond to severe public health emergencies (Timmis & Brüssow, 2020). The covid-19 pandemic has unfolded many aspects of health preparedness lacking in India (Mahmood et al., 2020). Moreover, lack of knowledge gap about the pandemic, less approachable to technology to seek information about coronavirus, the level of education to seek corona information, some sought of disability, people turning off from the information seeking sources due to fear of mental pressure, threat, terrorism and depression, fear of losing their loved ones, lack of safety measures like masks, gloves, sanitization and hygiene and unavailability of proper mediation are some of the major factors which directed to lack of covid preparedness among the people working in the corporates (Cvetković et al., 2020).

4. IMPACT OF SECOND IN THE CORPORATE SECTOR

Once the first wave curve started to flatten, many countries and India started to reopen their economic business activities. However, the government underestimating the risk of virus, people's causal behavior, and lack of proper planning and preparedness led the world to experience the crucial impact of coronavirus second wave. It has been more than one year since the first case of Covid-19 was registered in India, and the virus continued to disrupt people's lives worldwide. The studies suggest that the present strain is much more dangerous than the first strain of Covid-19. This time virus spread to rural areas that remained intact during the first wave and thus resulted in more deaths in India (Ranjan et al., 2021). The pandemic is rushing in many counties worldwide, highlighting the lack of preparedness (Xu & Li, 2020).

Figure 2. reason to increased covid cases

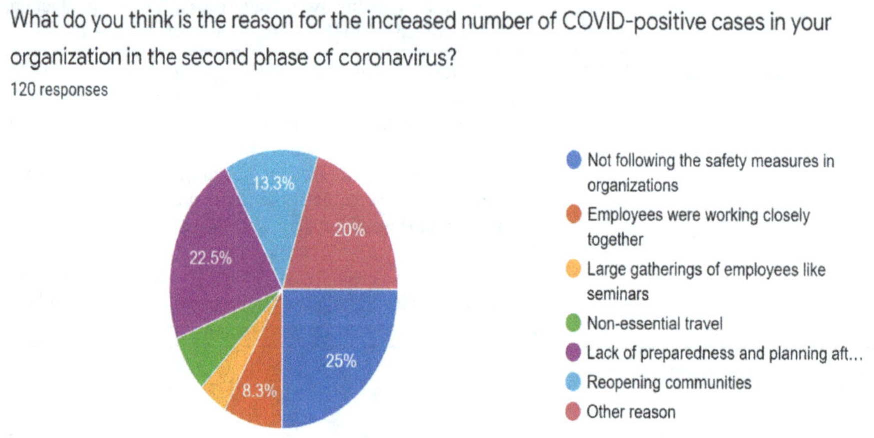

Figure-2 shows that as per our study, 25% of respondents have responded that not following the safety measures in the organization, and 22.5% of them responded that lack of proper planning and preparedness management were the two major factors that resulted in the second wave of coronavirus whereas 20% responded that employees working closely, 13.3% responded reopening communalities, 8.3% responded other reasons, 6.4% responded non-essential travel, and 4.5% responded large gathering of employees like seminars were the major factors for covid second wave.

Moreover, to reduce the risk of infection, there was a strict need to implement sanitary protocols for the smooth reopening of the organizations. However, the study suggests that implementing these protocols requires businesses to adjust their physical and operational infrastructure. Implementing these protocols in the organization requires a huge cost involvement, which was difficult for the businesses as they were already dealing with the first wave's losses, which was an extra burden on them in an already reduced demand. "An estimated cost of implementing these protocols for a five-worker firm is about $117,000 per worker, and it is about $39,000 if the company has 500 employees" (Janiak et al., 2021).

According to the (FICCI) Federation of Indian Chambers of Commerce and Industry survey, the second wave has highly impacted the Indian corporate sector as people's income level reduced due to job losses or the death of bread earners of the family. The reduced income is directed towards weakening their demand for commodities and services, thus impacting business activities. Similarly, the lockdown held during the first wave increased the prices of raw materials, which increased the cost of commodities and reduced the demand, which was yet another major factor behind the huge loss of businesses (Paliwal, 2021). Moreover, managing costs, reduced sales, and financial liquidity were creating hurdles in the operations of businesses.

The second wave of Covid-19 has resulted in a drop of 20 points in the overall index value. Vijay Kumar, a chief investment strategist, said that the emergence of the second-wave extended lockdown had raised concerns regarding the growth of GDP, and the business profit/earning seems to be lacking Foreign Portfolio Investment (FPI) outflow. This created a panic situation among investors and made them nervous and alert. "In May, foreign investors pulled Rs 6,452 crore from the Indian market. In mid of May, the foreign portfolio investors drew out Rs 6,427 crore from equities and Rs 25 crore from debt" (Business Today, 2021). Similarly, in April, Indian foreign trade weakened due to the second wave of covid-19 and impacted the country's economic activities. "Merchandise exports dropped from $34.4 billion in March to $30.21 billion in April. Merchandise imports dropped from $48.4 billion in March to $45.45 billion in April. Moreover, the value of jewelry, non-petroleum, and non-gem exports decreased from $27.42 billion in March to $23.51 billion in April. Similarly, Precious metal, non-oil, silver, and non-gold were reduced from $29.62 billion in March to $26.05 billion in April" (Nahata, 2021).

The new normal in the second phase of Covid-19 is an assortment of the essential need for improvisation, technology adoption, and association to enhance the effectiveness and efficiency of the organizations (Pinzaru et al., 2020). The highlighted facts suggest that returning to the traditional work culture without proper preparedness can lead to the third pandemic, worsening the economic condition. Therefore, following the required interventions and collaborating with digital solutions seems to be the only way out. However, adopting digital solutions and changing the business strategies and operational infrastructure will take time.

5. COVID-19 HEALTH RESPONSIBILITY (CHR)

The emergence of Covid-19 has enhanced the organization's concern for the health and safety of its employees. It is now the responsibility of the managers to create a work culture where the employees' health and safety are treated as one of the most important concerns. Organizations now are focusing more on remote working to ensure the safety of their employees. For instance, in a survey of CEO pulse review, after the second wave in April 2021, about 98% of the organizations are giving work from home to more than 90% of their employees (NASSCOM, 2021). The health and safety of employees are now the top priorities of most organizations. They are taking various steps to improve the work environment, which fosters the well-being of employees as a most important element for which they are forming the covid-19 management teams, arranging the beds for isolation and meditation, quickening employee's vaccination, and improving their insurance covers (NASSCOM, 2021).

5.2 Covid-19 Health Responsibility Management (CHRM)

Organizations should keep themselves prepared for the third wave of covid-19 and the possible extension of lockdown. The organizations need to learn from the experiences of the second wave and prepare a plan where they can communicate with the staff easily and ensure the appropriate availability of technical support required by the organization to continue their work. HRM should review the policies and practices of the organization for flexible working, which may be required shortly to work during the lockdown. Similarly, the policies related to remote working should be reviewed to ensure data safety and ensure the sufficiency of performance procedures to make the employees accountable for their work performance. Moreover, HRM should ensure the proper health checkups of employees and support them to avoid mental illness among employees. Therefore, CHRM regards the management of the overall risk of infection for the long-term success of the businesses and the health and well-being of employees. If the organization wants to eliminate the risk of virus, it must collaborate with both employer and employee. Therefore, organizations should frame the CHRM system for employees by ensuring implementation of practices recommended by WHO.

Table 2. Practices should be followed for Covid-19 health management system

5.1 Start from your workplace	• Even if the virus infection is reducing now, the chances of a third wave can increase if the situation continues to be the same. Therefore, we must start to stop the spread of viruses before it reaches your organization. • Keep proper maintenance of health and hygiene of workplace from surface to objects everything should be disinfected regularly to remove the contamination from everything employees and customers encountered. • Encourage the washing of hands of everyone in the organization regularly. Put sanitizers, exhibit handwashing posters, and use all communication measures to promote handwashing and make sure whoever comes into the organization is properly sanitized. • Encourage best respiratory hygiene among employees by exhibiting posters and other communication measures. Ensure the employees are wearing masks, especially when they are encountering colds and coughs. • Before going on any business trip, advise your employees to refer national travel for assistance. • Advice your employees that if your community is Covid-19 infected and has mild fever or cough, they must stay at home and take medicines. The employees should also be given sick leaves.
5.2 Safety during meetings and events	**Before meetings** • Take advice from higher authorities on where to hold the meeting. • Prepare a proper preparedness plan for preventing the infection during meetings. Try to replace face to face meetings with telecom meetings, involve fewer people, partner up with health care authorities for proper communication and ensure information, ensure proper supply of health kits (mask, sanitizers, etc.), monitor the symptoms of employees before the meeting, and share the participant's details with health care authorities to avoid infection. • Develop a response plan in case any employees show the symptoms of Covid-19 during the meeting. Keep ready an isolated room for infected employees, plan their safe reaching hospitals, plan what to do if any meeting member comes covid positive, and take advice from healthcare providers. **During meeting** • Communicate about the safety measures from all means of communication. Build trust, encourage using a safety kit, encourage the used tissue or cover the face while sneezing, and give the participants numbers of healthcare hotlines for medical advice. • Use an alcohol-based disinfectant for regular handwashing at the event place. • Do sitting arrangements at least 1 meter away from each other. • Keep the room validated by keeping the doors and windows open. • Suppose any participant feels not well, then follow the preparedness plan. Call the healthcare number, isolate the member in a separate room, offer a mask and send them for covid assessment. **After meeting** • Save the name and contact details of the members who attended the meeting for at least one month. This will help the health authorities find all the participants who attended the meeting if any members found positive after the meeting. • While meeting, if any participant is suspected to be positive, this should be informed to all the other members to isolate themselves for 14 days to avoid infection, if any. • If any participant has even minor symptoms, they should communicate this to their health care providers, giving them details about their symptoms and travel.

continued on following page

Table 2. Continued

5.3 Safety while traveling	**Before traveling** • Ensure that employees have information about the area they are traveling to know about the covid spread of that area. • Based on contamination of that area, make a proper plan, and assess the risk and benefits of the related travel. • Avoid sending those employees who have a serious illness and other diseases like diabetes. • Make sure that the person traveling should be briefed by any health care providers. • Ensure employees traveling should be equipped with the safety kit. **During traveling** • Encourage employees to properly use safety kits, keep sanitizing their hands, and maintain social distancing from everyone. • Make sure that employees take advice from health authorities if they get ill while traveling. • Ensure employees follow the instructions of local authorities of that area where they are traveling. **After traveling** • The employees who have traveled should be monitored for 14 days and should be given work from home. • If that employee has mild symptoms of covid infection, they should be isolated and keep taking advice from health care providers for meditation.
5.4 Safety if the virus reaches your community	• Develop a preparedness plan in the case any employee gets positive in your organization. Isolate the member for distancing, and take advice for health care provides immediately. • Prepare a report of employees who are at greater risk of infection and support them without any discrimination. These employees are seriously ill, have recently traveled anywhere, or come to work from contamination zones. • Encourage telecommuting among employees so that the health authorities may advise the employees to take necessary precautions. • Prepare your organization well for the possibility of cases that may occur. • Plan which can run business continuously with minimum contact and even when any member associated with the organization could not reach the workplace. • Communicate the plan with all the associated members of the company and tell them to strictly follow the safety measures guidelines. • Adress the mental health and social consequences in your plan and support the employees. • All businesses should partner up with health care providers in case of emergency.

Source: WHO, 2020

6. COVID-19 PREPAREDNESS MANAGEMENT SYSTEM

Covid-19 has raised concern in people's minds about their health and safety and of their loved ones. After the first wave of covid-19, the organizations opened for work on-site, and employees started to work, risking their lives to help the companies revive from the losses held during the first wave. However, a lack of preparedness management system spread the infection among employees, putting their lives at risk and many deaths. In this situation, employees are in dilemma whether they should quit the jobs as they fear losing their lives or continue to work where the threat of disease will create mental illness. Therefore, higher executives' responsibility is to create an environment of trust and care among employees and implement the policies and practices with the overall preparedness management system. Therefore, our paper proposes a conceptual framework for the companies to help them segregate the employees based on their health cards.

The concept of providing health cards supports the vaccination and regular health checkups of employees and generates their covid license and covid passport based on their existing health condition. As there are millions of people working in India, they must be segregated based on their present profession. The jobs which are essential and require human presence for serving the customers should be given covid license, and covid passports, the jobs which can be handled remotely should be given guidelines for remote working and the jobs which are non-essential and can be suspended for some time should be given guidelines for temporary closure as shown in fig-8. The employees who qualify for the covid license and covid passport will have to go through the medical checkups every month, and based on this concept, their health cards will be generated, and the eligibility for the type of covid license covid passport will be decided.

- The employees who are covid negative and are vaccinated will be given a green card.
- The employees who are covid negative but are not vaccinated will be given yellow cards.
- A few months ago, the employees who tested positive but have recently recovered will be given a blue card.
- The employees who tested positive will be quarantined and will be given red cards.

Figure 3. Conceptual framework for allocating primary health cards to employees

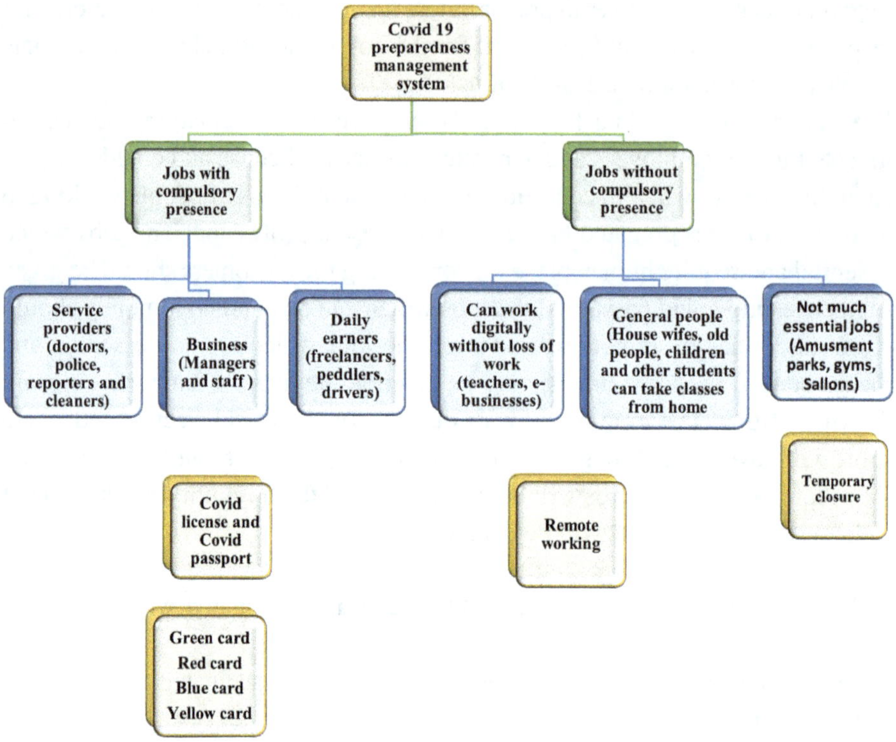

Covid license will be given to the employees who are working in India. The license will play a crucial role in making other employees aware of the existing health condition of their colleagues without even knowing them personally. Similarly, the covid passport will be given to the employees traveling outside India for some organizational work. The covid passport will be attached with the employee's vaccination and health checkups to help them, and other people be safe from the virus infection. This mechanism will help the employees stay alert and understand their colleague's present health condition. It will also be very helpful for the organizations as they can differentiate their employees easily and plan their reopening strategies accordingly. Besides, the organizations can arrange their shifts according to employees' health cards and quarantine them with red cards.

Figure 4. Responses for providing covid-19 passport and license to employees as a preparedness management system

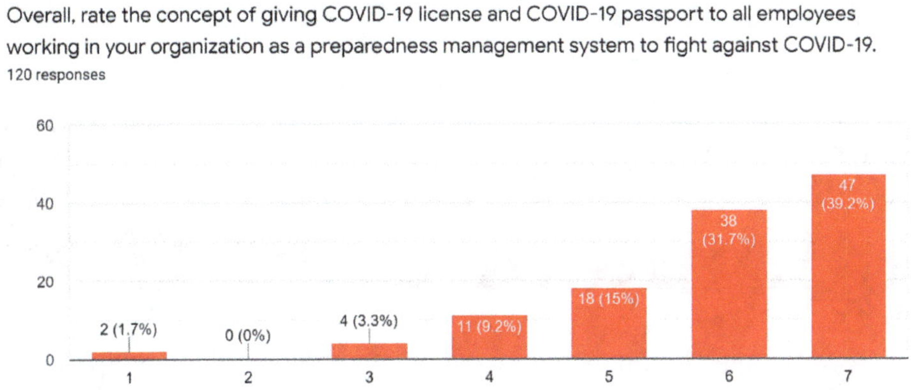

Figure-4 shows that as per our study, 39.2% of employees were highly satisfied, 31.7% were positively satisfied and 15% were somewhat satisfied with the concept of providing a covid license and covid passport to reduce the virus infection, whereas only 5% of the employees were dissatisfied, 2% were highly dissatisfied and 9.2% were neutral.

6.1 Covid-19 Medical Insurance

The health and safety of employees is yet another major concern for the companies which should be considered while planning the reopening strategies. Covid-19 medical insurance will motivate the employees to feel positive about the company and help them work for the organization enthusiastically. According to our study, most of the respondents have answered positively for having the covid-19 medical insurance.

Figure 5. Responses to covid-19 medical insurance

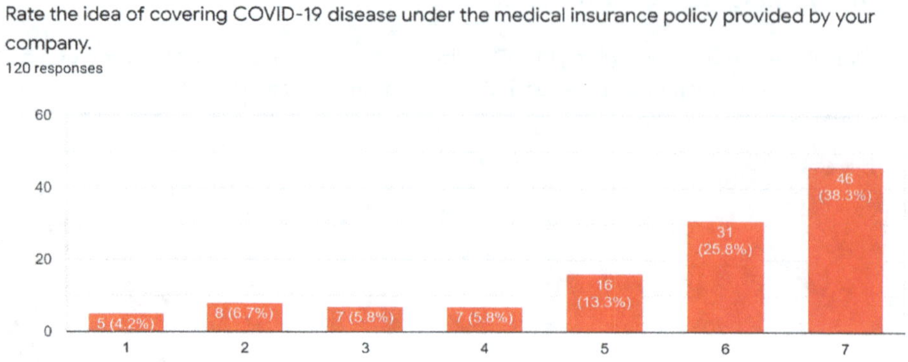

Figure 5 depicts that as per our study, 38.3% of the employees were highly satisfied, 25.8% were satisfied, and 13.3% were somewhat satisfied with the idea of covering covid-19 the disease under the medical insurance policy whereas 4.2 were highly dissatisfied, 6.7% were dissatisfied, 5.8% were somewhat satisfied and 5.8% were neutral for the same.

7. RESEARCH METHOD

It is a quantitative study based on random sampling that has been used to collect data for the study as this method of data collection helps the researcher to collect the data from the respondents, without any biasness and in this sampling method all the population has an equal chance of getting selected. The sample size of the study is the respondents who are working in the public and private enterprises as these enterprises are responsible for the economic stability of every country, and it is very much important that these organizations should be reopened with proper planning and preparedness management system to eliminate the risk of the virus among employees. The data was collected from the employees working in different states of India. The questionnaire was prepared with the help of a google form and was distributed among the respondents through email, for which a total of 120 responses were received. The survey measures were self-administrated, consisting of 15 questions, and divided into three goggle form categories. In the first category of the item based on organizational response to the first wave were placed, in the second category of items based on the organizational response to second-wave were placed and the final category of item was to know the responses of employees for providing the covid-19 license, passport, and medical insurance as a part of the covid-19 preparedness management system and the reliability of the scale was checked. The measurements of the survey item in this study are constructed on a seven-point Likert scale ranging from 1-7 in which one was for "highly disagree," 2 for "disagree," 3 for "somewhat disagree," 4 for "neutral," 5 for "somewhat agree," 6 for "agree" and 7 for "highly disagree." The total number of male respondents was 63.2%, whereas female respondents were 36.8%. The data collected with the help of a questionnaire was analyzed with the help of SPSS.

7.1 Objectives of the Study

The study attempted to:

- Assess the idea of providing covid-19 license, passport, and medical insurance to the employees as a part of the covid-19 preparedness management system.
- Identify the factors which were responsible for the second wave.
- Identify the factors for which the employees were positively concerned while preparing a preparedness plan for reopening.

8. DATA ANALYSIS AND RESULTS

8.1 Scale Reliability

Table 3 (a).

Case Processing Summary			
		N	%
Cases	Valid	119	99.2
	Excluded[a]	1	.8
	Total	120	100.0

a. Listwise deletion based on all variables in the procedure.

Table 3 (b).

Reliability Statistics	
Cronbach's Alpha	**N of Items**
.799	15

The table-3 (a) shows that the total number of respondents was 120, of which 119 responses were valid. To test the reliability of the scale, the Cronbach's Alpha value is being analyzed, and the table-3 (b) depicts that the items within the constructs and the overall score of Cronbach's alpha value were .799, which is higher than the preferable score value, which is 0.70 highlighting that the scale is reliable and consistent enough to measure the constructs (Cavana, et al., 2001).

8.2 Descriptive Statistics and Correlation Analysis

Table-4 shows the descriptive statistics includes the mean values, the standard deviation of the constructs. The respondents, on average, have confirmed positive concern about all the categories of items is placed in a construct. As per the mean values, they have shown the greatest positive concern about safety measures and the preparedness of the second wave. The other variables for which the respondents were much positive are the motivation for vaccination and organizations alerting employees. The variables about which the respondents were least positive are mandatory government apps as their mean value is below 3, i.e., 2.20.

Table 4 (a).

Descriptive Statistics			
	Mean	**Std. Deviation**	**Analysis N**
Alert employees	6.00	1.228	119
Safety measures	6.16	1.097	119
Nonessential travel	3.58	2.234	119
Motivated employees to follow safety rules	5.77	1.715	119
Colleagues' precautions	5.93	1.254	119
Preparedness second wave	6.09	1.000	119
Mandatory gov apps	2.20	2.331	119
Lessons countries	5.67	1.502	119
Motivation vaccination	6.00	1.353	119
Reason increased cases	4.12	2.285	119

continued on following page

Table 4 (a). Continued

Descriptive Statistics			
	Mean	**Std. Deviation**	**Analysis N**
Covid insurance	5.50	1.765	119
Health check-ups	5.70	1.554	119
Covid cards	5.92	1.176	119
Rate passport	5.57	1.516	119
Rate passport license	5.87	1.279	119

In the correlation analysis of the construct measures, the Pearson correlation is given in table-4 (b) depicts that majority of the hypothesized relations established were statistically significance at level p, 0.01 (2 tailed) as it is clear from the table that all values are more than (p, 0.001) which means that they all the items are correlated with each other. Moreover, the preferable score of the correlation coefficient should not be more than 0.8 to eliminate the risk of multicollinearity (Field, 2005). It is very much clear from the table below that the maximum correlation coefficient is 0.779, which is smaller than 0.8, suggesting that no multicollinearity exists in the scale.

Table 4 (b). Correlation Analysis between constructs

		Alert employees	**Safety measures**	**Non-essential travel**	**Encouraged employees**	**Colleagues precautions**	**Preparedness Second wave**	**Mandatory Gov apps**
Alert employees	Pearson Correlation	1	.640**	.083	.444**	.492**	.466**	-.091
	Sig. (2-tailed)		.000	.366	.000	.000	.000	.325
Safety measures	Pearson Correlation	.640**	1	-.008	.450**	.497**	.433**	-.095
	Sig. (2-tailed)	.000		.931	.000	.000	.000	.301
Nonessential travel	Pearson Correlation	.083	-.008	1	.129	.114	.033	-.117
	Sig. (2-tailed)	.366	.931		.159	.214	.719	.204
Motivated employees to follow safety rules	Pearson Correlation	.444**	.450**	.129	1	.398**	.342**	-.156
	Sig. (2-tailed)	.000	.000	.159		.000	.000	.088
Colleagues precautions	Pearson Correlation	.492**	.497**	.114	.398**	1	.321**	.259**
	Sig. (2-tailed)	.000	.000	.214	.000		.000	.004

continued on following page

Table 4 (b). Continued

		Alert employees	Safety measures	Non-essential travel	Encouraged employees	Colleagues precautions	Preparedness Second wave	Mandatory Gov apps
Preparedness Second wave	Pearson Correlation	.466**	.433**	.033	.342**	.321**	1	-.178
	Sig. (2-tailed)	.000	.000	.719	.000	.000		.051
Mandatory Gov apps	Pearson Correlation	.091	.095	-.117	-.156	.259**	-.178	1
	Sig. (2-tailed)	.325	.301	.204	.088	.004	.051	
Lessons countries	Pearson Correlation	.489**	.367**	.096	.366**	.385**	.573**	-.006
	Sig. (2-tailed)	.000	.000	.297	.000	.000	.000	.947
Motivation vaccination	Pearson Correlation	.418**	.354**	-.036	.384**	.335**	.413**	-.038
	Sig. (2-tailed)	.000	.000	.694	.000	.000	.000	.685
Reason Increased cases	Pearson Correlation	.063	.203*	-.033	.083	.238**	.246**	.106
	Sig. (2-tailed)	.495	.026	.719	.369	.009	.007	.249
Covid insurance	Pearson Correlation	.327**	.300**	.115	.182*	.263**	.403**	-.094
	Sig. (2-tailed)	.000	.001	.211	.047	.004	.000	.307
Health Check-ups	Pearson Correlation	.326**	.289**	.097	.178	.329**	.446**	-.099
	Sig. (2-tailed)	.000	.001	.293	.052	.000	.000	.282
	N	120	120	120	120	120	120	120
Covid cards	Pearson Correlation	.273**	.365**	.030	.212*	.202*	.338**	.176
	Sig. (2-tailed)	.003	.000	.748	.020	.027	.000	.055
Rate passport	Pearson Correlation	.185*	.341**	-.124	.163	.155	.293**	.233*
	Sig. (2-tailed)	.043	.000	.179	.076	.090	.001	.010
Rate Passport and license	Pearson Correlation	.266**	.346**	.046	.192*	.264**	.334**	.062
	Sig. (2-tailed)	.003	.000	.619	.036	.004	.000	.500

Table 4 (c). Correlation Analysis between constructs

Lessons countries	Motivation vaccination	Reason Increased cases	Covid insurance	Health Checkups	Covid cards	Rate passport	Rate Passport license
.489**	.418**	.063	.327**	.326**	.273**	.185*	.266**
.000	.000	.495	.000	.000	.003	.043	.003
.367**	.354**	.203*	.300**	.289**	.365**	.341**	.346**
.000	.000	.026	.001	.001	.000	.000	.000
.096	-.036	-.033	.115	.097	.030	-.124	.046
.297	.694	.719	.211	.293	.748	.179	.619
.366**	.384**	.083	.182*	.178	.212*	.163	.192*
.000	.000	.369	.047	.052	.020	.076	.036
.385**	.335**	.238**	.263**	.329**	.202*	.155	.264**
.000	.000	.009	.004	.000	.027	.090	.004
.573**	.413**	.246**	.403**	.446**	.338**	.293**	.334**
.000	.000	.007	.000	.000	.000	.001	.000
-.006	-.038	.106	-.094	-.099	.176	.233*	.062
.947	.685	.249	.307	.282	.055	.010	.500
1	.655**	.274**	.497**	.532**	.481**	.411**	.446**
	.000	.002	.000	.000	.000	.000	.000
.655**	1	.359**	.397**	.367**	.437**	.409**	.509**
.000		.000	.000	.000	.000	.000	.000
.274**	.359**	1	.152	.117	.334**	.349**	.263**
.002	.000		.097	.203	.000	.000	.004
.497**	.397**	.152	1	.779**	.320**	.351**	.395**
.000	.000	.097		.000	.000	.000	.000
.532**	.367**	.117	.779**	1	.453**	.425**	.462**
.000	.000	.203	.000		.000	.000	.000
120	119	120	120	120	120	120	120
.481**	.437**	.334**	.320**	.453**	1	.649**	.644**
.000	.000	.000	.000	.000		.000	.000
.411**	.409**	.349**	.351**	.425**	.649**	1	.748**
.000	.000	.000	.000	.000	.000		.000
.446**	.509**	.263**	.395**	.462**	.644**	.748**	1
.000	.000	.004	.000	.000	.000	.000	

**. Correlation is significant at the 0.01 level (2-tailed).

*. Correlation is significant at the 0.05 level (2-tailed).

8.3 Factor Analysis

8.3.1 KMO and Bartlett's Test

Table-5 shows that the Kaiser-Meyer-Olkin (KMO) value that highlights the sample adequacy is 0.816, which is more than the threshold value of 0.5, which means the sample size is adequate to conduct the study (Field, 2005). Bartlett's test value is .000, which is less than 0.05 (p-value < 0.05), which means that all the variables contribute significantly to the study.

Table 5.

KMO and Bartlett's Test		
Kaiser-Meyer-Olkin Measure of Sampling Adequacy.		**.816**
Bartlett's Test of Sphericity	Approx. Chi-Square	800.008
	df	105
	Sig.	.000

8.3.2 Communalities

The communalities table-6 tells that how much variance is explained by an individual variable. The variables having extraction values (variance) more than 0.5 are the important variables, and the Principal Component Analysis examined 15 items. As the results of the commonalities table show, the majority of the extracted values explain the variance more than 0.5, which means that all these values are important for the study.

Table 6.

Communalities		
	Initial	**Extraction**
Alert employees	1.000	.701
Safety measures	1.000	.646
Nonessential travel	1.000	.232
Motivated employees to follow safety rules	1.000	.525
Colleagues' precautions	1.000	.544
Preparedness second wave	1.000	.495

continued on following page

Table 6. Continued

Communalities		
	Initial	**Extraction**
Mandatory gov apps	1.000	.487
Lessons countries	1.000	.628
Motivation vaccination	1.000	.538
Reason increased cases	1.000	.319
Covid insurance	1.000	.740
Health check-ups	1.000	.806
Covid cards	1.000	.661
Rate passport	1.000	.765
Rate passport license	1.000	.670

Extraction Method: Principal Component Analysis.

8.3.3 Total variance

The total variance suggests that the total Initial Eigenvalue, more than one, will be extracted. In the table-7 there are three variables which are having values of more than one is 5.561, explaining the variance of 37.071%, 1.932 explaining the variance of 12.879%, and the last value 1.265 is explaining the variance of 8.433%; thus, these three values will be extracted from the table, and all the variables will be distributed among these three extracted factors. Tabachnick and Fidell (2013) factors with eigenvalue below one should be omitted, and factor loading less than 0.5 will be eliminated (Liao et al., 2007). The cumulative percent of the last value is 58.384% which is more than 50% and is good for management researches.

Table 7. Total Variance Explained

Component	Initial Eigenvalues			Extraction Sums of Squared Loadings			Rotation Sums of Squared Loadings		
	Total	% Of Variance	Cumulative %	Total	% Of Variance	Cumulative %	Total	% Of Variance	Cumulative %
1	**5.561**	**37.071**	**37.071**	5.561	37.071	37.071	3.480	23.203	23.203
2	**1.932**	**12.879**	**49.950**	1.932	12.879	49.950	3.258	21.722	44.925
3	**1.265**	**8.433**	**58.384**	1.265	8.433	58.384	2.019	13.459	58.384
4	.969	6.458	64.841						
5	.907	6.048	70.890						
6	.796	5.305	76.194						

continued on following page

Table 7. Continued

Component	Initial Eigenvalues			Extraction Sums of Squared Loadings			Rotation Sums of Squared Loadings		
	Total	% Of Variance	Cumulative %	Total	% Of Variance	Cumulative %	Total	% Of Variance	Cumulative %
7	.723	4.822	81.016						
8	.613	4.087	85.104						
9	.542	3.613	88.717						
10	.446	2.974	91.691						
11	.375	2.503	94.195						
12	.273	1.823	96.018						
13	.242	1.610	97.628						
14	.184	1.227	98.855						
15	.172	1.145	100.000						

Extraction Method: Principal Component Analysis.

8.3.4 Component matrix and Rotated matrix

The table-8 depicted the component matrix and rotated component matrix, which tells which variable will go with which factor loading extracted from the total variance table having a value more than 1. All 15 variables are distributed under three extracted variables, and both the tables are sorted by size. The possible values range from -1 to +1 because of correlation. The variables having the highest values in the line will go with that factor loading. In the rotated component matrix, the seven values of "Rate passport," "Covid cards, "Rate passport license," "Motivation vaccination," "Reason increased cases," "Lessons countries," "Mandatory gov apps" will go under the factor loading fir[st] which are 0.866, 0.780, 0.760, 0.527, 0.512, 0.489, 0.458 and will be given a nomenclature Covid-19 preparedness next five values of "Alert employees," "Safety measures," "Encouraged employees," "Colleagues' precautions," "Preparedness second wave" will go under the factor loading seco[nd] which are 0.820, 0.759, .0721, 0.702, .530 and will be given a nomenclature organizations safety measures and the last three values of "Health checkups," "Covid insurance," "non-essential travel" will go under the 3[rd-]factor loading which are 0.762, 0.759, 0.432 and will be given a nomenclature Measures avoiding the virus.

Table 8.

Component Matrix				Rotated Component Matrix			
	Component				Component		
	1	2	3		1	2	3
Lessons countries	.788	.024	.082	Rate passport	.866	.089	.086
Motivation vaccination	.720	.054	.127	Covid cards	.780	.196	.118
Rate passport license	.711	.402	.049	Rate passport license	.760	.177	.245
Health check-ups	.708	.060	.550	Motivation vaccination	.527	.481	.170
Covid cards	.684	.434	.074	Reason increased cases	.512	.195	.136
Preparedness second wave	.674	.197	.048	Lessons countries	.489	.478	.401
Safety measures	.663	.285	.354	Mandatory gov apps	.458	.292	.439
Rate passport	.659	.572	.063	Alert employees	.135	.820	.107
Covid insurance	.656	.008	.556	Safety measures	.262	.759	.023
Alert employees	.653	.436	.292	Encouraged employees	.046	.721	.058
Colleagues' precautions	.573	.445	.132	Colleagues' precautions	.059	.702	.217
Motivated employees to follow safety rules	.511	.430	.280	Preparedness second wave	.289	.530	.362
Reason increased cases	.404	.291	.267	Health check-ups	.446	.163	.762
Mandatory gov apps	.065	.623	.308	Covid insurance	.372	.160	.759
Nonessential travel	.078	.287	.380	Nonessential travel	.207	.057	.432
Extraction Method: Principal Component Analysis.				Extraction Method: Principal Component Analysis. Rotation Method: Varimax with Kaiser Normalization.			
a. three components extracted.				a. Rotation converged in 5 iterations.			

CONCLUSION, LIMITATIONS OF THE STUDY AND FUTURE SCOPE

In this study, the respondents were highly positive about the concept of providing a covid-19 license, passport, and medical insurance to them to reduce the risk of the virus. According to the primary data collected through questionnaires, 70.9% of the employees were highly satisfied and satisfied, and 15% of the employees were somewhat satisfied with the concept of covid-19 cards as a measure to prepare the organizations for reopening in the crucial time of the pandemic. Moreover, 25% of respondents have responded that not following the safety measures in the organization and 22% of them responded that lack of proper planning and preparedness management were the two major factors that resulted in the second wave of coronavirus, followed by 20% response for employees were working closely together and 13.3% for reopening commonalities as a reason for the second wave of covid-19.

Safety measures and preparedness during the second were the factors for which the employees were positively concerned while preparing a preparedness plan for the reopening of organizations. These factors were followed by motivation for vaccination and organizations alerting employees factors based on employees' positive concerns while making a preparedness plan. The variables about which the respondents were least positive were mandatory government apps as their mean value is below three, i.e., 2.20. The factors other than mandatory government apps have values of more than one, which means that all the factors taken for study are significant for measuring the covid-19 preparedness management system. The current study is a preliminary study analyzed with the help of factor analysis.

The study is a small-scale study where the sample size is 120 respondents, and a similar sought of study can be conducted taking a larger sample size in the future. Moreover, the same study can be conducted in other countries by adding more constructs to the scale for more accurate results.

REFERENCES

Acharya, R. M. & N. (2021). Indian stock market's reaction to Covid-19 crisis is surprisingly muted. Business Standard India. https://www.business-standard.com/article/markets/indian-stock-market-in-line-with-global-peers-even-during-covid-crisis-121051400159_1.html

ADB. (2020). *Asian Development Outlook 2020 Highlights: What Drives Innovation in Asia?* Asian Development Bank., DOI: 10.22617/FLS200120-3

Ali, I. (2020). COVID-19: Are We Ready for the Second Wave? *Disaster Medicine and Public Health Preparedness*, *1–3*(5), e16–e18. Advance online publication. DOI: 10.1017/dmp.2020.149 PMID: 32379015

Andrews, M. A., Areekal, B., Rajesh, K. R., Krishnan, J., Suryakala, R., Krishnan, B., Muraly, C. P., & Santhosh, P. V. (2020). First confirmed case of COVID-19 infection in India: A case report. *The Indian Journal of Medical Research*, *151*(5), 490–492. DOI: 10.4103/ijmr.IJMR_2131_20 PMID: 32611918

Bartik, A., Bertrand, M., Cullen, Z. B., Glaeser, E. L., Luca, M., & Stanton, C. (2020). How Are Small Businesses Adjusting to COVID-19? Early Evidence from a Survey. Harvard Business School Working Paper. 20(102):1–37. https://hbswk.hbs.edu/item/how-are-small-businesses-adjusting-to-covid-19-early-evidence-from-a-survey

Bartik, A. W., Bertrand, M., Cullen, Z. B., Glaeser, E. L., Luca, M., & Stanton, C. T. (2020). How Are Small Businesses Adjusting to COVID-19? Early Evidence from a Survey (Working Paper No. 26989; Working Paper Series). National Bureau of Economic Research. w26989. DOI: 10.3386/w26989

Beland, L.-P., Fakorede, O., & Mikola, D. (2020). Short-Term Effect of COVID-19 on Self-Employed Workers in Canada. *Canadian Public Policy*, *46*(S1, Suppl 1), S66–S81. DOI: 10.3138/cpp.2020-076 PMID: 38629981

Bohn, S., Mejia, M. C., & Lafortune, J. (2020). The Economic Toll of COVID-19 on Small Business. Public Policy Institute of California. Retrieved December 13, 2021, from https://www.ppic.org/blog/the-economic-toll-of-covid-19-on-small-business/

Breier, M., Kallmuenzer, A., Clauss, T., Gast, J., Kraus, S., & Tiberius, V. (2021). The role of business model innovation in the hospitality industry during the COVID-19 crisis. *International Journal of Hospitality Management*, *92*, 102723. DOI: 10.1016/j.ijhm.2020.102723 PMID: 36919038

Business Today. (2021a). Businesses bear the brunt in COVID fight. Business Today. Retrieved December 13, 2021, from https://www.businesstoday.in/latest/corporate/story/businesses-bear-the-brunt-in-covid-fight-296444-2021-05-19

Business Today. (2021b). COVID-19 second wave: FPIs net sellers of Indian stocks worth Rs 6,452 crore in May so far. Business Today. https://www.businesstoday.in/markets/stocks/story/covid-19-second-wave-fpis-net-sellers-of-indian-stocks-worth-rs-6452-cr-in-may-so-far-296216-2021-05-16

Cavana, R., Delahaye, B., & Sekeran, U. (2001). Applied Business research: Qualitative and Quantitative Methods. John Wiley & Sons. https://eprints.qut.edu.au/10523/

. Chowdhury, Md. T., Sarkar, A., Paul, S. K., & Moktadir, Md. A. (2020). A case study on strategies to deal with the impacts of COVID-19 pandemic in the food and beverage industry. Operations Management Research, 1–13. DOI: 10.1007/s12063-020-00166-9

Cvetković, V. M., Nikolić, N., Radovanović Nenadić, U., Öcal, A., & Noji, K., E., & Zečević, M. (. (2020). Preparedness and Preventive Behaviors for a Pandemic Disaster Caused by COVID-19 in Serbia. *International Journal of Environmental Research and Public Health*, *17*(11), 4124. DOI: 10.3390/ijerph17114124 PMID: 32527056

Donthu, N., & Gustafsson, A. (2020). Effects of COVID-19 on business and research. *Journal of Business Research*, *117*, 284–289. DOI: 10.1016/j.jbusres.2020.06.008 PMID: 32536736

Eklund, M. A. (2021). The COVID-19 lessons learned for business and governance. *SN Business & Economics*, *1*(1), 25. DOI: 10.1007/s43546-020-00029-2 PMID: 34778816

Fairlie, R. (2020, October). (2020). The impact of COVID-19 on small business owners: Evidence from the first 3 months after widespread social-distancing restrictions. *Journal of Economics & Management Strategy*, *29*(4), 727–740. Advance online publication. DOI: 10.1111/jems.12400 PMID: 32904856

Field, A. (2005). *Discovering Statistics Using SPSS* (2nd ed.). Sage.

GEM. (2020). Entrepreneurship in India. GEM Global Entrepreneurship Monitor. https://www.gemconsortium.org/economy-profiles/india

India Today. (2021). A tragedy of errors: 10 reasons behind India's catastrophic Covid crisis—Cover Story News—Issue Date: May 17, 2021. https://www.indiatoday.in/magazine/cover-story/story/20210517-a-tragedy-of-errors-10-reasons-behind-india-s-catastrophic-covid-crisis-1799925-2021-05-07

Jalali, S. F., Ghassemzadeh, M., Mouodi, S., Javanian, M., Akbari Kani, M., Ghadimi, R., & Bijani, A. (2020). Epidemiologic comparison of the first and second waves of coronavirus disease in Babol, North of Iran. *Caspian Journal of Internal Medicine, 11*(Suppl 1), 544–550. DOI: 10.22088/cjim.11.0.544 PMID: 33425273

Janiak, A., Machado, C., & Turén, J. (2021). Covid-19 contagion, economic activity and business reopening protocols. *Journal of Economic Behavior & Organization, 182*, 264–284. DOI: 10.1016/j.jebo.2020.12.016 PMID: 33390632

Jones, L., Palumbo, D., & Brown, D. (2020). Coronavirus: A visual guide to the economic impact - BBC News. Diplomatic Academy. Retrieved December 13, 2021, from https://www.unic.ac.cy/da/2020/05/08/coronavirus-a-visual-guide-to-the-economic-impact-bbc-news/

. Kuckertz, A., & Brändle, L. (2021). Creative reconstruction: A structured literature review of the early empirical research on the COVID-19 crisis and entrepreneurship. Management Review Quarterly. 1–27. DOI: 10.1007/s11301-021-00221-0

Liao, C. H., Tsou, C. W., & Huang, M. F. (2007). Factors influencing the usage of 3G mobile services in Taiwan. *Online Information Review, 31*(6), 759–774. DOI: 10.1108/14684520710841757

Luisetto, M., & Latiyshev, O. Y. (2020). (PDF) Covid-19 Pandemic and the Management Strategy for Business and Economy #BUSINESS #ECONOMY #FINANCIAL #CRISIS #PANDEMIA #COVID-19 #CORONAVIRUS, #PANDEMIA#LOCKDOWN #FEAR #EMOTIONAL APPROACH,#LIBRARY,#SCHOOL,#LONDON,#NEW YORK,#PARIS,#ROME,#G OGLE. ResearchGate. https://www.researchgate.net/publication/341642216_Covid_-19_Pandemic_and_the_Management_Strategy_for_Business_and_Economy_BUSINESS_ECONOMY_FINANCIAL_CRISIS_PANDEMIA_COVID-19_CORONAVIRUS_PANDEMIALOCKDOWN_FEAR_EMOTIONAL_APPROACHLIBRARYSCHOOLLONDONNEW_Y

Mahmood, S., Hasan, K., Colder Carras, M., & Labrique, A. (2020). Global Preparedness Against COVID-19: We Must Leverage the Power of Digital Health. *JMIR Public Health and Surveillance, 6*(2), e18980. DOI: 10.2196/18980 PMID: 32297868

McKinsey, C. (2020). COVID Promo—Video Desktop. https://ceros.mckinsey.com/coronavirus-promo-video-desktop

Miocevic, D. (2021). Investigating strategic responses of SMEs during COVID-19 pandemic: A cognitive appraisal perspective. *BRQ Business Research Quarterly, 23409444211005780*. Advance online publication. DOI: 10.1177/23409444211005779

Nahata, P. (2021). India Foreign Trade Weakens in April Amid Second Wave of Covid-19. Retrieved December 13, 2021, from https://www.bloombergquint.com/business/indias-foreign-trade-weakens-in-april-amid-second-wave-of-covid-19

NASSCOM. (2021, May 25). CEO Pulse Review: Impact of COVID-19 2nd Wave on India Tech Sector. NASSCOM. https://nasscom.in/knowledge-center/publications/ceo-pulse-review-impact-covid-19-2nd-wave-india-tech-sector

Paliwal, A. (2021). Explained: Why Indian businesses are worried about 2nd Covid wave. India Today. https://www.indiatoday.in/business/story/explained-how-2nd-covid-wave-has-impacted-indian-businesses-1809014-2021-05-31

Paul, S. K., & Chowdhury, P. (2020). Strategies for Managing the Impacts of Disruptions During COVID-19: An Example of Toilet Paper. *Global Journal of Flexible Systems Managment*, *21*(3), 283–293. DOI: 10.1007/s40171-020-00248-4 PMID: 38624462

. Pinzaru, F., Zbuchea, A., & Anghel, L. (2020, October 16). The Impact of The COVID-19 Pandemic on Business. A Preliminary Overview. ResearchGate.

Prasad, S., Su, H.-C., Altay, N., & Tata, J. (2015). Building disaster-resilient micro enterprises in the developing world. *Disasters*, *39*(3), 447–466. DOI: 10.1111/disa.12117 PMID: 25546436

Ranjan, R., Sharma, A., & Verma, M. (2021). Characterization of the Second Wave of COVID-19 in India. DOI: 10.1101/2021.04.17.21255665

Salamzadeh, A., & Dana, L. P. (2021). The coronavirus (COVID-19) pandemic: Challenges among Iranian startups. *Journal of Small Business and Entrepreneurship*, *33*(5), 489–512. DOI: 10.1080/08276331.2020.1821158

Shafi, M., Liu, J., & Ren, W. (2020). Impact of COVID-19 pandemic on micro, small, and medium-sized Enterprises operating in Pakistan. *Research in Globalization*, *2*, 100018. DOI: 10.1016/j.resglo.2020.100018

Shirey, M. R. (2021). Leadership Lessons From COVID-19 and the Path Forward. *Journal for Healthcare Quality*, *43*(1), 1–2. DOI: 10.1097/JHQ.0000000000000290 PMID: 33394837

Tabachnick, B. G., & Fidell, L. S. (2013). *Using multivariate statistics* (6th ed.). Pearson.

Timmis, K., & Brüssow, H. (2020, June). (2020). The COVID-19 pandemic: Some lessons learned about crisis preparedness and management, and the need for international benchmarking to reduce deficits. *Environmental Microbiology*, *22*(6), 1986–1996. Advance online publication. DOI: 10.1111/1462-2920.15029 PMID: 32319151

UNCTAD. (2020). Impact of the coronavirus outbreak on global FDI | UNCTAD. Retrieved December 13, 2021, from https://unctad.org/press-material/impact-coronavirus-outbreak-global-fdi

Vijayraghavan, K., & Philip, L. (2021). Concerned about Covid wave, but better prepared than last year: Companies. The Economic Times. https://economictimes.indiatimes.com/news/company/corporate-trends/concerned-about-covid-wave-but-better-prepared-than-last-year-companies/articleshow/82138179.cms

WHO. (2020). https://www.who.int/docs/default-source/coronaviruse/getting-workplace-ready-for-covid-19.pdf

Wikipedia. (2021). COVID-19 vaccination in India. In Wikipedia. https://en.wikipedia.org/w/index.php?title=COVID-19_vaccination_in_India&oldid=1027104660

Wikipedia. (2021a). COVID-19 pandemic in India. In Wikipedia. https://en.wikipedia.org/w/index.php?title=COVID-19_pandemic_in_India&oldid=1027006555

World Population Review. (2020). India Population 2021 (Demographics, Maps, Graphs). https://worldpopulationreview.com/countries/india-population

Worldometer. (2021). India COVID: 28,809,339 Cases and 346,784 Deaths - Worldometer. https://www.worldometers.info/coronavirus/country/india/

Xu, S., & Li, Y. (2020). Beware of the second wave of COVID-19. *Lancet*, *395*(10233), 1321–1322. DOI: 10.1016/S0140-6736(20)30845-X PMID: 32277876

Chapter 9
Occupational Health and Safety in the Age of Industry 4.0:
Legal and Regulatory Responses to Digital Risk Environments

Rajiv Gurugopinath
http://orcid.org/0000-0003-3845-6778
Ramaiah College of Law, India

Ramya Krishnappa
http://orcid.org/0000-0001-8900-454X
Christ University, India

ABSTRACT

The rapid diffusion of artificial intelligence, automation, data analytics, and platform-based work has fundamentally transformed occupational risk profiles, challenging industrial-era frameworks of occupational health and safety (OHS). While traditional OHS law focused on physical and mechanical hazards, digitally mediated workplaces generate new forms of harm—psychosocial stress, cognitive overload, algorithmic bias, and privacy erosion—that remain inadequately regulated. The study highlights structural gaps, regulatory fragmentation, and emerging best practices. The chapter proposes a reform agenda emphasizing inclusive protection for gig and informal workers, algorithmic accountability, psychosocial risk recognition, and anticipatory regulation, positioning digital occupational safety as a cornerstone of equitable and sustainable future-of-work governance.

DOI: 10.4018/979-8-3373-6370-7.ch009

1. INTRODUCTION

The fourth industrial revolution, now alternatively called Industry 4.0, is the massive reorganization of production processes and workplaces based on the incorporation of cyber-physical systems, artificial intelligence (AI) and automation, the Internet of Things (IoT) and analytics that are based on data. These technologies have changed the nature, organizational form and even content of work itself in addition to industrial efficiency. Occupational health and safety (OHS) that has always been reserved to bodily threats like mechanical injury, exposure to chemicals and ergonomic dangers is challenges in this changing environment more than ever before (Obasi & Benson, 2025). The digitalized workplaces are further subjecting human Labour to work out of direct physical interactions to monitoring, inter, and decision-making in the framework of systems controlled by algorithms, thus allowing the definition of the occupational risk to extend beyond the physical harm to cognitive, psychological, and informational vulnerability. This change requires that OHS be reinterpreted as a holistic concept that can counteract the real and imaginary harms in work-related digital environments (Moraru, 2024) and argues that such risks require a jurisprudential and regulatory reimagining of OHS grounded in dignity, autonomy, and constitutional labour protections rather than narrow compliance-oriented safety norms.

The appearance of automation and AI-controlled working forces has added another set of so-called techno-psychosocial risks within the scope of which the current regulatory framework is poorly equipped. The algorithmic management systems have taken over the task allocation and work intensity, performance appraisals and payment in the sectors of manufacturing and logistics to platform-level and gig work. At the same time, wearable devices, biometric gadgets and productivity trackers allow tracking the actions, motions, and physiological parameters of employees at all times. Even though the existence of such systems is frequently explained by the need to enhance efficiency, precision, and predictive safety, these systems erase the borders between supervision and surveillance, diminish worker autonomy, and established power and accountability imbalance (Sebastian et al., 2025). These digital practices are an empirical study that is gradually adding stress, anxiety, burnout, and job insecurity to the list of consequences, especially those related to service-based and remote work-based economies, like the Indian IT-enabled sectors. The shift of the focus of the mechanical hazards to the cognitive and emotional strain posed by digital technology, therefore, is a paradigmatic one in the nature of the occupational hazard, where the concern with algorithmic strain, digital fatigue, and harm related to surveillance should be viewed as a legitimate one, in terms of OHS (Rahmi et al., 2025).

The digitalization of Indian industry happens in the context of a changing worldwide regulation of algorithmic governance and digital work. Though Indian reforms are marked by the willingness of modernizing occupational regulation, it is still prevalent as they are still quite informed by traditional paradigms of safety and inadequately responsive to AI controlled working reality (Sun & Chen, 2025). Conversely, European regulatory practices focus on order, human control, reasonableness, and responsibility in robotic systems, putting the worker-focused protective measures against the threats of robotic bias and unrestricted surveillance. The necessity to incorporate the psychosocial risk management and digital well-being into the national OHS strategies is also strengthened by international organizations. These are the comparative reflections on why adaptive legal pluralism and the blending of Labour law, data protection and ethical AI governance could respond to the challenge of multiple risks through Industry 4.0 workplaces (Obasi & Benson, 2025).

1.1 Theoretical Framework: Understanding Digital Occupational Risk

This chapter adopts an interdisciplinary socio-legal framework drawing from labour jurisprudence, surveillance studies, and political economy to interpret emerging digital occupational risks. Michel Foucault's theory of surveillance and disciplinary power explains how algorithmic monitoring and wearable technologies normalize continuous visibility and behavioural compliance in digital workplaces. Shoshana Zuboff's concept of surveillance capitalism situates employee data extraction within broader economic logics of value creation, revealing how workplace monitoring is not merely managerial but structurally embedded in platform-driven business models. Braverman's labour process theory further illuminates the phenomenon of digital Taylorism, where managerial control is intensified through algorithmic task allocation, performance scoring, and automated discipline, often resulting in deskilling and loss of worker autonomy.

From a legal standpoint, the traditional employer's duty of care—originating in common law and later codified in labour statutes—has historically focused on preventing physical injury. This chapter extends that doctrinal foundation to argue for a recalibrated duty encompassing psychosocial, cognitive, and informational harms arising from algorithmic governance. Together, these theoretical lenses enable a deeper understanding of why existing OHS frameworks fail to capture the lived realities of Industry 4.0 workplaces and why legal reform must address power, control, and dignity alongside safety.

1.2 Review of Literature

The contemporary literature on Industry 4.0 and occupational health reflects a fundamental tension between technological optimism and labour protection skepticism. One body of scholarship highlights the potential of AI, predictive analytics, and robotics to enhance workplace safety, reduce accidents, and optimize ergonomic interventions. These studies frame digitalization as a tool for proactive risk prevention and efficiency-driven safety management.

In contrast, a growing critical literature documents the psychosocial consequences of algorithmic management, including stress, burnout, loss of autonomy, and privacy erosion. Scholars adopting sociological and critical legal perspectives argue that digital monitoring systems intensify managerial control and shift occupational risk from acute physical injury to chronic cognitive and emotional harm. Methodologically, while quantitative studies focus on correlations between monitoring intensity and health outcomes, qualitative and socio-legal research emphasizes lived experiences of precarity, invisibility, and contestation in platform and remote work.

Despite this growing body of work, the literature remains fragmented. Most studies examine technology, labour law, or data protection in isolation, with limited engagement across doctrinal, ethical, and regulatory domains. Furthermore, Global South perspectives—particularly Indian legal analyses—remain underdeveloped, especially regarding the integration of constitutional labour rights with digital governance. This gap necessitates a holistic legal framework that reconceptualizes digital surveillance and algorithmic control as occupational hazards warranting preventive regulation.

1.3 Research Gap and Original Contribution

Existing scholarship inadequately addresses three critical gaps. First, it fails to integrate occupational safety law with data protection and AI governance, resulting in siloed regulatory analyses. Second, it marginalizes psychosocial and informational harms by treating them as secondary managerial concerns rather than core occupational risks. Third, there is limited doctrinal analysis contextualized within Indian labour law and constitutional jurisprudence.

This chapter contributes to the field by advancing an integrated digital OHS framework that reconceptualizes algorithmic surveillance, data extraction, and techno-stress as legally cognizable workplace hazards. Through a comparative analysis of Indian and European regulatory approaches, it proposes a rights-based, human-centered model of occupational safety grounded in dignity, autonomy, and proportionality. In doing so, the chapter bridges doctrinal labour law, data protection principles, and ethical AI governance, offering a coherent regulatory pathway for digitally transforming economies.

1.4 Research Methodology

Methodologically, the chapter employs doctrinal legal analysis combined with comparative regulatory study and socio-legal interpretation. This approach is essential to address Industry 4.0 risks that are diffuse, algorithmically mediated, and transnational in nature, and which cannot be adequately captured through traditional industrial safety paradigms alone.

1.5 Problem Statement: Regulatory Lapses to Digital Workplace Hazards

Even though the modern work environment is speedily being digitized, modern OHS and Labour law framework continue to be largely based in the industrial-era assumptions that focus on the physical safety, rather than the psychosocial and informational well-being of the work environment. For Example, in India there is an encouragement of a shift to consolidation and modernization, although existing frames persist with marginalizing mental health, data protection and accountability of algorithms in the regulatory frame of governance in the workplace through contemporary legislative changes. Current laws are too weak to share concerns related to automated decision-making, constant surveillance of the workers, and performance-based metrics that are current and use AI technology. This lag of regulation results in a structural imbalance between the technologically innovative workplaces and the legal protection that remains outdated, making the workers vulnerable to the unregulated threats of digital workplaces. The lack of clear norms in the area of surveillance at work, the openness of algorithms in managing employees, as well as the responsibility of employers in psychosocial damage, highlight the importance of the urgent regulatory rebalancing of Industry 4.0.

The same tendency is observed in many countries: digital workplace transformations have become quicker than the occupational health and labour legislations. Even in the United States, the Occupational Safety and Health Act continues to pay attention to the apparent risks, such as machinery and physical protection. The psychosocial problems that are reported more often include algorithmic work pressure, electronic monitoring stress, and mental burnout, yet the legal approach to them is exceptionally tender and dependent on the will of employers, rather than binding prescribes. Protecting data at workplaces is discontinuous whenever there are sector-specific laws, therefore, areas of algorithmic decision-making in hiring, productivity points, and terminations have minimal regulations.

In particular, worker dignity and data rights are more safeguarded by the laws of the European Union through the General Data Protection Regulation. However, the occupational safety in the Framework Directive is still based on the old industrialistic premises, where the notion of risk is considered exclusively as physical and ergonomic. Such work-related pychosocial risks as surveillance fatigue, algorithmic bias, and AI-induced work pressure are cited in the policy, but they are not implemented equally in Member States. The new Artificial Intelligence Act has a fresh chance at dealing with the issue of algorithmic accountability although it has not reached full adherence to occupational health norms yet.

The United Kingdom continues to prioritize the elimination of physical risks in Health and Safety at Work Act, although platform-mediated and digitally-monitored work is growing. Although the Health and Safety Executive has issued advice on work-related stress, these are not compulsory and proactive but rather advisory. Management systems are often dysfunctionally situated, and there are gaps in mental-well-being and algorithmic transparency of grey and gig jobs.

Occupational health and safety regulations in Canada are largely provincial and continue to deal with the traditional industrial hazards, even as small and remote work, online surveillance, and artificial intelligence-based performance management are currently ascending at a pace greater than before. Workplace safety is increasingly discussed in the context of mental health, but there is a lack of legal obligations regarding the subject and enforcement is fragile. The privacy legislation is not very effective against intrusive surveillance, and the relationships between data regulation, algorithmic choice, and the welfare of the workers are not highly developed.

The only commonality to all these countries is that labour and occupational safety law continues to be based on a physical risk mental-state; the psychosocial harm, information, and algorithmic control are addressed in a fragmented, somewhat irresponsible manner. Such mismatch demonstrates the necessity to make OHS and labour law fit in digital workplaces, where the danger is becoming increasingly data-based, surveillance-based, and, crucially, automated, than a visible industrial accident.

1.6 Objectives of the Chapter

The chapter will have four objectives that are interrelated:

1. To critically discuss how the emerging technologies that are related to Industry 4.0 are transforming the environment of occupational health and safety hazards.
2. To determine the suitability of prevailing OHS and Labour law measures to deal with technology-psychosocial and data-oriented occupational hazards.
3. To understand the crossroads between data rights and Labour rights by comparing them in the case of India and the European Union.

4. To present a normative and a regulatory framework that would be able to reconcile technological innovation with worker protection, dignity, and autonomy.

1.7 Value and Importance of the Chapter

Placing occupational health and safety in the wider contexts of digital transformation, this chapter adds to the under-investigated but progressively important field of legal and policy analysis. It makes a case of how OHS governance in the era of Industry 4.0 needs to go beyond compliance-focused regulation to prospective, human-focused legal development. Following the principle of alignment of the technological advancement with the constitutional and human rights laws of dignity, independence and egalitarianism in the workplace, the chapter places OHS as a significance plot in the equitable and sustainable industrial development. By so doing, it provides a conceptual and regulatory framework in the redesign of workplace safety in digital risk working conditions, especially in quickly digitizing economies of the Global South.

2. THE DIGITAL WORKPLACE: TRANSFORMATION AND EMERGING OHS CHALLENGES- KEY FEATURES OF INDUSTRY 4.0: IoT, AI, ROBOTICS, BIG DATA, AND PREDICTIVE ANALYTICS

Industry 4.0 is a system that unites digital and physical systems and is marked with the use of the Internet of Things (IoT), artificial intelligence (AI), robotics, big data analytics, and predictive modeling. All these technologies allow the creation of the so-called smart workplace, in which the sensors are interconnected, the autonomous machines operate, and the process of optimizing production and delivering services to users is performed continuously by the self-learning algorithms (Nesterak et al., 2024). The IoT infrastructures enable immediate transmission of data between machines and human operators, whereas the AI systems process those data streams to anticipate failures and determine the safety level of an object, as well as determine the best overall efficiency of the workflow.

Robotics and collaborative robots do more duties in hazardous or repetitive surroundings and big data analytics assists with evidence-based occupational health intervention. Predictive analytics specifically enables employers to foresee accidents, track fatigue and indicate signs of ergonomic strain ahead of any injury occurring (Saxena, 2024). On one hand, these innovations can provide more security and efficiency, but on the other hand, they introduce different types of vulnerabilities due to the option of their use to collect a large amount of data in real-time, profile behaviors and interconnect electronic and physical aspects, which is how the acceptance of the specified innovations is redefining occupational risks and not abolishing them.

2.1 Change in Occupational Risk Profiles: A Physical Injury to Digital and Cognitive Damage

Occupational risks have been transformed by the digitalization of work in nature, temporality, and visibility. The classical OHS systems were not suited to address acute and visible dangers like mechanical accidents or chemical exposure by using engineering measures and safety precautions (Doval, 2024). Conversely, the workplaces of Industry 4.0 cause a stream of relentless and progressive risks that are associated with cognitive overworking, algorithmic controls, and ongoing digital interaction. Employees are now being obliged to react to real-time performance indicators, algorithmic feedback, and work at an output up to the borders of work, rest and monitoring (Bowdler et al., 2025). This change is a shift of event based risk to an ambient risk, in which damage is acquired progressively in the process of mental fatigue, stress and loss of autonomy but never directly by physical harm. Based upon risk society theory, such development highlights the necessity of extensions of protection to OHS regimes that address bodily safety, as well as psychological and digital health, which continue to reside as fringe concerns in the majority of labour laws, including the one in India. The modern concept of safe work thus has to include human work intensity, reasonable work cognitive and sense of data-driven coercion protection (Pisarczyk, 2025).

2.2 Digital Work Control and Algorithms of Management

The feature of Industry 4.0 is that tasks will be performed during algorithmic management, where the automated system will conduct the distribution of tasks, evaluation of performance, scheduling, and discipline with the human factor being very minimal. Proprietary algorithms in digital platforms and massive companies are increasingly dependent on Labour process regulation, generating obscure kinds of managerial control commonly referred to as digital Taylorism (Huram, 2025). Although they are more efficient in their operations, such systems compromise transparency, due process, and worker agency because workers are seldom, if ever, asked about what criteria are used to make an algorithmic decision about them, whether in terms of pay, workload, or termination.

Digital surveillance also adds to eroding a boundary between proper supervision and intrusive surveillance into creating a sense of 24-hour visibility that undermines a atmosphere of trust and psychological safety (Munn, 2024). On the regulatory front, the concept of algorithmic management makes it difficult to enforce traditional OHS, which is based on apparent attribution of employer liability. In cases where injuries are caused by the functionality of automated scheduling, or erroneous data analytics, responsibility becomes watered down among employers, software developers, and the platforms or intermediaries, revealing vast gaps in the current Labour and safety jurisprudence (Новиков, 2025).

2.3 Instances of AI-Based Performance Monitoring in Warehouses, Gig Work Platforms, and Remote Work Settings

Digital risk environments have some implications that are manifested in various sectors. Artificial intelligence powered surveillance systems are used in warehousing and logistics to monitor worker flows, speed, and efficiency using wearables and computer vision, usually rewarding unsustainable and excessive performance rates leading to musculoskeletal trauma and stress-associated diseases (Nilsson et al., 2025). Gig workers in platform economy are regulated by incomprehensible algorithms which assign work, price and penalise through rating schemes, which generates an exaggerated stress, working long hours and no avenue to complain, thereby resulting in grievance being redressable. Their no pursuit of formal employment categories means that they lack protection of OHS. Remote and hybrid work environments bring with them further psychosocial hazards due to productivity-monitoring software, keypad nailing, and webcam tracking that push manager oversight into the domestic environment (Andrade et al., 2025). The practices enhance mental exhaustion, loss of privacy, and work life imbalance, and have disproportionate effects on women and caregivers. Altogether, the circumstances portray that digitalization not only influences the working process but also the power dynamics and medical consequences in the workplace (Gnugesser et al., 2025).

2.4 Documented Case Studies of Algorithmic Harm in Digital Workplaces

While the general patterns of digital risk are well documented, recent high-profile case studies illustrate how algorithmic management and performance-tracking technologies translate into tangible occupational health and safety harms. In large-scale warehousing operations, the deployment of wearable tracking devices and AI-driven productivity metrics has attracted sustained regulatory and judicial scrutiny. Investigative reports and litigation relating to productivity quotas in automated warehouses reveal that real-time monitoring systems incentivize unsafe work speeds, contributing to elevated rates of musculoskeletal injuries, chronic fatigue, and stress-related disorders. Workers have reported limited discretion to pause, adjust pace, or seek ergonomic relief without triggering automated penalties, effectively subordinating bodily well-being to algorithmic efficiency (Nilsson et al., 2025).

Similarly, platform-based ride-hailing and delivery services offer documented examples of algorithmic management producing psychosocial harm. Protests and legal challenges by drivers and couriers across multiple jurisdictions highlight opaque rating systems, automated deactivations, and income volatility generated by proprietary algorithms. These systems regulate access to work, remuneration, and continued employment without meaningful transparency or procedural safeguards. The resulting uncertainty, constant performance anxiety, and lack of human oversight have been associated with heightened stress, sleep disorders, and economic precarity, raising serious concerns regarding the adequacy of existing occupational safety protections for platform workers (Sander et al., 2025).

Remote work environments present a third documented site of digital harm. The increased use of keystroke logging, webcam surveillance, and productivity software has blurred the boundary between professional supervision and private life. Empirical studies and labour complaints demonstrate that such practices intensify work-life imbalance, contribute to burnout, and disproportionately affect women and caregivers. These cases collectively illustrate that digital technologies do not merely reorganize work processes but actively reshape risk distribution, power relations, and health outcomes in contemporary workplaces (Daneshvar & Otterbach, 2025).

2.5 Digital Occupational Hazards: Towards an Idea Framework

The accumulating impact of digitalization requires identification of the so-called digital occupational hazards as a new type of occupational risk. Such risks are a loss of privacy by over data gathering, mental exhaustion due to the continuous use of the brain, and techno-stress related to learning to use complex and intrusive digital technologies. Digital risks, unlike traditional ones, tend to be invisible, personal and temporally sanguine and transpire across their exposure duration as opposed to acute events (Frielingsdorf et al., 2025). Privacy loss compromises informational autonomy and dignity and techno-stress leads to anxiety, burnout and lack of job satisfaction especially in employees that have little control over technological systems. Mental fatigue is widespread but not highly considered in the statutory OHS structures. The idea of these harms as the occupational risk makes it necessary to replace the reactive approach to regulation with the anticipatory one, ensuring the principles of privacy-by-design, human control, and ethical responsibility are directly adopted as aspects of digital work architecture. With Industry 4.0, occupational safety thus needs to be conceptualized to ensure that it encompasses beyond the physical protection, cognitive wellness, emotional stability, and informational integrity (Obasi & Benson, 2025).

One of the defining strengths of contemporary occupational health and safety (OHS) analysis lies in its interdisciplinary reach. Digital workplace risks cannot be adequately understood through technological or legal analysis alone; rather, they emerge at the intersection of technological design, legal governance, and psychological experience. For instance, the deployment of computer vision systems in warehouses and offices illustrates this convergence clearly. From a technological perspective, computer vision enables real-time productivity tracking and safety monitoring. Legally, however, such systems raise serious concerns regarding informational privacy, consent, and proportionality under data protection regimes. Psychologically, continuous visual monitoring contributes to surveillance anxiety, heightened stress responses, and behavioural self-censorship among workers (Sum et al., 2025).

By linking these dimensions, the chapter demonstrates that occupational risk in Industry 4.0 is not a linear cause–effect phenomenon but a layered interaction of systems. Surveillance technologies may be legally permissible yet psychologically harmful; efficiency-enhancing tools may comply with safety standards while undermining autonomy and dignity. Recognizing this convergence strengthens analytical depth by revealing how regulatory blind spots emerge when disciplinary silos persist. An interdisciplinary approach thus becomes not merely descriptive but diagnostic, allowing OHS governance to respond to the full spectrum of harm experienced in digitally mediated work (Obasi & Benson, 2025). This integrated perspective is crucial for developing holistic mitigation strategies that address both the manifest and latent consequences of digitalization on worker well-being (Singh, 2023).

3. LEGAL DIMENSIONS OF OCCUPATIONAL HEALTH IN THE DIGITAL ERA: EXPANDING THE LEGAL DEFINITION OF "WORKPLACE HAZARDS" IN THE CONTEXT OF AI AND DIGITAL SYSTEMS

The conceptual bases of occupational health law have superseded by the fast dispersion of artificial intelligence (AI), data analytics, and automated management systems. Historically, the concept of work place hazards had a very simplistic meaning of physical, chemical, biological, and ergonomic hazards as a result of industrial activity. However in digitally mediated workplaces harm is produced in more and more aspects in the form of algorithmic decision-making, ubiquitous surveillance and mental overload. Employees no longer have to deal with obvious hazards like heavy machinery or poisonous exposure but are exposed to unremitting control, information-yellowed assessment, and psychosocial pressure factors eroding personal autonomy, respect, and psychological wellbeing. This change requires a jurisprudential broadening of the concept of hazard, a materialist conception of the physical risk to a socio-technical model, which includes informational, psychological, and algorithmic harms (Chapman et al., 2025).

Online occupational risks are both physical (screen-induced warning signs and musculoskeletal injuries) and psychological (invasion of privacy, algorithmic discrimination, and algorithmic stress). Although the international bodies are recognizing the emergent digital risks of information overload, and cognitive strain, the Indian occupational safety law is still pegged on the sense of hazard based on the industrial era. Lack of statutory legitimacy of psychosocial and data-intensive risks restrains regulatory ability to subjugate precautionary duties within the digitally intensive workplaces. It is hence imperative that the scope of the law is extended to incorporate algorithmic and informational harms taking into account the need to make occupational health governance meet the modernities of work (Mbare et al., 2024).

3.1 Digital Right to Health, Dignity, and Safe Working Conditions

The entitlement to healthy work conditions and health is solidly rooted in the constitutional and international Labour jurisprudence. The interpretation of right to life in India by the judiciary has incorporated health, dignity and humane working conditions in Article 21 accompanied by the Directive Principles, which is focused on the welfare of workers, and social justice. On the international leg, legislation identifies safe and healthy work as a basic human right (key international covenants on Economic, Social and Cultural rights and core Labour conventions), and European regulations specifically define dignity, privacy and safety as elements of a Labour protection standard (Beridze, 2024).

In order to transfer these guarantees into the digital workplace, it is important to realize that informational and psychological damages are harmful just as the physical ones. Unremitting surveillance, the transparent algorithmic assessment, and data mining harm individual autonomy and dignity in the manner that explicitly violates the constitutional right to health. As a result, occupational health jurisprudence should be changed in such a way that the algorithmic bias, techno-stress, and privacy intrusion are not discussed as side-by-side managerial concerns anymore. The identification of digital health as part of occupational health enhances the normative base of work involving the regulation of AI-mediated workplaces (Akki & Ouddasser, 2024).

3.2 The Duty of Care of Employer in Algorithms Mediated Workplaces

The occupational health and safety law places emphasis on the duty of care undertaken by the employer and it should undergo a re-calibration to fit the workplaces which are under algorithmic control. Employers have assumed a new role as physical space custodians, but also as data lords and system deployers that influence a workplace and employment outcome. This increased responsibility includes technical accountability to build algorithmic accuracy and security, procedural fairness in the form of transparency and employee notification, and psychosocial accountability to avoid over-monitoring and unrealistic performance measures (Milanez et al., 2025).

Data-protection regimes are putting increasing obligations like these on them, by mandating transparency, a reduction of data, and protection against purely automated decision-making with a substantial employment impact. In comparison, the Indian Labour law does not explicitly discuss the algorithmic governance, and the legal provisions in data-protecting do not have standards in the workplace. Acknowledging a digital duty of care would permit Labour law to combat injuries due to flaws, prejudiced, or obscure algorithms to put the concept of employer liability into the context of technologically mediated danger (Abraha, 2025).

3.3 Automated Decision-Making, Consent, and Surveillance at the Workplace

Recruitment, performance assessment and workforce management automation present complicated legal issues that allude to consent, proportionality and accountability. Even in hierarchical workplaces, the agreement to the processing of data in employment relationships is not consistently voluntary or even at times, nominal and the workers may agree to be monitored obtrusively because of economic coercion. Legal systems should thus no longer stick at the aspect of formal consent but necessitate the use of necessity based reasons in the collection of data at work places (Fairfield & Reilly, 2025).

On the same note, the practice of surveillance should be proportional by ensuring that the amount of surveillance should be restricted to that which is significantly important to achieve the right goals. Such over tracking creates surveillance anxiety and distrust, especially when there is no transparency of the algorithmic system in operation. The perplexing nature of machine-learning models also complicates due process because employees can not comprehend or challenge the decisions made about their remuneration, labor, or dismissal. To be certain that automation does not overtake human rights and legal control, it is necessary to integrate the issue of transparency, the desire of explainability, and the mechanisms of accountability into the Labour regulation (Doellgast et al., 2025).

3.4 Predictive Risk Systems: Due Process and Algorithmic Opacity

What comes with the application of predictive analytics in the area of safety management, discipline, and workforce optimization is the issue of due-process. The operation of algorithmic systems is usually a black box operation that is hard to audit or challenge and generates results. This obfuscation is a betrayal of the primary provisions of natural justice, such as the right to hear and the need to make a rational decision-making. The judicial review in government workforce might involve exposing algorithmic reasoning in order to meet the fairness in its procedures whereas in the case of privative employment, the regulatory oversight should be employed to close gaps in accountability (Mariani & Lozada, 2023).

Attempts to regulate AI systems in relation to the employment are undergoing comparative regulatory practices that employ high-risk categories to embrace risk assessment, documenting tasks, and enforcing human intervention. The process of adapting comparable standards would would enhance procedural justice in the digitally mediated workplaces by making sure that automated tools complement and not supplant human judgment. Placing these precautions in to occupational and Labour law would help to alleviate the due-process shortfall caused by predictive technologies (Adams-Prassl et al., 2024).

The legal framework of occupational health should move towards a more reactive and compliance-focused approach to a more anticipatory one that is able to concern itself with algorithmic and digital environments of risk. The broadening of the concept of workplace hazard, restating the right of digital comfort and re-exploring the responsibility of care presented by the employer are critical milestones on the way of ensuring the serviceman dignity in the time of automated management. Facilitating the presence of transparency, proportionality, and due process in algorithmic decision-making is not only an issue in data-governance but a principle in the modern Labour law (Aloisi et al., 2025).

The high rate of digitalization in India creates a chance of creating a rights-based approach to the digital occupational health jurisprudence that balances innovation with the constitutional values. The future workplace can be technologically sophisticated, yet still retaining its human nature by introducing the principles of accountability, explainability and the ethical supervision in the Labour and safety regulation (Oyekunle et al., 2024).

3.5 Cybersecurity Failures as Occupational Health and Safety Risks

The occupational health implications of cybersecurity failures remain an underexplored dimension of digital work environments. Cyberattacks, data breaches, and system failures pose both direct and indirect threats to worker safety, psychological well-being, and professional responsibility. In healthcare settings, ransomware attacks that encrypt patient records have forced healthcare professionals to operate without access to critical clinical information, increasing cognitive load, moral distress, and the risk of medical error. The psychological burden on healthcare workers during such incidents—characterized by anxiety, fear of liability, and ethical conflict—constitutes a significant occupational hazard (Aloisi et al., 2025).

In industrial settings, cyberattacks targeting supervisory control and data acquisition (SCADA) systems can disrupt safety-critical operations in factories, power plants, and chemical facilities. Loss of system integrity may result in machinery malfunction, emergency shutdowns, or hazardous material exposure, placing workers at immediate physical risk. Even in the absence of physical injury, the uncertainty and loss of control generated by cyber incidents contribute to heightened stress and long-term psychosocial harm (Oyekunle et al., 2024).

These scenarios demonstrate that cybersecurity is not merely an information-technology concern but a core component of occupational safety governance. Integrating cybersecurity resilience, system redundancy, and incident response protocols into OHS frameworks is essential for addressing the evolving risk landscape of digitally mediated workplaces.

4. INDIA'S REGULATORY LANDSCAPE: FRAGMENTED BUT EVOLVING- THE CODE ON OCCUPATIONAL SAFETY, HEALTH AND WORKING CONDITIONS, 2020: STRENGTHS AND SILENCES

The Code on Occupational Safety, Health and Working Conditions, 2020 is a major advancement in consolidating Labour in India, which involves consolidating several industry-specific statutes into one common law. Its goals are to streamline compliance, harmonize safety standards and make workplace protections available outside of the factory environment. The duties of the employers, inspection, and safety committees are progressive provisions that establish an institutional framework that can tackle the occupational risks in a larger proportion. On procedural level, the Code sets the direction of modernization and administrative consistency of occupational safety governance (Beridze, 2024).

Substantively, though, the Code is still pegged on an industrial age way of understanding workplace risk. Its regulative emphasis remains to be placed on physical, chemical, and ergonomic risks where the dominance of digitally mediated work settings associated with psychosocial, cognitive, and informational risks are not given much attention. The key terminologies like its meaning of hazard, establishment, and worker fail to reflect algorithmic management, data-based monitoring, and unconventional types of work like gig work and remote work (Bråten et al., 2024). The dependence of the Code on a delegated rule-making process also poses a threat of fragmentation between the regulatory systems of the states even in cases where digital work is transboundary. As a result, although the OSHWC Code offers a general outline, it cannot be considered substantially adapted to the new Industry 4.0 realities (Obasi & Benson, 2025).

4.1 The Marginalization of Platform Work and Mental Health and the Labour Codes

The larger Labour code structure in India, including wages, industrial relations, social security and occupational safety, is seen to be an attempt of rationalizing, rather than reimagining. Despite the fact that platform workers and gig workers enjoy limited status within the benefits of the social security, they do not enjoy the coverage of occupational safety and health benefits (Au-Yeung et al., 2024). The resulting regulatory paradox, which is brought about by this selective inclusion, is that digital workers are recognized as actors in the economy, but they are not granted access to safe and dignified working conditions (Abraha, 2025).

It is also important that mental health and psychosocial risk recognition has not been addressed in Labour statutes. Even with increasing literatures on work-related stress, burnout and cognitive fatigue among digitally intensive industries, the current occupational health in Indian Labour law still assumes that it is mainly physical. Such gap is compounded due to the absence of integration between Labour law and digital governance (Rahmi et al., 2025). Data protection laws and information technology exist beyond occupational safety standards, creating an effect of siloed regulation that cannot combat risks compounded in relation to algorithmic oversight, data exposure and techno-stress. Such disintegration negatively affects the holism of protecting workers in the digital economy.

4.2 Digital Personal Data Protection Act, 2023: Privacy Rights in the Absence of Occupational Assurances

The Digital Personal Data Protection Act, 2023 sets a broad roadmap on the lawful process of data processing according to permission, restriction of purpose, and minimization of data. It is not Labour specific but carries serious consequences on the surveillance of the work place and human resource management by artificial intelligence. Data fiduciaries (employers) must use employee data to carry out lawful activities, but in the context of employment relations, there is no significant consideration of the structural power disparity(Anjumanwarshaik et al., 2025).

Far-reaching exceptions where data processing may be done with reasonable purposes in employment give employers a lot of discretion, which may justifiably be use of intrusion monitoring. There is no clear regulatory protection against over-surveillance, algorithmic profiling, or inferred data analytics which leads to regulation vacuum. Furthermore, the Act fails to impose algorithmic clarity and workforce effect characteristics, which exposes the employees to unreasonable and unexplainable decision-making (Abraha, 2025). Owing to this, though the DPDPA is a positive step towards enhancing the digital privacy of the general, there is a risk of pacifying the intrusive data gathering measures regarding employment affairs unless the act is complemented with labor-specific regulations.

4.3 Sectoral OHS Frameworks and the Digital Regulation Gap

The sector occupational safety systems in India are either unequal or outdated considering the digital transformation. Regulations concerning information technology, e-business, logistics and home-based services still focus on physical infrastructure and ergonomics with very little consideration to psychosocial health, algorithmic pacing, or stresses due to surveillance (Dollard & Potter, 2025). Platform-mediated industries and knowledge intensive, even though large in scale and economical, lie outside a formal OHS control.

There are no common digital OHS standards that span across the sectors to protect workers who still face uncontrolled risks associated with AI-powered data monitoring and performance metrics, or cyber-physical systems (Bråten et al., 2024). This regulatory gap is in opposition to new global policies including digital fatigue, data ethics, and algorithmic fairness as the primary occupational health and safety issues. Having industry-specific digital OHS guidelines would enable India to deal with these gaps and still have the flexibility to do so within context.

4.4 Developments and Policy Signals in Judicial Developments

Court decisions have established constitutional grounds to solve the digital occupational hazards, especially by developing privacy and informational self-determination as inherent to the individual freedom. Necessity, proportionality, and legitimate purpose constitute the principles that are conceptualized by constitutional courts and can be applied to workplace surveillance and algorithmic governance. Yet, these principles have not been operationalized or applied to the disputes of automated employment decisions directly in Labour law.

The policy initiatives indicate a growing realization of worker-focused AI regulation, albeit mostly on a merely advisory basis. Without constitutional codification, the constitutional guarantees will only provide a partial reprieve to employees who are the victims of online surveillance and algorithmic oppression. What emerges thus is more regulatory ecosystem, where normative aspiration is strong, but decisional weakness prevails.The regulatory reaction in India towards the digitalized operations of work place is still diverse in the areas of Labour law, data protection and technology policy.

Although new reforms point to the transition to modernization, they are insufficient to decrease the synergistic threat of algorithmic management, monitoring in the workplace, and techno-psychosocial injury. To close these loopholes, a specific legal framework that would focus on addressing digital exposure, data management, and mental health as essential aspects of occupational safety is needed (Saxena, 2024). This framework must account for the dynamic emergence of OSH risks in algorithmic management, moving beyond a sole focus on "safety by design" to encompass both ex ante and ex post assessments (Cefaliello et al., 2023).

In fact, an integrated framework of occupational health law in conjunction with data protection and ethical AI management would enable India to correspond the technological innovation up to the constitutional principles of dignity, fairness, and proportionality. By drawing such evolution, the regulatory tools would not just be refreshed but occupational safety, as a whole, would be re-contextualized to fit the requirements of Industry 4.0. Such a comprehensive approach would involve not only robust legislative measures but also the establishment of clear enforcement mechanisms and continuous adaptation to emerging technologies and their implications for worker well-being (Cefaliello et al., 2023).

5. EU AND GLOBAL PERSPECTIVES: BENCHMARKING DIGITAL OHS REGULATION- OVERVIEW OF THE EU'S AI ACT AND GENERAL DATA PROTECTION REGULATION (GDPR)

European Union has become a global leader in the way digital technologies impact Labour in employment with the right and human-based approach to their regulation. EU digital occupational health and safety (OHS) is shaped by two primary policies, namely the General Data Protection Regulation (GDPR) and an upcoming Artificial Intelligence (AI) Act (Adikwu et al., 2024). The GDPR is based on an extensive data protection framework based on human dignity, transparency, the limitation of purpose, and minimization of data. These principles directly limit the way employers gather, work with and use employee data, especially in a setting where there is widespread surveillance, as well as entrust as algorithms are involved (Gaudio, 2024).

A major protection is the fact that no decision can be made on automated basis that impacts a lot on the individuals unless given the right protective means and legal mandate. The protection would especially apply to employment matters where the automated systems affect the hiring, performance appraisal, or termination. In addition to the GDPR, there is an AI Act in place, which adds a horizontal regulatory framework based on risk to the classification of AI systems (Lin, 2024). The AI applications applied in workforce management and employment are categorized as high-risk which results in the necessity to conduct mandatory risk assessments, transparency documentation, and human-in-the-loop controls. Raising the notion of occupational safety beyond the physical injury to include informational safety, mental freedom, and against robotic injustice, these tools make the idea bigger (Callari et al., 2024).

5.1 Worker-Centric Guidelines: Openness, Human Supervision, and Algorithms Fairness

The Labour legislative framework of the EU enhances the digital safety with a constellation of new directives incorporating the occupational health, data governance, and procedural fairness. The Framework Directive on Safety and Health at Work places an obligation on the employer that is not delegable to determine the risk posed by new technologies, which is increasingly understood to encompass psychosocial as well as digital risk. This prevention method presupposes the employers to foresee the harm during design and implementation levels of the technological systems. Moreover, the GDPR further mandates principles like lawfulness, fairness, and transparency for processing personal data, especially when AI algorithms influence HR decisions, requiring employers to inform employees about data usage (Hoxhaj et al., 2023).

The Transparent and Predictable Working Conditions Directive also enhances the protection of the workers by providing it to disclose the critical terms of working, such as automated decision making and algorithmic management tools. Above all, the Platform Work Directive is a direct appeal to the concept of algorithmic governance, which demands that automated decision parameters be disclosed, that access to human review should be possible, and that no discrimination possibilities exist.

All these measures make transparency, accountability and participation of workers institutionalized organizational part of digital OHS governance, and reflect a vision of a model of the digital Labour constitutionalism that incorporates AI ethics into employment policy. Furthermore, the European regulatory landscape, particularly with the AI Act, mandates stringent risk assessments for "high-risk" AI systems used in employment, such as those for recruitment or performance evaluation, necessitating conformity assessments and adherence to guidelines promoting transparency, robustness, accuracy, and traceability (Moore, 2023).

5.2 WHO and ILO Normative Guidance on Digital Fatigue and Psychosocial Risk

On the international level, the digitalization has become more and more a determinant of occupational health identified by international organizations. According to joint WHO-ILO research findings, longer working hours, digital exhaustion, and psychosocial stress are among the key factors that cause work illness. The newest policy guidance specifically recognizes technostress, work from home stress, and continuous connectedness as two new occupational hazards and suggests regulatory actions to be implemented to address them, which include rest policies, restrictions on after hours online communication, as well as institutional acceptance of mental health as an issue in the workplace.

These recommendations underscore the necessity for employers to proactively manage the psychosocial impacts of digital transformation, moving beyond traditional physical safety paradigms to encompass mental well-being in digital work environments (Oyekunle et al., 2024).

The ILO guide to digitalization and occupational safety is about the dual character of technology: AI and data analytics can help to improve predictive safety and compliance surveillance, but at the same time, over-surveillance and automation may harm psychological and job satisfaction. To this end, the world standards are starting to embrace the tendency of combining digital risk control with psychosocial health interventions, introducing human control in the automated systems, and providing employees with participatory roles in the implementation of technology.

These norms consolidate the discourse that digital well-being is a key part of the modern OHS policy. The emphasis on human-centered economics further highlights the need for policies that protect workers from the potential negative impacts of AI tools and digital work on their well-being, including issues related to data privacy and psychological strain (Lindholm et al., 2020; Samans, 2023).

5.3 European Union National Proceeds: Germany, France and Spain

The Member States of the EU have operationalized these principles in different but convergent national ways. Germany puts the democratic control in the focus of codetermination, where the psychological risk assessment needs to take into consideration the digital stressors as well as providing the worker councils with the power to regulate the implementation of AI systems. France has made the right to disconnect institutional, committing employers to establish parameters on digital communication and the affirmation of proportionality in monitoring of the workplace. Spain has implemented a clear digital rights regime that ensures algorithmic decision-making is transparent and platform workers are also protected, which requires sharing of principles defining how work is allocated and remunerated.

These and other national experiences are examples of a triangle of accountability to the European people, founded on transparency, worker participation and ethical accountability. They show that successfully governing OHS digitally is a blend of legal obligations and social dialogue, as well as institutional review, which provides flexible approaches to jurisdictions in the process of rapid digitalization. This diverse approach across member states highlights the importance of tailoring digital OHS strategies to specific national contexts while adhering to overarching European principles (Fabrellas, 2022; Malenfer et al., 2023).

5.4 International Soft Law/ Ethical AI: Indian Relevance

In addition to controlling authority, the international soft law has been instrumental in the development of ethical AI regulation possibly leading to Labour protection. International standards of AI focus on human values, transparency, responsibility, and appreciation of human rights of the algorithmic systems. Occupationally, these schemes embrace ethical risk assessment which factors in the psychological, social and dignity-based consequences of the workers. These frameworks emphasize the imperative for AI systems to be designed and deployed in a manner that prioritizes worker well-being, acknowledging potential negative impacts such as informal overtime and psychosocial risks (Avogaro, 2022).

In the case of India, playing with such norms offers an enabler through which the domestic Labour law can be harmonized with the new international standards. Even without formal obligatory rules, soft law instruments may provide normative advice that may guide the forms of law reform, judicial construction, and policy formulation. The application of ethical AI principles to the national occupational safety policies would aid gaps in bridging the receurring differences in states conduct between Labour laws and data management, creating uniformity in constitutional promises to dignity and equity in the virtual work-place. Such an integration would require a comprehensive reassessment of existing legal frameworks to explicitly incorporate provisions for ethical AI deployment and data protection in occupational health and safety (Saxena, 2024).

It is proved by the practice of European and other Western nations that occupational safety, confidentiality of personal data, and responsibility of AI providers cannot be completely secured without an elaborate regulatory framework that unites these three key elements in a unified manner. Transparency, human involvement and participatory governance are integrated into the legal framework of the EU algorithmic systems, whereas the guidelines of WHO and ILO emphasize the role of psychosocial health as OHS component.

In the case of India, the models can give it a roadmap through which technological modernization and the respect of workers will integrate. By incorporating algorithmic fairness, moral risk evaluation, and institutional responsibility in national OSHA policies, one would guarantee that the digital revolution contributes to not only the efficiency but also equity and human happiness. This includes a focus on predictive analytics as a cornerstone of OHS strategy, utilizing artificial intelligence and machine learning to anticipate and mitigate workplace hazards proactively (Saxena, 2024).

The comparative analysis between India and the European Union underscores that regulatory effectiveness is deeply contingent on institutional, economic, and cultural contexts. While the EU's regulatory model emphasizes precaution, worker participation, and rights-based oversight, India's labour ecosystem is shaped by informality, platform-based employment, and resource constraints. A direct transplantation of European regulatory mechanisms into the Indian context risks formal compliance without substantive protection.

This comparison highlights the necessity of contextual adaptation rather than regulatory imitation. India's approach to digital OHS must account for informal labour, fragmented enforcement capacity, and developmental priorities, while still aligning with global norms of dignity and fairness. By foregrounding context-sensitive regulatory pluralism, the chapter strengthens the credibility of its policy recommendations and avoids the pitfalls of "copy-paste" governance in complex socio-economic environments.

6. TOWARDS AN INCLUSIVE AND EQUITABLE DIGITAL OHS FRAMEWORK IN INDIA: BRIDGING THE GAP BETWEEN FORMAL AND INFORMAL DIGITAL LABOUR

India has the potential to develop safer and more resilient work environments by implementing innovative strategies that integrate cutting-edge technology with traditional safety protocols, offering a nuanced approach to occupational health and safety (Saxena, 2024). The industrial and digital economies have taken a middle ground in the workforce of India. Although some parts of the formal sector are quickly transforming to be automated, artificial intelligence (AI), and remote working patterns, almost ninety percent of the Indian workers are informal and can hardly enjoy technological coverage and Labour rights. The emergence of digital informal labor, workers who work in mobile applications and online platforms as well as crowdsourcing systems but are not provided with a formal contract has challenged the traditional beginnings and ends of formality and informality. These employees are involved in cognitively intensive and data mediated work that is completed in dispersed settings with risks that are psychological, emotional and informational rather than physical.

Occupational health and safety (OHS) regimes that existed in India had a traditional state that was modeled to the workplace that was fixed, the employer was identifiable, and the working hours were standardized. Digitalization has broken this infrastructure which has created disjointed and frequently invisible workplaces that exist in and ultimately are directed by algorithms but not human supervisors. Workers in the realm of platforms delivery, online tutors, content moderators, and micro-workers live in regulatory grey spaces, which lacks any health insurance, data protection, and legal consideration (Ramesh, 2025). To address this gap, a workplace based regulatory framework has to be replaced by a work-based regulatory framework- one that acknowledges Labour in both physical and digital workplaces and in hybrid spaces. Expanding digital OHS coverage to informal and platform workers would bring constitutional guarantees of equality and dignity to life and enhance the progress of the Indian pledges under sustainable development goal 3 and 8 (Krishna, 2025).

6.1 Adding Gig Workers, Platform Workers and Remote Freelancers

The growing popularity of the gig economy is an indication of the need to regulate Labour inclusively. Even though the new social security laws recognize gig and platform workers, they only limit them to welfare and not occupational safety and mental health protection measures, as well as data right restrictions. These types of work should be clearly encompassed in a current digital framework of OHS by statutory means and joint responsibility. Platform aggregators must share accountability of workers safety and data management in cases whereby workload, remuneration, and evaluation are set by an algorithm system (Bowdler et al., 2025).

Inclusion should be multi-layered in nature: recognition of digital workplaces in the OHS law; the development of a portable safety right: a platform-funded insurance system; the development of ergonomic and psychosocial norms of working remotely; and the legalization of digital fatigue, technostress, and algorithmic overwork. Considering that many digital Labour platforms are transnational, coordinated regulation by labor, technology, and external affairs institutions is the solution to avoid regulatory arbitrage and loss of worker rights during the cross-border digital labor (I. et al., 2025).

6.2 Harmonizing Labour Law, Data Protection and AI Governance

The regulatory framework of work, data, and technology in India is still disorganized. Physical risk is covered by Labour laws, personal information is covered by the data protection law, and the mapping of AI policy is conducted all in part by means of soft-law tools, which create systemic blind spots in the employer-employee relationships meditated with algorithms. The harmonization of cross-sectoral equitable digital OHS framework needs to be present. Ethics of data should be incorporated into Labour laws so that the surveillance, tracking and assessment of algorithms should be done in accordance with the acceptable parameters of consent, necessity and proportionality (Abraha, 2025).

The applicability of occupational safety law has to be extended to cover cognitive, psychological, and informational harms as hazard and compulsory digital risk and psychosocial audits in the facilities with AI-intensive working places. It is also important that institutional coordination should take place. Harmonized standard-setting, monitoring, and enforcement in digital Labour ecosystems would be possible with a centralized regulatory mechanism having multidisciplinary competence (Mehta et al., 2025). This integration would bring the domestic system of India on the international standards of ethical AI and employee protection, which would mark the shift to the rights-based digital labour jurisprudence rather than the reactive, industrial-era control.

6.3 Trade Union, Society and Cooperative Models Role

The administration of digital work cannot be based on the statutory intervention only; it must be influenced by the democratic involvement and social discussion as well. A digital shift to the unionism is also depicted with trade unions repositioning themselves as champions of data rights, algorithmic transparency, as well as psychosocial safety. The civil society groups and research institutions have been critical to predeterminists of algorithmic responsibility, privacy and well-being of workers in policy discussions as well as provide avenues to participatory rule-making and educating workers (Doellgast et al., 2025). An alternative model of governance, empowering ethical norms of OHS through digital platform design, is offered in the worker cooperatives and platform cooperativism. The increment of worker welfare can be achieved through the decentralization of ownership and decision making, which continues to benefit technological architecture and cooperative platforms in other areas. The culture of cooperation is the reason why India provides a promising market within which such models can be tested in the context of delivery services and freelancing, as well as distance education within the context of remote work, allowing to institutionalize the principle of worker safety and data control in digital Labour markets (Chaudhary et al., 2025).

6.4 The Development of Ethics Training Concerning Compliance with Ethics in SMEs and Startups

The IIAD The key driver of the Indian digital economy are small and medium enterprises and startups, which are likely unable to deploy ethical OHS and data governance frameworks. The compliance needs to be addressed with sustainable practice such as capacity building (via legal literacy training programs, where available), provision of compliance mechanisms, and external audit of ethics (Sarker et al., 2025). The development of awareness about psychosocial safety and data rights is possible through the process of digital OHS literacy programs, whereas the realization of self-assessment and a real-time compliance could be helped by technology-facilitated regulatory measures applied to enterprises. Digital OHS checks (such as algorithmic fairness, proportionality of surveillance, and mental health threats) should be carried out once every few years in addition to conventional safety inspections (Adikwu et al., 2024). Voluntary compliance can also be motivated by using incentives, which can be certification and financial rewards. The collaboration between academics and industries will be able to lead to the creation of evidence-based standards of psychosocial risk assessment, which makes the digital ethics an inherent part of the entrepreneurial education and innovation environment (Hall et al., 2025).

6.5 Ethical Dilemmas in Digital OHS: Competing Frameworks and Practical Resolution

Although concepts such as "human-centeredness" and "ethical AI" are frequently invoked in regulatory discourse, their operationalization requires engagement with underlying ethical theories. A utilitarian approach prioritizes efficiency, productivity, and aggregate welfare, often justifying extensive data collection and surveillance if overall safety or output is improved (Sarker et al., 2025). However, such an approach risks legitimizing invasive monitoring and psychological harm when individual worker dignity is subordinated to organizational gains.

By contrast, deontological ethics emphasize inviolable rights, including privacy, autonomy, and informed consent, regardless of efficiency gains. From this perspective, constant surveillance and opaque algorithmic decision-making are ethically impermissible even if they marginally enhance productivity or safety metrics. Theories of justice further demand attention to distributive and procedural fairness, particularly where digital risks are disproportionately borne by precarious, platform-based, or informal workers.

This section advances a rights-based ethical framework that reconciles these tensions by grounding digital OHS governance in constitutional principles of dignity, proportionality, and accountability. Ethical deployment of workplace technology must therefore satisfy necessity, minimal intrusion, transparency, and meaningful human oversight. Such a framework ensures that efficiency gains do not come at the expense of fundamental labour rights and psychological well-being.

6.6 Digital Literacy and Skill Development in Occupational Health Protection

Regulatory reform alone is insufficient to address digital occupational risks without parallel investment in digital literacy and skill development. Workers and managers must possess the capacity to understand, question, and safely interact with algorithmic systems. Limited awareness of data practices, automated decision-making, and surveillance technologies exacerbates vulnerability, reduces reporting of harm, and weakens informed consent.

Digital OHS literacy programs should therefore form an integral component of occupational safety strategies. These programs should include training on data rights, algorithmic transparency, cybersecurity hygiene, and psychosocial risk recognition.

For managers and supervisors, ethical technology governance training is essential to prevent misuse of monitoring tools and to promote health-conscious deployment of digital systems (Adikwu et al., 2024). Embedding digital literacy within OHS frameworks enhances preventive capacity, empowers workers, and complements legal protections by fostering informed, participatory, and resilient digital workplaces.

7. POLICY RECOMMENDATIONS/ LEGAL REFORM: WORKPLACE SAFETY CODE TOWARDS UNITY IN DIGITAL OCCUPATIONAL HEALTH AND SAFETY

The unintegrated legal system in India requires the creation of a single Digital Occupational Health and Safety Code that would incorporate Labour rights, data protection, and AI ethics. This code must acknowledge that workplaces are digital-physical ecosystem and provides protection on mental, emotional and informational aspects of work. Its key goals should cover protection of Labour rights in all types of employment, data-right safeguarding of working places and institutionalization of analogousness, visibility, and people authority. The integrated Digital OHS Code would also result in regulation certainty and normative guidance, and on the other hand, guarantee that technological innovation promotes the welfare of workers as opposed to working against it through unifying the scattered laws and other legal standards on the subject into a consistent framework built around the principles underlying the constitution and the global norms/standards.

7.1 The Compulsory Algorithmic Risk Impact Evaluations

AI implementation in the employment field must undergo obligatory Algorithmic Risk Impact Assessment before use. Such appraisals should consider fairness, transparency, proportionality, psychological impact, and data governance controls, and should be a preventive governance tool that is similar to environmental impact assessments. Digital Labour governance would be improved by making it more trustworthy and legitimate through a periodic review and accountability. The job hazard has to be redefined to incorporate psychological distress, digital follow-up, algorithmic bias and the exploitation of data. Identification of these harms in OHS law would mean they would perform psychosocial measures, set up grievance mechanisms, offer mental health help, and broadcastment of the stress-related disorders. Incorporating these injuries in compensation and insurance plans would confirm the fact that occupational safety will involve the maintenance of dignity and psychological comfort.

Implementation will require a special governmental body to oversee digital Labour regulation practices and do so appropriately with interdisciplinary skills to establish standards and uphold control and dispute resolution procedures. Simultaneous collaboration with foreign allies, especially the European Union, may help in sharing knowledge, conducting research and aligning AI and digital work ethics. This cooperation would strengthen the regulatory capacity of India and its global trust as well as advance a rights-based digital transformation model. The technification of the workplace raises an issue of overturning the assumptions of the occupational safety law. India has to act holistically, proactively and based on constitutional morality. It is necessary to bridge formal-informal gaps, afford protection to gig and remote employees, bring about a balance between data and Labour regulation and to make investments in institutional and capacity-building changes to protect dignity in the digital workplace. By taking the initiative and loading reform with a sense of ethics, India will be in a position to make its movement to the Industry 4.0 not only effective and competitive but, as well, fair, inclusive, and human-centered in its essence.

7.2 Looking Ahead: The Future of Digital OHS Regulation Toward 2030

The timeliness of current regulatory developments which includes the enactment of India's Digital Personal Data Protection Act, 2023, the European Union's AI Act, and evolving international guidance on psychosocial risk and also offers a critical foundation for forward-looking analysis. If existing trends continue, occupational health and safety regulation by 2030 is likely to undergo three major transformations.

First, OHS frameworks will increasingly adopt anticipatory and preventive governance, mandating algorithmic risk assessments and psychosocial audits before digital systems are deployed. Second, digital well-being—encompassing mental health, informational autonomy, and cognitive load—will emerge as a core component of statutory workplace safety, rather than a peripheral concern. Third, regulatory convergence between labour law, data protection, and AI governance will intensify, replacing fragmented oversight with integrated compliance models.

For India, this trajectory presents both opportunity and risk. Without proactive reform, digital labour may outpace legal protection. Conversely, embedding constitutional values of dignity, proportionality, and social justice into future OHS reforms could position India as a normative leader among rapidly digitizing economies. Anticipating these shifts strengthens the chapter's policy relevance by ensuring that regulatory recommendations are not merely reactive but future-ready.

8. CONCLUSION: REAFFIRMING THE HUMAN-CENTERED VISION OF OHS IN THE DIGITAL AGE - REAFFIRMING THE HUMAN-CENTERED VISION OF OHS IN THE DIGITAL AGE

The development of the law on occupational health and safety (OHS) reflects the larger direction taken by fluctuations in technology and the economy. Although previous regulatory frameworks were already created to respond to the physical risks involved in the production process in the industries, the modern digital work environment creates an entirely different threat structure (Obasi & Benson, 2025). The harms in Industry 4.0 spaces also become more prone to algorithmic decision making, ubiquitous data monitoring, cognitive overload, and psychosocial stress instead of physical injury being seen. This change requires a shift in the paradigm of OHS governance to a more humanistic oriented one- where dignity, autonomy, and well being are central to the technological advances (Li et al., 2025). This vision has long since lost its aspirational quality in the digital era and today, when it is important to affirm that inclusive, just, and democratic labor systems should be maintained, not only in the digital era.

This chapter set out to examine how the rapid diffusion of Industry 4.0 technologies reshapes the nature of occupational health and safety (OHS) risks, challenges the adequacy of existing regulatory frameworks, and demands a reorientation of labour protection grounded in human dignity, autonomy, and well-being. The analysis demonstrates that these questions cannot be addressed through traditional industrial safety paradigms alone. Instead, digitally mediated workplaces generate a qualitatively different risk environment, the one that is characterized by algorithmic management, continuous surveillance, cognitive overload, psychosocial stress, and informational vulnerability. By tracing these transformations across technological, legal, psychological, and policy dimensions, the chapter answers the central concern raised in the introduction: existing OHS regimes, designed for visible and event-based physical harm, are structurally ill-equipped to govern the diffuse, continuous, and often invisible risks of digital work.

The chapter further establishes that emerging digital harms are not speculative or abstract. Through concrete case analyses involving warehousing, platform labour, remote work, and cybersecurity failures, it illustrates how technologies such as computer vision, wearable tracking devices, opaque rating algorithms, and vulnerable digital infrastructures directly shape worker safety, mental health, and professional responsibility. These cases reveal that occupational risk in the digital age is produced at the intersection of technological design, legal permissibility, and psychological experience. In doing so, the chapter substantiates its core claim that algorithmic surveillance, data extraction, and system failures must be recognized as legally cognizable occupational hazards rather than peripheral managerial concerns.

From a normative perspective, the chapter answers the question of *how* law should respond to these challenges by advancing a rights-based, human-centered framework for digital OHS governance. Drawing on surveillance theory, labour process theory, and ethical frameworks, it demonstrates that efficiency-driven approaches grounded solely in utilitarian logic risk legitimizing intrusive monitoring and psychological harm. By contrast, a deontological and justice-oriented approach that is anchored in dignity, proportionality, transparency, and accountability which offers a principled basis for reconciling technological innovation with worker protection. This ethical analysis moves beyond aspirational references to "ethical AI" and operationalizes human-centeredness through concrete legal and institutional safeguards.

The comparative analysis between India and the European Union further clarifies that regulatory solutions cannot be mechanically transplanted across jurisdictions. While EU frameworks emphasize precaution, worker participation, and integrated data governance, India's labour ecosystem is shaped by informality, platform-mediated employment, and uneven enforcement capacity. The chapter demonstrates that effective reform therefore requires context-sensitive regulatory pluralism rather than a copy-paste approach. By situating Indian regulatory challenges within global normative trends—while remaining attentive to domestic constitutional values—the chapter contributes a Global South–informed perspective to digital OHS discourse.

Importantly, the chapter also responds to the forward-looking dimension of OHS governance by projecting how current trends may shape regulation toward 2030. It argues that future frameworks are likely to evolve toward anticipatory regulation, mandatory algorithmic risk assessments, recognition of psychosocial and informational harms, and deeper integration between labour law, data protection, and AI governance. In this trajectory, digital literacy and skill development emerge as indispensable complements to regulation, empowering workers and managers to engage safely and ethically with technological systems.

Further, the chapter reframes Industry 4.0 not merely as a technological shift, but as a normative test for contemporary labour law and social policy. Whether digital transformation deepens precarity or advances dignity depends on how law, ethics, and institutional design respond today. By offering an integrated analytical framework, concrete case evidence, and future-oriented policy pathways, this chapter contributes to reimagining occupational health and safety as a foundational pillar of equitable and sustainable digital development.

Finally, the digital age of occupational safety lies in the ability of the law to reconcile the power of technology and human frailty. With work becoming how more and more of a data subject and algorithmically controlled, the law should protest against the down-grading of workers to the status of data subjects and reaffirm that they are rights-bearing subjects. Legal systems can create mechanisms to enable technological advancement to improve human well-being instead of undermining it through human-focused OHS structures, participative governance and ethical regulation. The issue of Industry 4.0 is, therefore, more a regulatory challenge, but also an opportunity- to recycle occupational health and safety as the foundation of human, inclusive, and sustainable digital development.

REFERENCES

Abraha, H. H. (2025). Navigating workers' data rights in the digital age. In *ILO eBooks*. DOI: 10.54394/MLUH5441

Adams-Prassl, J., Rakshita, S., Abraha, H. H., & Silberman, M. S. (2024). Towards an International Standard for Regulating Algorithmic Management: A Blueprint. SSRN *Electronic Journal*. DOI: 10.2139/ssrn.4684947

Adikwu, F. E., Esiri, A. E., Aderamo, A. T., Akano, O. A., & Erhueh, O. V. (2024). Leveraging digital technologies for health, safety, and environmental (HSE) management in industrial operations. *International Journal of Applied Research in Social Sciences*, *6*(11), 2712–2726. DOI: 10.51594/ijarss.v6i11.1722

Akki, S., & Ouddasser, A. (2024). Gesundheit und Sicherheit am Arbeitsplatz im digitalen Zeitalter: Theoretische Untersuchung der Auswirkungen der digitalen. *Transformation (Durban)*. Advance online publication. DOI: 10.5281/zenodo.12998523

Aloisi, A., Paliokaitė, A., Moore, P., & Potocka-Sionek, N. (2025). Study Exploring the Context, Challenges, Opportunities, and Trends in Algorithmic Management. SSRN *Electronic Journal*. DOI: 10.2139/ssrn.5198804

Andrade, M. D., Martínez-Córcoles, M., Fialho, P., & Guimaraes, M. (2025). The effects of being under watch: The impact of electronic monitoring on remote workers' psychological safety. *Journal of Management & Organization*, *1*, 1–18. Advance online publication. DOI: 10.1017/jmo.2025.10039

Anjumanwarshaik, S., Shaikh, S., Desai, P. S., & Ym, G. (2025). Artificial intelligence in human resource management and employee well-being. *International Journal of Applied Research*, *11*(10), 219–224. DOI: 10.22271/allresearch.2025.v11.i10c.12943

Au-Yeung, T. C., Chan, C. K., Ming, C. K. K., & Tsui, W. Y. A. (2024). The gig economy, platform work, and social policy: Food delivery workers' occupational welfare dilemma in Hong Kong. *Journal of Social Policy*, *1*. Advance online publication. DOI: 10.1017/S0047279423000673

Avogaro, M. (2022). A holistic approach to remote digital work: From the inconveniences suffered by employees to a comprehensive 'fair compensation.'. *IDP Revista de Internet Derecho y Política*, *36*(36). Advance online publication. DOI: 10.7238/idp.v0i36.400857

Beridze, G. (2024). "Unexpected" Europeanization: The Impact of EU-Georgia Association Agreement on Occupational Safety and Health Policy in Georgia. *TalTech Journal of European Studies/TalTech. Journal of European Studies, 14*(1), 196–232. DOI: 10.2478/bjes-2024-0010

Bowdler, M., Lahti, H., Jelenko, M., & Valtonen, T. (2025). Algorithmic management and psychosocial risks at work: An emerging occupational safety and health challenge. *Scandinavian Journal of Work, Environment & Health.* Advance online publication. DOI: 10.5271/sjweh.4270 PMID: 41353777

Bråten, M., Rasmussen, S., & Oppegaard, S. M. N. (2024). The Working Environment of the Future. In *TemaNord*. DOI: 10.6027/temanord2024-531

Callari, T. C., Segate, R. V., Hubbard, E., Daly, A., & Lohse, N. (2024). An Ethical Framework for Human-Robot Collaboration for the Future People-Centric Manufacturing: A Collaborative Endeavour with European subject-matter experts in Ethics. *Technology in Society, 78*, 102680. DOI: 10.1016/j.techsoc.2024.102680

Cefaliello, A., Moore, P., & Donoghue, R. (2023). Making algorithmic management safe and healthy for workers: Addressing psychosocial risks in new legal provisions. *European Labour Law Journal, 14*(2), 192–210. DOI: 10.1177/20319525231167476

Chapman, K., Smith, G., Klabacka, K., Winslow, H., Das, S., Wiśniewski, P., Knijnenburg, B. P., & Page, X. (2025). Beyond the Legal Lens: A Sociotechnical Taxonomy of Lived Privacy Incidents and Harms. *arXiv (Cornell University)*. DOI: 10.48550/2511.20791

Chaudhary, K., Manchanda, N., Timilsina, R. R., Rahut, D. B., & Sonobe, T. (2025). *From Platforms to Paychecks: Comparative Insights into Wages of Low-Skilled Platform and Nonplatform Workers in India*. DOI: 10.56506/LOHF9401

Daneshvar, E., & Otterbach, S. (2025). Workplace stressors and burnout among healthcare professionals: Insights from the pandemic and implications for future public health crises. *Scientific Reports, 15*(1), 42015. DOI: 10.1038/s41598-025-21540-2 PMID: 41298499

Doellgast, V., Appalla, S., Ginzburg, D., Kim, J., & Thian, W. L. (2025). Global case studies of social dialogue on AI and algorithmic management. In *ILO eBooks*. DOI: 10.54394/VOQE4924

Dollard, M. F., & Potter, R. (2025). Managing Psychosocial Risks at Work Through National Policy and Regulation. *Occupational Health Science, 9*(4), 865–869. Advance online publication. DOI: 10.1007/s41542-025-00236-z

Doval, E. (2024). Workplace Health and Safety Management in the Digitalization and Artificial Intelligence Era. *American Journal of Biomedical Science & Research, 23*(6), 700–705. DOI: 10.34297/AJBSR.2024.23.003133

Fabrellas, A. G. i. (2022). How to ensure workers well-being in the digital age? *IDP Revista de Internet Derecho y Política, 35*(35), 1–15. DOI: 10.7238/idp.v0i35.392944

Fairfield, J., & Reilly, A. (2025). Home—The Final Frontier: Why Privacy Means Protecting Workers' Rights to Time and Space. SSRN *Electronic Journal*. DOI: 10.2139/ssrn.5245735

Frielingsdorf, H., Fomichov, V., Rystedt, I., Lindstrand, S., Korhonen, L., & Henriksson, H. (2025). Associations of time spent on different types of digital media with self-rated general and mental health in Swedish adolescents. *Scientific Reports, 15*(1), 993. DOI: 10.1038/s41598-024-83951-x PMID: 39762280

Gaudio, G. (2024). Litigating the Algorithmic Boss in the EU: A (Legally) Feasible and (Strategically) Attractive Option for Trade Unions? *International Journal of Comparative Labour Law and Industrial Relations, 40*(1), 91–130. DOI: 10.54648/IJCL2024002

Gnugesser, E., Jöllenbeck, M., Schlenger, W., & Ochsmann, E. (2025). Navigating the digital shift: Working conditions and employee health in digital social care. *Journal of Innovation & Knowledge, 10*(3), 100706. DOI: 10.1016/j.jik.2025.100706

Gonzalez Vazquez, I., Fernandez Macias, E., Wright, S., & Villani, D. (2025). Digital monitoring, algorithmic management and the platformisation of work in Europe.

Hall, S., Hastings, O., Prentice, K., Brown, B. J., Andrews, J. A., Marner, S., Woodcock, R., Martin, J., & Hall, C. L. (2025). Principles of Industry-Academic Partnerships Informed by Digital Mental Health Collaboration: Mixed Methods Study. *JMIR Mental Health, 12*, v12i3e77439. Advance online publication. DOI: 10.2196/77439 PMID: 40929728

Hoxhaj, O., Halilaj, B., & Harizi, A. (2023). ETHICAL IMPLICATIONS AND HUMAN RIGHTS VIOLATIONS IN THE AGE OF ARTIFICIAL INTELLIGENCE. *Balkan Social Science Review, 22*(22), 153–171. DOI: 10.46763/BSSR232222153h

Huram, K. (2025). Algorithmic Management and the Future of Human Work: Implications for Autonomy, Collaboration, and Innovation. *arXiv (Cornell University)*. DOI: 10.48550/2511.14231

Krishna, N. R. (2025). Occupational Safety, Health and Working Conditions (OSH) Code, 2020 and Interstate Women Migrants: A Gender Integrative Evaluation. *The Indian Journal of Public Administration, 71*(2), 406–411. DOI: 10.1177/00195561251338111

Li, K., Xia, M., & Bentley, T. (2025). The paradoxical effects of emerging mining technologies on psychosocial work factors: An integrated framework and research agenda. *Technological Forecasting and Social Change, 222*, 124391. DOI: 10.1016/j.techfore.2025.124391

Lin, H. (2024). Ethical and Scalable Automation: A Governance and Compliance Framework for Business Applications. *arXiv (Cornell University).* DOI: 10.48550/2409.16872

Lindholm, M., Reiman, A., & Väyrynen, S. (2020). On Future Occupational Safety and Health Challenges. *International Journal of Occupational and Environmental Safety, 4*(1), 108–127. DOI: 10.24840/2184-0954_004.001_0009

Malenfer, M., Sarrey, M., Clerté, J., Héry, M., Biéri, M., Braunschweig, B., Chatellier, R., Fatès, N., Halluin, S., de Jouvenel, F., Mandinaud, V., Munoz, J. H., Olympio, A., Silvestre, T., & Soupizet, J.-F. (2023). *Artificial intelligence in the service of health and safety at work: Perspectives and challenges from now to 2035 - A prospective study*. Qeios., DOI: 10.32388/NRENI6

Mariani, K., & Lozada, F. V. (2023). The Use of AI and Algorithms for Decision-making in Workplace Recruitment Practices. *Journal of Student Research, 12*(1). Advance online publication. DOI: 10.47611/jsr.v12i1.1855

Mbare, B., Perkiö, M., & Koivusalo, M. (2024). Algorithmic management, wellbeing and platform work: Understanding the psychosocial risks and experiences of food couriers in Finland. *Labour & Industry, 1*(4), 386–411. Advance online publication. DOI: 10.1080/10301763.2024.2423442

Mehta, P. G., Reshma, C. U., Polisetty, A., & Suraj, P. (2025). A systematic review on decent work through automation [Review of *A systematic review on decent work through automation*]. *Discover Sustainability, 6*(1). Springer Nature. DOI: 10.1007/s43621-025-01720-w

Milanez, A., Lemmens, A., & Ruggiu, C. (2025). Algorithmic management in the workplace. In *OECD artificial intelligence papers*. DOI: 10.1787/dee339a8-en

Moore, P. (2023). Workers' right to the subject: The social relations of data production. *Convergence (London), 30*(3), 1076–1098. DOI: 10.1177/13548565231199971

Moraru, R. I. (2024). Challenges and Opportunities in Occupational Health and Safety Digitalization. In *Intelligent systems reference library* (p. 189). Springer Nature., DOI: 10.1007/978-3-031-55952-5_10

Munn, L. (2024). Expansive and Invasive: Mapping the "Bossware" Used to Monitor Workers. *Surveillance & Society*, *22*(2), 104–119. DOI: 10.24908/ss.v22i2.16179

Nesterak, J., Szelągowski, M., & Radziszewski, P. (2024). Workplace performance measurement: Digitalization of work observation and analysis. *Journal of Intelligent Manufacturing*. Advance online publication. DOI: 10.1007/s10845-024-02419-x

Nilsson, K. H., Bodin, T., Strauss, P., Matilla-Santander, N., Badarin, K., Brulin, E., & Håkansta, C. (2025). Algorithmic management is associated with psychological distress, musculoskeletal pain, and occupational accidents: A cross-sectional study in logistics. *International Archives of Occupational and Environmental Health, 98*(9-10), 929–942. Advance online publication. DOI: 10.1007/s00420-025-02180-5 PMID: 41251732

Obasi, I. C., & Benson, C. (2025). The Impact of Digitalization and Information and Communication Technology on the Nature and Organization of Work and the Emerging Challenges for Occupational Safety and Health. *International Journal of Environmental Research and Public Health, 22*(3), 362. Multidisciplinary Digital Publishing Institute. DOI: 10.3390/ijerph22030362

Oyekunle, D., Boohene, D., & Preston, D. (2024). Ethical Considerations in AI-Powered Work Environments: A Literature Review and Theoretical Framework for Ensuring Human Dignity and Fairness [IJSRM]. *International Journal of Scientific Research and Management, 12*(3), 6166–6178. DOI: 10.18535/ijsrm/v12i03.em18

Pisarczyk, Ł. (2025). Bezpieczeństwo i higiena pracy w stosunkach pracy z wykorzystaniem nowoczesnych technologii. *Acta Universitatis Lodziensis Folia Iuridica, 112*, 39–60. DOI: 10.18778/0208-6069.112.03

Rahmi, K. H., Fahrudin, A., Supriyadi, T., Herlina, E., Rosilawati, R., & Ningrum, S. R. (2025). Technostress and cognitive fatigue: Reducing digital strain for improved employee well-being: A literature review [Review of Technostress and cognitive fatigue: Reducing digital strain for improved employee well-being: A literature review]. *Multidisciplinary Reviews, 8*(12), 2025380. DOI: 10.31893/multirev.2025380

Ramesh, D. (2025). Affect, Aspirations, and Accountabilities: Informing Human-Centered AI through Interpretivist Empirical Inquiry. *Deep Blue (University of Michigan)*. DOI: 10.7302/26812

Samans, R. (2023). *Human-Centred Economics*. DOI: 10.1007/978-3-031-37435-7

Sander, D. J., Jane, H., Tzu-Sheng, K., & Moberg, J. R. Niels, van B., & Haiyi, Z. (2025). Understanding, Demystifying and Challenging Perceptions of Gig Worker Vulnerabilities. *arXiv (Cornell University)*. DOI: 10.48550/2511.00273

Sarker, I. H., Janicke, H., Mohsin, A., & Μαγλαράς, Λ. (2025). SME-TEAM: Leveraging Trust and Ethics for Secure and Responsible Use of AI and LLMs in SMEs. *arXiv (Cornell University)*. DOI: 10.48550/2509.10594

Saxena, R. R. (2024). Applying Predictive Analytics to Occupational Health and Safety in India. *arXiv (Cornell University)*. DOI: 10.48550/2412.16038

Sebastian, R. A., Ehinger, K., & Miller, T. (2025). Do we need watchful eyes on our workers? Ethics of using computer vision for workplace surveillance. *AI and Ethics*, *5*(4), 3557–3577. Advance online publication. DOI: 10.1007/s43681-025-00726-4

Singh, M. P. (2023). Impact of Industry 4.0 on Occupational Safety and Health. *International Journal for Research in Applied Science and Engineering Technology*, *11*(11), 1687–1692. DOI: 10.22214/ijraset.2023.56941

Sum, C. M., Shi, C. M., & Fox, S. (2025). "It's Always a Losing Game": How Workers Understand and Resist Surveillance Technologies on the Job. *Proceedings of the ACM on Human-Computer Interaction*, *9*(2), 1–32. DOI: 10.1145/3710902

Sun, C., & Chen, X. (2025). Destined for Balance? Centralized and Decentralized Approaches to AI Governance. *Politics and Governance*, *13*, 10197. Advance online publication. DOI: 10.17645/pag.10197

Chapter 10
The Role of AI and Machine Learning in Predictive Safety Analytics

P. Selvakumar
http://orcid.org/0000-0002-3650-4548
Department of Science and Humanities, Nehru Institute of Technology, Coimbatore, India

Undabatla Rambabu
http://orcid.org/0000-0003-2731-1610
Lakireddy Bali Reddy College of Engineering, India

V. M. R. Hiremath
Rao Bahadur Y. Mahabaleswarappa Engineering College, India

Vijay Uprikar
http://orcid.org/0000-0003-2546-8307
Datta Meghe Institute of Management Studies, India

R. Kalai Selvi
Kuppam Engineering College, India

Ninad Rithe
http://orcid.org/0009-0004-5729-6081
Datta Meghe Institute of Management Studies, India

ABSTRACT

Predictive safety analytics in healthcare represents a transformative approach to managing patient care, clinical risks, and operational efficiency by using data-driven insights to anticipate safety issues before they arise. Unlike traditional reactive methods, which respond to adverse events after they occur, predictive analytics proactively identifies patterns and risk factors that signal the potential for harm. By leveraging predictive safety analytics enables healthcare systems to forecast patient deterioration, reduce medical errors, prevent hospital-acquired conditions, and optimize resource allocation—ultimately fostering a safer, more efficient health-

DOI: 10.4018/979-8-3373-6370-7.ch010

care environment. Predictive safety analytics refers to real-time data anticipate in healthcare settings. Encompasses the collection, structured, and ranging electronic health (EHRs), laboratory test results, and vital signs to clinician notes, wearable device data, and even social determinants of health. Patients are categorized based on their risk levels, allowing clinicians to prioritize interventions.

UNDERSTANDING PREDICTIVE SAFETY ANALYTICS IN HEALTHCARE

Predictive safety analytics in healthcare represents a transformative approach to managing patient care, clinical risks, and operational efficiency by using data-driven insights to anticipate safety issues before they arise. Unlike traditional reactive methods, which respond to adverse events after they occur, predictive analytics proactively identifies patterns and risk factors that signal the potential for harm. By leveraging predictive safety analytics enables healthcare systems to forecast patient deterioration, reduce medical errors, prevent hospital-acquired conditions, and optimize resource allocation—ultimately fostering a safer, more efficient healthcare environment. Predictive safety analytics refers to real-time data anticipate in healthcare settings. Encompasses the collection, structured, and ranging electronic health (EHRs), laboratory test results, and vital signs to clinician notes, wearable device data, and even social determinants of health. Patients are categorized based on their risk levels, allowing clinicians to prioritize interventions. Clinical Decision Support (CDS): Predictive insights are embedded into the clinician workflow, offering timely alerts and recommendations at the point of care. Feedback Loops: Systems continuously learn from new data and clinician input, improving predictive accuracy over time. Operational and System-Level Benefits Beyond individual patient care, predictive safety analytics improves the overall safety and efficiency of healthcare systems. It enables: Resource Optimization: Hospitals can forecast demand for ICU beds, staffing needs, or equipment usage, improving readiness and reducing bottlenecks.

Technological Enablers: The rise of predictive analytics in healthcare has been enabled by several technological advancements. The digitization provides a rich data source for analytics. Cloud Computing: Enables large-scale data storage and real-time analysis without the need for extensive local infrastructure. Allows for useful data sources like discharge summaries. Interoperability Standards: Frameworks like Wearable IoT Devices: Continuous data from smartwatches, monitors, and biosensors feeds predictive models with real-time insights into patient status. Understanding predictive safety analytics is essential for navigating the future of healthcare. As the industry shifts toward value-based, patient-centered care, the ability to prevent harm rather than respond to it becomes a critical differentiator in delivering quality outcomes. By anticipating adverse events before they happen, predictive analytics can transform the healthcare landscape—enhancing safety, improving efficiency, and ultimately saving lives. However, its success depends on high-quality data, ethical implementation, clinician buy-in, and continual refinement. When these factors align, predictive safety analytics stands as one of the most powerful tools in the quest for safer, smarter, and more equitable healthcare.

THE ROLE OF AI AND MACHINE LEARNING IN ENHANCING SAFETY: ROLE

Numerous sectors, but their role in enhancing safety, transportation, manufacturing, cybersecurity, and public safety is particularly profound. As the digital landscape grows increasingly complex, traditional safety mechanisms are often inadequate for velocity, and variety systems must process in real time (Anastasi et al., 2021). AI and ML fill this gap by enabling intelligent, predictive, and adaptive safety solutions that can detect, analyze, and mitigate risks faster and more accurately than human systems alone. From predicting clinical deterioration in hospitals to monitoring machinery for early signs of failure, AI and ML offer revolutionary improvements to both reactive and proactive safety protocols. Understanding Safety Contexts: Artificial Intelligence is a broad domain involving the simulation of human intelligence by machines. It includes subfields like natural language processing (NLP), computer vision, and robotics. Machine Learning, a subset of AI, involves algorithms that enable systems to learn from data, identify patterns, and improve over time without being explicitly programmed. In safety applications, models can process massive data streams to recognize anomalies, anticipate failures, and issue warnings, often before a human operator would even perceive a problem.

These technologies are increasingly embedded into decision support systems, sensors, surveillance infrastructure, and autonomous control systems to boost both situational awareness and response speed. AI and ML in Healthcare Safety: impactful applications in healthcare safety. Powered by ML, can identify patients at risk for sepsis, cardiac arrest, or lab results, and real-time vitals. AI-driven tools assist clinicians by providing early alerts and treatment recommendations, thus preventing adverse events (Arjmandnia & Alimohammadi, 2024). Clinical decision support systems (CDSS) use ML models to detect prescribing errors, contraindications, and potential allergic reactions, significantly reducing the rate of medication errors. Furthermore, AI enhances radiology and pathology by detecting anomalies in imaging studies or biopsy results, many diagnostic. In Industrial and Occupational Safety: In industrial environments, worker safety is a critical concern, especially in sectors such as construction, mining, and manufacturing. AI-powered computer vision systems monitor live feeds from CCTV cameras to detect unsafe behaviors like the absence of protective gear or risky proximity to heavy machinery. Predictive maintenance systems, driven by ML algorithms, analyze equipment performance and environmental data to anticipate malfunctions or breakdowns before they lead to hazardous conditions. Wearable technologies enhanced with AI can monitor workers' physiological or fatigue levels— issue real-time alerts if safety thresholds are breached, thereby preventing accidents caused by exhaustion or stress (Fallah Madvari, 2022).

Enhancing Cybersecurity and Digital Safety: In the digital realm, cybersecurity is paramount for both personal and organizational safety. AI and ML technologies are central to next-generation cybersecurity solutions, as they can identify malware, phishing attempts, and intrusions based on learned behavior rather than predefined rules. ML models can detect patterns in network traffic that signal potential cyberattacks, enabling rapid response and containment (Rehan, 2024) . Moreover, AI supports the development of intelligent firewalls, automated patching systems, and user authentication mechanisms (such as biometric identification) that adapt to evolving threats in real time. AI also plays a role in content moderation, helping to detect and remove harmful or illegal content on digital platforms, thereby promoting safer online environments. Public Safety and Disaster Response AI and ML are also proving invaluable in enhancing public safety and disaster response. In emergency management, AI systems, social media posts, and damage after disasters resource deployment. Predictive models help forecast wildfires, floods, and earthquakes by analyzing environmental and geophysical data. In law enforcement, AI aids in crime prediction (predictive policing), facial recognition for missing persons, and real-time surveillance analysis to detect suspicious activity. However, these applications must be implemented with caution, as issues of algorithmic bias and privacy intrusion can pose ethical concerns. Across domains is both expansive and transformative (Chib et al., 2025). These technologies allow for the anticipation, detection, and mitigation of risks with a level of speed and precision previously unattainable. Whether it is in predicting patient outcomes in hospitals, ensuring occupational safety in factories, defending against cyber threats, or managing public crises, AI and ML are becoming foundational components of modern safety ecosystems. Ensuring use, eliminating bias, safeguarding privacy, and fostering human-AI unlocking AI-driven safety. As innovation continues, safety protocols will save lives and redefine how we perceive and manage risk in the digital age.

PREDICTING ADVERSE EVENTS WITH MACHINE LEARNING ALGORITHMS: ADVERSE EVENT PREDICTION

Adverse events (ML) algorithms risk, especially in domains like healthcare, pharmaceuticals, industrial safety, and aviation. Traditional systems for identifying potential hazards have relied on reactive or rule-based models, which, although effective to a degree, often fail to process the enormous volume and complexity of data generated in real-time environments. Machine learning, with its capacity to learn from historical and real-time datasets, uncover subtle patterns, and update its understanding over time, introduces a proactive and highly adaptive method for forecasting harmful incidents before they occur (Abegaz et al., 2023). From preventing patient deterioration in hospitals to forecasting equipment failures in factories, ML-based adverse event prediction systems are reshaping how organizations approach safety, accountability, and operational efficiency. Understanding Adverse Event Prediction: Adverse events refer to unexpected, undesirable outcomes that can cause harm or negatively impact health, productivity, or safety. In healthcare, adverse events, sepsis, and unexpected patient deterioration. Manufacturing and aviation encompass equipment breakdowns, structural failures, and operational accidents. ML algorithms can be trained on vast datasets—comprising structured and unstructured data—to recognize precursors to these incidents. By identifying patterns and correlating risk factors, these models offer predictive alerts that allow stakeholders to intervene early, thereby mitigating or even avoiding the harm entirely Various methods are employed to predict adverse events, reinforcement learning. Decision trees (SVM), trained on labeled datasets to classify or likelihood known outcomes, or can identify unknown or outliers that may correspond to hidden risks or previously unidentified event precursors. Reinforcement Learning: This approach, though less common in current predictive applications, is used in dynamic decision-making environments, learning optimal strategies through trial and error. Complex (e.g., physiological signals) or image-based information (e.g., X-rays or equipment scans), improving predictive performance in high-stakes scenarios (Deimazar & Sheikhtaheri, 2023). Perhaps the most high-impact application of ML in adverse event prediction is in the healthcare sector. Clinical settings generate vast amounts of data—vital signs, lab results, medication history, clinician notes, and imaging studies. ML models analyze this data to predict: Sepsis Onset: Algorithms trained on electronic health records (EHRs) can detect subtle trends indicating early sepsis hours before it is clinically obvious. Patient Deterioration: Real-time monitoring combined with ML predictions helps identify patients at risk of cardiac arrest, respiratory failure, or other life-threatening conditions. Medication Errors: ML can cross-reference prescriptions, diagnoses, allergies, and lab values to flag potential adverse drug events (ADEs). Hospital Readmission: Predictive models assess whether a discharged patient is likely to be readmitted within 30 days, enabling tailored follow-up care. For instance, Google's DeepMind developed a model with high accuracy—giving clinicians a critical window to prevent irreversible damage. Adverse Event Prediction in Pharmacovigilance: In the pharmaceutical industry, adverse drug reactions (ADRs) can result in serious harm, regulatory action, or product withdrawal (D'Ascenzo et al., 2021). ML algorithms are increasingly used in pharmacovigilance to analyze spontaneous reporting systems (e.g., FAERS), clinical trial data, and real-world evidence from social media or EHRs. Natural language processing (NLP) techniques extract adverse event mentions from text-based records, while ML models evaluate drug-event relationships and detect rare side effects. This automated surveillance accelerates detection, improves drug safety profiles, and supports regulatory compliance.

Industrial and Occupational Safety Applications: Manufacturing environments pose multiple risks—from machinery malfunctions to exposure to hazardous substances. Predictive maintenance systems, powered by ML, monitor equipment sensors and operational data to anticipate breakdowns before they result in safety incidents. For example, Vibration Analysis: ML models analyze abnormal vibration patterns to predict failures in rotating machinery (Rana & Shuford, 2024). Worker Behavior Monitoring: Wearables and cameras, interpreted by ML algorithms, can predict fatigue-related accidents or unsafe actions. Process Deviations: ML detects deviations from normal process conditions in chemical plants, reducing the likelihood of explosions or toxic leaks. These predictive systems minimize downtime, enhance worker safety, and reduce liability. ML algorithms improve situational awareness for air traffic controllers, predicting potential mid-air collisions or runway incursions (Shashank et al., 2025). In road transport, ML is integrated into autonomous driving systems and traffic management platforms to predict collisions, pedestrian incidents, or road hazard exposures based on real-time conditions. Machine learning's role in predicting adverse events is a transformative advancement across industries. By enabling early identification of risks through intelligent pattern recognition, these systems can significantly reduce harm, save costs, and improve human well-being. However, responsible implementation demands attention to data ethics, user trust, interpretability, and contextual adaptation. As ML models become more sophisticated, their ability to detect, anticipate, and prevent adverse outcomes will increasingly redefine the standards of safety, making reactive strategies obsolete and paving the way for a predictive, precision-based approach to risk management.

IDENTIFYING SAFETY SIGNALS WITH AI-POWERED ANALYTICS: SAFETY SIGNAL IDENTIFICATION

In today's data-rich environments, the identification of safety signals—early indicators of potential harm associated with products, treatments, or operations—has become sectors, pharmaceuticals, transportation, manufacturing, and consumer goods. Traditionally, safety signals have been identified through manual review, rule-based algorithms, and retrospective analysis of adverse event reports, but these methods are increasingly being replaced or enhanced by artificial intelligence (AI)-powered analytics (Kalla, 2022) . AI, particularly through, enables faster, more accurate, and proactive detection of signals by analyzing vast and complex datasets that human analysts could not feasibly process on their own. AI-powered safety signal detection marks a transformative shift toward real-time risk surveillance, improving not only safety outcomes but also regulatory compliance, public health, and consumer confidence. Understanding Safety: A safety signal is defined as information that arises from one or multiple sources—such as clinical data, pharmacovigilance reports, or consumer feedback—that suggests a new or known adverse event may be associated with a particular product or process and warrants further investigation. Identifying these signals early allows stakeholders to investigate, mitigate, and respond to risks before they cause widespread harm. However, in complex systems, signals may be weak, obscured by noise, or emerge only under specific conditions. Limitations of Traditional Safety Signal Detection: Historically, safety signal detection relied on spontaneous reporting systems, disproportionality analysis, and periodic safety updates (Ozobu et al., 2025). While effective in many contexts, these methods face several limitations: Latency: Manual signal detection is often delayed, allowing harm to persist before action is taken. Bias and Underreporting: Human-reported data (e.g., in healthcare or pharmaceuticals) may be incomplete or biased. Lack of Scalability: Analysts struggle to review the growing volume of safety data from diverse and unstructured sources. Inability to Capture Complex InAteractions: Traditional methods often miss signals involving multiple factors, such as comorbidities or environmental variables. AI-powered analytics addresses these gaps by bringing speed, scalability, and pattern recognition capabilities to the forefront of safety monitoring. AI Techniques in Safety Signal Detection: AI systems for identifying safety signals employ several advanced techniques: Machine Learning (ML): ML models to or anomalies associated with adverse outcomes. Meaningful sources include clinical notes, patient narratives, social media posts, and regulatory databases. —process sequential and contextual information, enhancing prediction and classification of rare or hidden signals. Anomaly Detection: Unsupervised models flag deviations from normal behavior or statistical baselines, which may indicate emerging risks. Time-Series Analysis: AI tracks trends over time, identifying signals that escalate, diminish, or cyclically reappear (Zafer, 2025). These technologies enable a dynamic, data-driven approach that continually learns and adapts to changing conditions. Applications in Pharmacovigilance and Drug Safety: One of the most developed applications of AI-powered safety signal detection is in pharmacovigilance—the monitoring of adverse drug reactions (ADRs) and drug safety. Regulatory agencies and pharmaceutical companies deploy AI to: Monitor Spontaneous Reports: AI algorithms analyze millions of individual case safety reports (ICSRs) from databases like FAERS (FDA) and VigiBase (WHO), identifying disproportional reporting patterns. Integrate: AI processes insurance claims and patient forums to detect safety signals in real-world settings.

Safety Signal Detection in Clinical Settings: In hospitals and clinical environments, AI plays a growing role in detecting safety signals in patient care and treatment protocols. Real-Time Monitoring: AI systems process real-time data from patients or medical errors (Fazle et al., 2023). Adverse Event Prediction: and live anticipate events like sepsis, allergic reactions, or surgical complications. Clinical Documentation Analysis: NLP scans medical records to surface subtle indications of adverse outcomes not coded in structured formats. These AI-enhanced systems provide clinicians with alerts, enabling faster interventions and improving patient safety outcomes. 6. Industrial Safety and Manufacturing: In industrial environments, safety signals may relate to machine failure, environmental hazards, or unsafe human behavior. AI supports signal detection by: Sensor Data Analysis: AI interprets data from IoT sensors monitoring machinery temperature, pressure, vibration, or movement, identifying precursors to equipment failures. Behavioral Monitoring: Vision-based AI systems analyze worker movements and PPE (personal protective equipment) compliance, flagging unsafe practices. Predictive Maintenance: ML forecasts or becomes unsafe, preemptive (Malik et al., 2025). These systems help reduce downtime, improve compliance with occupational safety standards, and minimize accidents. AI-powered analytics are redefining how organizations identify, evaluate, and act on safety signals. By automating the detection of risks from vast and varied datasets, these tools enable faster, more accurate, and proactive responses. While challenges remain in terms of data quality, model transparency, and regulatory alignment, the trajectory of AI-driven safety signal identification is undeniably transformative. As industries are an indispensable ally in safeguarding health, operational integrity, and societal trust.

PERSONALIZED RISK ASSESSMENT WITH MACHINE LEARNING

Personalized risk assessment, particularly when empowered by machine learning (ML), represents a transformative advancement in predictive analytics across domains such as healthcare, finance, insurance, cybersecurity, and environmental safety. Unlike traditional risk assessment methods that rely on population-level averages or static rule-based models, personalized risk assessment utilizes individual-level data to deliver tailored predictions and recommendations (Jagatheesaperumal et al., 2023). Through learn complex high-dimensional datasets enables dynamic, granular, and real-time risk stratification that can significantly enhance decision-making, prevention strategies, and resource allocation. By integrating behavioral, demographic, genetic, lifestyle, and contextual information, ML models can produce individualized risk scores, detect early signs of adverse outcomes, and support targeted interventions—thus efficiency in risk management. Foundations of Personalized Risk Assessment: Risk assessment aims to estimate the likelihood of a specific negative outcome occurring, such as disease onset, system failure, financial default, or data breach. Traditionally, risk has been quantified using actuarial tables, which work well on average but often lack accuracy at the individual level. These models typically use predefined inputs. Personalized risk assessment, however, recognizes that risk is not uniform across individuals. Factors such as genetic predispositions, co-existing conditions, social determinants, or behavioral patterns can dramatically influence individual risk profiles. Machine learning's ability to uncover nonlinear interactions, adapt to new data, and process large-scale inputs makes it a powerful tool to advance personalized risk prediction (Taylor et al., 2024). Machine Learning Techniques for Risk Stratification: Several machine learning methods are applied in personalized risk assessment, each with unique strengths: Supervised Learning (SVM), random forests, gradient boosting machines, and LightGBM. Commonly used to classify individuals into risk categories based on labeled historical data. Feedforward architectures, unstructured (e.g., images, time-series, free text), are used to discover complex patterns. DBSCAN) identify hidden subgroups within populations that share similar risk profiles. Ensemble Methods: By combining multiple models, ensemble approaches improve robustness and accuracy. LIME, attention mechanisms offer interpretability, showing how specific features contribute to an individual's risk score—an essential requirement for clinical and regulatory use. These algorithms can be trained on various data types—structured tabular data, electronic health records (EHRs), images, sensor data, genomic sequences, and natural language inputs—to produce highly personalized and actionable insights. Personalized Risk in Healthcare: Healthcare is perhaps the most significant beneficiary of personalized risk assessment using ML (Pieszko et al., 2023). From predicting disease onset to guiding treatment plans, the potential applications are vast: Chronic Disease Prediction: patients assess diabetes, hypertension, cardiovascular, and other chronic illnesses based on lifestyle, lab tests, family history, and wearable sensor data. Cancer Risk Stratification: Personalized genomic data combined with clinical features enables risk assessment for cancers like breast, prostate, or colon cancer, guiding decisions on screening frequency and preventive measures. Hospital Readmission Prediction: ML predicts the likelihood that a discharged patient will be readmitted, helping allocate follow-up care and reduce costs.

Figure 1. Predictive Analytics with Machine Learning

Applications Beyond Healthcare: Machine learning-based personalized risk assessment extends into other critical domains: Finance and Credit Scoring: Banks use ML to personalize creditworthiness evaluations, considering transactional behavior, spending patterns, and Insurance: Personalized risk scores help insurers offer tailored premiums and services, assessing risk based on driving habits (in auto insurance), lifestyle (in health insurance), or property characteristics (in home insurance) (Wong et al., 2024). Cybersecurity: User-specific behavioral patterns (e.g., keystroke dynamics, access times) help detect anomalies and predict which users are most vulnerable to phishing, malware, or insider threats. Workplace Safety: AI systems monitor individual workers' behaviors, environmental exposures, and fatigue levels to assess risk of injury in real time. Disaster Preparedness: ML helps personalize evacuation strategies and communication during natural disasters by analyzing demographics, mobility data, and vulnerability indices. These applications improve outcomes, operational efficiency, and user satisfaction by aligning actions with individual risk profiles. Data Sources and Integration: The effectiveness of personalized risk assessment richness and. ML models leverage diverse data sources: Electronic Records: Health, financial, legal, or academic records provide structured historical data. Wearables and IoT Devices: Continuous physiological and environmental data enhance real-time risk prediction. Genomic and Biomarker Data: Used in precision medicine and personalized health risk modeling. Natural Language Data: NLP techniques extract risk factors from unstructured clinical notes, news reports, or user-generated content. Remote Sensing and GIS Data: Used in environmental and public health risk models (Mehta et al., 2025). Social Determinants: Housing, education, income, and community safety influence personal risk and can be modeled with appropriate data sources. Combining these datasets through data fusion, federated learning, or multimodal AI creates a holistic and nuanced view of individual risk. Personalized risk assessment powered by machine learning represents a paradigm for risk management, data-driven, and individualized. By leveraging advanced algorithms and diverse data streams, organizations can predict adverse events with greater accuracy, intervene early, and allocate resources efficiently. In healthcare, finance, cybersecurity, and beyond, this approach supports more intelligent systems and informed decision-making. However, to fully realize its potential, personalized risk assessment must be implemented with careful attention to ethics, fairness, privacy, and transparency. As technologies mature and governance frameworks evolve, personalized ML-based risk models will become indispensable tools in creating safer, more efficient, and more equitable systems.

THE IMPORTANCE OF DATA QUALITY IN PREDICTIVE SAFETY ANALYTICS

In the rapidly evolving field of predictive safety analytics, the accuracy and reliability of outcomes are directly dependent on the quality of the data underpinning the models and decision-making processes. Predictive safety analytics, which leverages statistical models, artificial intelligence (AI), and machine learning (ML) to anticipate adverse events and improve safety outcomes, can only be as effective as the data it is built (Rangineni et al., 2023). Poor-quality data—characterized by inaccuracies, inconsistencies, incompleteness, or outdated information—can compromise model performance, misinform stakeholders, and ultimately lead to suboptimal or even dangerous decisions. As organizations across healthcare, manufacturing, aviation, transportation, and other safety-critical industries increasingly rely on predictive analytics to prevent harm and manage risk, the importance of data quality cannot be overstated. Ensuring high data quality is not merely a technical task—it is a strategic imperative that demands rigorous governance, continuous monitoring, and organizational commitment. Defining Data Quality in the Context of Predictive Safety Analytics: Data quality encompasses multiple dimensions, each of which plays a vital role in predictive safety analytics: Accuracy: Data should reflect real-world conditions or events as closely as possible. Inaccurate data may result in false alarms or failure to detect real threats. Completeness: All relevant fields and records should be present; missing data can introduce bias or render models ineffective. Consistency: Uniformity across datasets ensures that data from multiple sources align properly; inconsistent data undermines integration and model reliability. Timeliness: Data must be current, especially for real-time or near-real-time analytics used in safety-critical contexts. Validity: Data should conform to the required formats and rules; invalid data can cause technical errors or skew analyses. Integrity: Relationships between data elements must be maintained accurately, ensuring the cohesiveness of linked records. Relevance: Data used should be contextually appropriate and suitable for the intended analysis or prediction. These quality attributes collectively determine the trustworthiness and utility of the data being analyzed. Why Data Quality Matters in Predictive Safety Analytics: The implications of poor data quality in predictive safety analytics are profound: Model Inaccuracy: Algorithms trained on flawed data yield unreliable predictions, which can increase the likelihood of adverse events rather than preventing them. Increased False Positives/Negatives: Low-quality data can result in safety alerts being missed (false negatives) or unnecessarily triggered (false positives), leading to either oversight or alarm fatigue (Elouataoui W., et al, 2023). Wasted Resources: Efforts spent acting on inaccurate predictions drain time, personnel, and financial resources. Regulatory and Legal Risks: In regulated industries like healthcare or aviation, inaccurate predictions based on poor data can lead to compliance violations and legal consequences. Loss of Stakeholder Trust: Decision-makers may lose confidence in predictive systems if outcomes are frequently incorrect or inconsistent. Ethical and Equity Concerns: Biases in low-quality or unrepresentative data can perpetuate systemic inequalities, particularly in safety outcomes for marginalized populations. In high-stakes environments, the cost of poor data quality is measured not just in inefficiencies, but in lives, reputations, and public trust. 3. Sources and Types of Data Used in Safety Analytics. Predictive safety analytics draws from a diverse array of data sources, including Electronic Health Records (EHRs) and clinical and administrative data used in patient safety analytics. Sensor and IoT Data: Equipment usage, temperature, vibration, and movement data from industrial machinery or vehicles. Incident Reports and Safety Logs: Documentation of past safety events or near misses. Wearable Devices: Biometric data from health trackers used in occupational safety or remote patient monitoring. Environmental and Geospatial Data: Weather, terrain, or facility layout information relevant to risk exposure. Behavioral and Human Factors Data: Worker fatigue levels, compliance with protocols, or training effectiveness. Surveillance and Video Analytics: Real-time monitoring of environments for potential safety breaches. Each of these sources introduces unique challenges for ensuring data quality, from human error in data entry to sensor calibration errors or integration mismatches.

Identifying and mitigating these sources of poor data quality is essential to realizing the full value of predictive safety analytics. Strategies for Ensuring High Data Quality: To safeguard predictive analytics systems from the dangers of poor data, organizations should implement a multi-layered strategy: Establish Data Quality Standards: Define clear guidelines for what constitutes high-quality data across all relevant dimensions (Bauskar, 2024). These standards should be specific, measurable, and enforceable. Implement Data Validation and Cleansing: Use automated tools and algorithms to identify and correct anomalies, fill missing values, standardize formats, and flag suspicious entries. Leverage Metadata and Data Lineage: Document where data originates, how it is transformed, and how it is used. Metadata improves traceability and accountability, which are crucial for audits and investigations. d. Monitor Data Quality in Real-Time: Deploy real-time dashboards and monitoring tools that alert administrators when quality thresholds are breached, enabling timely intervention. e. Train Staff and Promote a Data-Driven Culture: Educate data entry personnel, analysts, and decision-makers on the importance of data accuracy and their role in maintaining it. f. Foster Interdepartmental Collaboration: Encourage collaboration between IT, safety, clinical, and operational teams to align on data governance and share best practices. g. Use AI to Assist in Data Cleaning: Machine learning can automate anomaly detection, pattern recognition, and resolution of ambiguous or conflicting records (Wu et al., 2021). In the realm of predictive safety analytics, data quality is foundational. No matter how advanced the algorithms or sophisticated the infrastructure, the insights derived from analytics can only be as good as the data they rely on. Poor-quality data introduces risk, while high-quality data enables foresight, precision, and prevention. Organizations must therefore invest in robust data quality frameworks, tools, and cultures to ensure that predictive safety analytics fulfills its promise: to prevent harm before it happens, protect lives and assets, and build a safer, smarter future.

SUPERVISED LEARNING FOR PREDICTING SAFETY OUTCOMES: SUPERVISED LEARNING

Supervised learning, a fundamental branch of machine learning (ML), has emerged as a transformative force in predicting safety outcomes across various industries, including healthcare, manufacturing, transportation, energy, and public safety. By training algorithms on labeled historical data—where inputs are associated with known outputs—supervised learning enables systems to identify complex patterns and relationships that are often invisible to human analysts (Kaushik et al., 2022). These insights empower organizations to anticipate and prevent accidents, adverse events, equipment failures, and health complications before they occur. In the era of big data and AI, supervised learning offers the precision, scalability, and adaptability necessary for developing intelligent safety systems that enhance decision-making and reduce harm. However, its successful implementation requires a deep understanding of model selection, data quality, domain-specific nuances, and continuous monitoring. Overview of Supervised Learning in the Context of Safety Supervised learning involves training a model using a dataset that contains both input features (predictors) and corresponding output labels (targets). The goal is to learn a function that maps inputs to outputs so that the model can accurately predict outcomes on new, unseen data. In the context of safety, the outputs often represent binary or multiclass labels such as These models can also predict continuous values (regression tasks), such as estimating the time before a safety-critical failure or the probability of an accident under certain conditions (Lin Y., et al, 2023). Common Algorithms Used for Safety Prediction Several supervised learning algorithms are commonly employed to predict safety outcomes, each with its strengths and trade-offs: Logistic Regression: Effective for binary classification problems such as predicting whether a patient will experience a complication based on clinical features. Decision Trees and Random Forests: Useful for interpreting complex interactions and handling missing or categorical data. Random forests reduce overfitting and are widely used in industrial safety systems.

Support Vector Machines (SVMs): Ideal for classification in high-dimensional spaces, useful for anomaly detection in security and quality control. Gradient Boosting Machines (GBM): Including XGBoost, LightGBM, and CatBoost, these models are known for high predictive accuracy and are extensively applied in safety-related competitions and real-world deployments. Neural Networks: Particularly deep learning architectures, which can model highly nonlinear relationships, are increasingly used in areas like image-based safety inspections or wearable sensor data analysis. K-Nearest Neighbors (KNN): Simple yet effective in situations with dense, high-quality labeled data. The choice of algorithm depends on the specific application, data characteristics, need for interpretability, and computational constraints. Applications of Supervised Learning in Predicting Safety Outcomes: Supervised learning has diverse applications across safety-critical domains: Healthcare and Patient Safety: Predicting hospital readmissions or adverse drug reactions. Forecasting patient deterioration (e.g., sepsis onset or cardiac arrest) using vital signs and EHR data. Identifying individuals at high risk of falls or pressure ulcers. Manufacturing and Industrial Safety: Forecasting equipment failure (predictive maintenance) using historical sensor data. Predicting worker fatigue or ergonomic risk using video and posture analysis (Westerveld et al., 2021). Assessing product quality to preempt defects that may cause harm. Transportation and Road Safety: Predicting the likelihood of vehicular accidents based on driver behavior, environmental conditions, and vehicle telemetry. Classifying dangerous driving patterns from dashcam or in-cabin monitoring systems. Energy Sector and Utilities: Detecting early signs of pipeline leakage, gas explosions, or transformer failure. Identifying areas at risk of wildfires based on historical fire data, weather, and vegetation density. Public Safety and Security: Classifying suspicious activities or individuals from surveillance footage or behavioral data. Predicting locations or periods with an elevated risk of crimes or natural disasters. Each of these applications leverages historical safety incidents to teach supervised models what "unsafe" looks like, enabling proactive intervention. Data Requirements and Challenges in Supervised Safety Model: High-quality, labeled data is the foundation of any successful supervised learning model. In safety analytics, obtaining such data presents several challenges: Imbalanced Classes: Safety incidents are often rare, leading to datasets where the "unsafe" class is underrepresented.

Models trained on outdated data may become less reliable. Sensor and Human Error: Inaccuracies in sensor readings or inconsistencies in manual data entry can compromise model learning. These challenges necessitate robust data governance strategies and thoughtful preprocessing pipelines to ensure models are learning from representative, accurate, and relevant data. Evaluating Supervised Models for Safety Predictions. Because the stakes in safety prediction are high, evaluation metrics must go beyond overall accuracy to include performance on minority classes and false positive/negative rates. Commonly used metrics include: Precision: The proportion of true positive predictions among all positive predictions—important to avoid false alarms. Recall (Sensitivity): The ability of the model to correctly identify all true safety incidents—critical for avoiding missed detections (Mahadevkar et al., 2022). Human-in-the-loop validation, where experts review model predictions before deployment, adds a layer of safety. Supervised learning has revolutionized the ability to predict and prevent safety incidents across numerous industries. By harnessing labeled historical data and advanced algorithms, organizations can move from reactive risk management to proactive harm prevention. Whether predicting patient deterioration in hospitals, preventing industrial accidents, or enhancing public security, supervised learning models empower stakeholders with actionable foresight. However, the successful application of these models requires high-quality data, rigorous validation, ethical vigilance, and ongoing model maintenance. As supervised learning continues to evolve and integrate with other AI paradigms, its role in advancing global safety outcomes will only grow more prominent—driving a future where intelligent systems and human expertise work together to safeguard lives, infrastructure, and environments.

DEEP LEARNING APPLICATIONS IN PREDICTIVE SAFETY ANALYTICS

Deep learning, a subset of artificial intelligence (AI) rooted in artificial neural networks, has emerged as a transformative technology in the realm of predictive safety analytics. By mimicking the human brain's capacity to process information, recognize patterns, and learn from complex data, deep learning offers unmatched capabilities in modeling non-linear relationships, processing high-dimensional data, and uncovering subtle indicators of safety risks (Mazhar et al., 2023). Its power lies in its ability to automatically extract features from raw data, removing the need for extensive manual preprocessing or expert-defined variables. As the volume, velocity, and variety of safety-related data continue to grow — including sensor outputs, video feeds, electronic health records, and unstructured incident reports — deep learning provides a scalable, adaptable framework for enhancing hazard detection, incident prediction, and decision support. This makes it indispensable for industries where safety is critical, such as healthcare, aviation, manufacturing, energy, and transportation. 1. Introduction to Deep Learning in Safety Contexts Deep learning models consist of multiple layers of interconnected neurons that transform input data into meaningful outputs through hierarchical feature learning (Jamal et al., 2021). These models can identify intricate patterns and learn abstract representations from vast amounts of unstructured or structured data. In predictive safety analytics, this translates into the ability to detect early warning signs of accidents, equipment failure, adverse health outcomes, or cybersecurity breaches — often before any human would recognize the signals. Unlike traditional machine learning models, which require manual feature engineering, deep learning architectures such as convolutional neural networks (CNNs), recurrent neural networks (RNNs), and transformers can process raw inputs like images, time series, text, and audio. This capability is especially important in safety analytics, where multimodal data is prevalent and the cost of oversight is high. Key Deep Learning Architectures for Safety Applications: a. Convolutional Neural Networks (CNNs): CNNs are particularly effective at processing spatial data such as images and video. In safety analytics, they are used to detect unsafe behaviors or environments through surveillance camera footage (e.g., detecting workers without helmets or slippery surfaces). Identify defects in manufacturing components via automated visual inspection. Analyze medical imaging (e.g., CT scans, MRIs) to detect anomalies indicative of future health risks. By automating and enhancing visual recognition, CNNs reduce human error and accelerate response times in safety-critical scenarios. Recurrent Neural Networks (RNNs) and Long Short-Term Memory Networks (LSTMs): RNNs and their advanced forms like LSTMs are tailored for sequential or time-dependent data: Monitor time-series sensor data in machinery to predict malfunctions or failures. Track patient vital signs to anticipate critical health events such as cardiac arrest. Analyze log data from vehicles, industrial systems, or IT networks to detect sequences leading to safety breaches. LSTMs can retain memory of past events, making them ideal for identifying trends and dependencies that evolve — a cornerstone of predictive safety. Autoencoders: Autoencoders are neural networks trained to reconstruct their input, often used for: Anomaly detection, where high reconstruction error indicates unusual or unsafe conditions (e.g., abnormal heart rhythms, machine vibration patterns). Dimensionality reduction to simplify complex safety datasets while preserving their structure (Badawy et al., 2023). Autoencoders are particularly effective in environments where labeled incidents are rare but early anomaly detection is critical. Transformers and Attention Mechanisms. Originally developed for natural language processing, transformers have revolutionized the analysis of text and sequence data: Extract patterns from unstructured incident reports, safety logs, and maintenance notes. Model complex relationships in health records, such as temporal associations between medications and adverse events. Enable explainable AI through attention layers that highlight which inputs contributed most to risk predictions. Transformers like BERT and GPT can be fine-tuned for domain-specific safety analysis tasks, significantly boosting predictive accuracy and interpretability.

Transportation and Automotive Safety: Deep learning enhances the safety of vehicles and infrastructure: Autonomous driving systems use CNNs to recognize road signs, lanes, and pedestrians. RNNs forecast driver fatigue or distraction based on facial cues and driving patterns. Deep reinforcement learning models optimize route planning while minimizing risk exposure (Hou L., et al, 2021). Advanced Driver Assistance Systems (ADAS) rely heavily on deep learning to provide collision warnings, emergency braking, and lane-keeping support. Energy and Utilities: In high-risk environments like oil rigs or power plants, deep learning is employed to Detect abnormal thermal patterns via infrared imaging (using CNNs). Monitor pipeline pressure and flow data to predict leaks (using LSTMs). Analyze textual maintenance records to anticipate system degradation (using NLP transformers). These insights contribute to safer, more resilient infrastructure. Cybersecurity and Digital Safety: Deep learning algorithms monitor complex network activity to detect threats: Autoencoders and GANs (Generative Adversarial Networks) uncover subtle anomalies in system behavior. NLP models interpret logs and emails to identify phishing or insider threats. LSTMs model user behavior sequences to identify deviations that may signal a breach. As digital systems are increasingly targeted, these capabilities ensure preemptive defenses. This holistic view of safety makes deep learning particularly useful in environments where context and coordination are critical (Kaur et al., 2025). Deep learning represents a paradigm shift in predictive safety analytics, offering powerful tools to anticipate, detect, and mitigate risks across diverse domains. With its capacity to process complex data, uncover hidden patterns, and enable real-time decision-making, deep learning supports a proactive approach to safety that goes far beyond traditional reactive models. From predicting patient deterioration and preventing industrial accidents to securing cyber infrastructure and enhancing autonomous systems, deep learning is unlocking new possibilities for saving lives, protecting assets, and fostering resilient systems. As technology continues to advance, integrating deep learning into safety practices will not only improve outcomes but also empower organizations to navigate an increasingly complex and interconnected world with confidence and intelligence.

BENEFITS OF AI AND MACHINE LEARNING IN PREDICTIVE SAFETY ANALYTICS: BENEFITS

The integration of Artificial Intelligence (AI) and Machine Learning (ML) into predictive safety analytics has revolutionized the way organizations identify, evaluate, and mitigate risks across diverse industries such as healthcare, manufacturing, transportation, energy, and cybersecurity. These technologies go far beyond traditional safety analysis by enabling real-time insights, dynamic risk assessments, and proactive interventions (Erinjogunola et al., 2025). By learning from vast and complex datasets, AI and ML can uncover hidden patterns, predict adverse events before they occur, and offer recommendations that enhance decision-making and resource allocation. Their benefits range from improving operational efficiency to saving lives, making them indispensable tools in any modern safety strategy. 1. Real-Time Risk Detection and Response: One of the foremost benefits of AI and ML in predictive safety analytics is the ability to monitor systems continuously and provide real-time alerts. Traditional safety systems often rely on historical data or periodic audits, which may miss emerging risks or time-sensitive incidents. AI-enabled systems can analyze streaming data from sensors, machines, and human interactions, flagging deviations or anomalies instantly. For example, in industrial settings, machine learning algorithms can monitor equipment performance in real time, predicting failures or hazardous conditions such as overheating, pressure imbalances, or mechanical wear. In healthcare, AI can alert clinicians to deteriorating patient conditions by analyzing vital signs, lab results, and historical health data, enabling timely interventions that prevent complications or fatalities. Early Prediction of Adverse Events: AI and ML excel at forecasting potential safety threats before they materialize. Predictive models can be trained on historical incident data, maintenance records, environmental factors, and human behavior to recognize patterns that precede accidents or system failures (Saxena, 2024). This predictive capability allows for preventive action rather than reactive damage control. In aviation, for instance, ML can predict aircraft maintenance needs based on sensor data, thus avoiding mid-flight failures. In construction, AI can assess site conditions, worker fatigue, and equipment usage to forecast accident probabilities. In hospitals, algorithms can anticipate patient falls, infections, or medication errors by integrating structured and unstructured clinical data. Data-Driven Decision Making: AI and ML support evidence-based decision-making by providing deep insights derived from vast, heterogeneous datasets. Safety decisions that once depended on intuition or anecdotal evidence can now be grounded in quantifiable predictions, simulations, and risk assessments. Leaders can use AI-generated dashboards to monitor key safety metrics, track progress, and identify high-risk areas. For example, logistics companies can decide when to replace fleet vehicles based on predictive maintenance analytics, while hospital administrators can allocate staff and equipment where AI indicates the greatest safety risks. This enhances the objectivity, transparency, and accountability of safety-related decisions (Rathod et al., 2024). Reduction in Human Error: While human expertise remains vital in safety-critical roles, human error remains a leading cause of accidents and failures across domains. AI and ML help reduce this risk by automating routine or complex tasks, offering decision support, and continuously monitoring for signs of error or fatigue.

Cost Savings and Operational Efficiency: Predictive safety analytics driven by AI and ML can lead to significant cost savings by preventing costly accidents, reducing downtime, and minimizing waste. Organizations can shift from reactive maintenance to predictive maintenance, extending the life of equipment and avoiding unplanned outages. For example, an energy company can use machine learning to optimize the operation of turbines or pipelines, reducing both safety risks and maintenance costs. Hospitals can reduce avoidable adverse events, thereby cutting legal liabilities, patient readmissions, and extended stays. These efficiencies not only reduce expenditures but also improve service delivery and stakeholder trust. Scalability and Automation: AI and ML systems are highly scalable and can handle increasing volumes of data without a proportional increase in human oversight. This makes them ideal for large organizations with complex operations, multiple sites, and massive data flows (Xu & Saleh, 2021). A global logistics provider can deploy machine learning models across its entire fleet to monitor driver behavior, vehicle condition, and road conditions — all in real-time. Similarly, a multinational manufacturing company can automate safety audits across plants using AI-powered image recognition and anomaly detection. The ability to scale safety analytics while maintaining quality and consistency is a crucial advantage. The benefits of AI and ML in predictive safety analytics are multifaceted and far-reaching. From real-time monitoring and early warning systems to personalized risk assessments and cost savings, these technologies are reshaping the way organizations approach safety. They not only enhance accuracy, responsiveness, and efficiency but also foster a culture of continuous improvement and proactive prevention. As AI and ML continue to evolve, their role in predictive safety analytics will expand, enabling smarter, safer, and more resilient systems across all sectors. Embracing these technologies is no longer optional — it is a strategic imperative for organizations committed to excellence in safety and risk management.

ADDRESSING BIAS AND FAIRNESS IN MACHINE LEARNING MODELS: BIAS AND FAIRNESS

As machine learning (ML) systems become integral to decision-making across industries—from healthcare and finance to criminal justice and education—issues of bias and fairness have emerged as critical concerns. These systems, though rooted in mathematical precision, do not exist in a vacuum; they are products of the data they are trained on, the objectives defined by developers, and the environments in which they operate (Ghani et al., 2023). When left unchecked, ML models can replicate or even exacerbate existing societal inequalities, leading to decisions that disadvantage certain individuals or groups based on race, gender, socioeconomic status, age, disability, or other sensitive attributes. Ensuring that ML models are fair and free of harmful bias is not just a technical necessity—it is an ethical imperative. Addressing bias in ML is a multidimensional task that requires understanding the sources of bias, developing mitigation strategies, ensuring accountability, and fostering transparency and inclusivity in the development lifecycle. Understanding Bias in Machine Learning: Bias in ML refers to systematic errors that lead a model to consistently produce prejudiced results against certain groups. Bias can enter at various stages of the ML pipeline, including: Data Collection Bias: This arises when the training dataset does not accurately represent the population the model is intended to serve. For example, a facial recognition system trained mostly on lighter-skinned individuals may perform poorly on darker-skinned faces, as seen in real-world deployments. Labeling Bias: Human annotators bring their assumptions and stereotypes into labeling processes. If a dataset used to train a hiring algorithm includes biased judgments about candidate qualifications, the model may learn and replicate these prejudices (Venkatasubbu & Krishnamoorthy, 2022). Measurement Bias: This occurs when proxies are used to represent outcomes or inputs. For instance, using arrest records as a proxy for criminal activity can perpetuate existing biases in policing. Algorithmic Bias: Even with balanced datasets, models may optimize for objectives that unintentionally disadvantage specific groups. For example, optimizing for overall accuracy may lead to higher error rates for minority subgroups if their representation is low. Deployment Bias: This emerges when an ML system is used in ways that deviate from its design assumptions, such as applying a model trained on urban hospital data to a rural clinic with different patient demographics. Types of Fairness in Machine Learning: Fairness in ML is a nuanced concept, and researchers have proposed various mathematical definitions, each with trade-offs. Common fairness criteria include: Demographic Parity: The model's outcomes are independent of protected attributes (e.g., race or gender). Equalized Odds: The model has equal true positive and false positive rates across groups. Predictive Parity: The model's predicted probabilities are equally accurate across groups. Individual Fairness: Similar individuals should receive similar predictions. In practice, it is often impossible to satisfy all fairness criteria simultaneously—a phenomenon known as the "impossibility theorem" in fairness. Therefore, addressing fairness involves carefully selecting definitions that align with the ethical and legal values of the context (Pagano et al., 2023).

Strategies for Mitigating Bias: Bias mitigation can occur at several stages of the ML pipeline: Pre-processing (Data Level): Re-sampling: Balance datasets by over-sampling underrepresented groups or under-sampling overrepresented ones. Data Augmentation: Create synthetic data to improve diversity in training sets. Reweighting: Assign weights to data samples to equalize representation. Fair Representations: Transform the input data into a representation that obfuscates sensitive attributes while preserving predictive power. In-processing (Algorithm Level): Fairness-Constrained Optimization: Incorporate fairness constraints into the model's objective function. Adversarial Debiasing: Train a model while simultaneously training an adversary that tries to predict sensitive attributes; the main model is penalized if the adversary succeeds. Regularization Techniques: Add regularization terms that penalize disparity in model outcomes across groups. Post-processing (Prediction Level): Threshold Adjustment: Apply different decision thresholds for different groups to equalize outcomes. In uncertain cases, defer decisions to human reviewers to prevent potential discrimination. Each approach has its benefits and limitations (Mehrabi et al., 2021). Pre-processing is often easiest to implement but may limit model performance. In-processing offers deep integration of fairness but requires access to model internals. Post-processing is flexible but may raise concerns about transparency and legitimacy. Bias and fairness in machine learning models are not peripheral issues—they strike at the heart of how we use AI to make decisions that affect people's lives. Ensuring fair outcomes requires more than statistical adjustments; it demands a deep understanding of context, stakeholder engagement, robust technical interventions, and ongoing monitoring. As ML systems become more powerful and pervasive, the responsibility to address bias grows. By proactively identifying sources of unfairness and implementing thoughtful mitigation strategies, organizations can build AI systems that are not only accurate but also equitable, transparent, and trustworthy. This work is essential to fostering public trust, meeting regulatory expectations, and ultimately delivering technology that serves all members of society fairly.

CONCLUSION

Artificial Intelligence (AI) and Machine Learning (ML) are revolutionizing the healthcare landscape by offering advanced, data-driven tools to improve patient safety. In an environment where medical errors remain a significant source of harm, the application of intelligent algorithms holds immense promise for transforming safety protocols and outcomes. Patient safety, a cornerstone of quality healthcare, can benefit from AI/ML through more accurate risk prediction, early detection of clinical deterioration, reduction of diagnostic and medication errors, real-time monitoring, and personalized interventions. By leveraging the vast amounts of health data generated in modern clinical environments, AI and ML enable healthcare systems to shift from reactive responses to proactive and preventive care strategies. Real-Time Risk Detection and Early Intervention: AI-powered systems excel at processing large datasets and detecting subtle patterns that may elude human observation. In patient safety, this capability is vital for real-time monitoring and early detection of critical conditions. For example, ML algorithms can continuously assess vital signs, laboratory results, and clinical notes to detect early signs of sepsis, hypoglycemia, or cardiac arrest, often hours before they become clinically evident. Such predictive alerts empower healthcare providers to intervene swiftly, improving patient outcomes and preventing escalation of complications. By reducing delays in diagnosis and treatment, these tools significantly decrease the likelihood of adverse events and enhance the overall safety of care. Minimizing Medication Errors: Medication errors are a major contributor to preventable harm in healthcare. AI and ML can help by supporting medication reconciliation, flagging potential drug interactions, identifying incorrect dosages, and ensuring compliance with prescribing guidelines. Natural language processing (NLP) can extract data from unstructured physician notes to cross-check prescriptions against patient-specific factors such as allergies, renal function, or prior adverse reactions. Decision support tools integrated with AI can suggest alternative medications or dosing based on real-time analytics and patient history, thereby reducing the incidence of medication-related errors and adverse drug events. Improved Diagnostic Accuracy: Misdiagnosis and delayed diagnosis are among the leading threats to patient safety. AI tools, particularly those using deep learning, have shown remarkable potential in enhancing diagnostic accuracy across a range of conditions, including cancer, cardiovascular diseases, and neurological disorders. In radiology, AI can analyze imaging data with high precision, highlighting anomalies that may be missed by the human eye. In pathology, ML models are being developed to interpret biopsy slides with accuracy comparable to expert pathologists. By augmenting human decision-making, these systems reduce diagnostic errors, shorten time to diagnosis, and ensure timely initiation of treatment. Predictive Analytics for Patient Deterioration: Patient deterioration in hospitals often goes unrecognized until it becomes critical. ML algorithms can monitor patient data continuously, identifying those at risk of rapid decline. These predictive models analyze diverse inputs, including heart rate, oxygen saturation, lab values, and nursing assessments, to forecast risks like respiratory failure or septic shock. Such early warnings allow care teams to deploy rapid response interventions, thereby preventing ICU transfers, emergency surgeries, or even mortality. These tools also support triage decisions, resource allocation, and step-down care planning, leading to safer, more efficient clinical pathways. A Safer Future Through Intelligent Innovation: AI and ML are not just tools—they are catalysts for a cultural shift in how healthcare approaches patient safety. By enabling earlier interventions, reducing errors, personalizing care, and supporting clinicians, these technologies are redefining safety from being reactive to being intelligent and preventive. As healthcare systems worldwide continue to digitize and innovate, the thoughtful and responsible integration of AI/ML into clinical practice will play a pivotal role in achieving the long-standing goal of harm-free, high-quality care. Ultimately, harnessing AI and machine learning for enhanced patient safety promises not just improved outcomes, but a healthcare environment that is smarter, safer, and more compassionate.

REFERENCES

Abegaz, T. M., Baljoon, A., Kilanko, O., Sherbeny, F., & Ali, A. A. (2023). Machine learning algorithms to predict major adverse cardiovascular events in patients with diabetes. *Computers in Biology and Medicine*, *155*, 106556. Advance online publication. DOI: 10.1016/j.compbiomed.2023.107289 PMID: 37557056

Anastasi, S., Madonna, M., & Monica, L. (2021). Implications of embedded artificial intelligence-machine learning on safety of machinery. *Procedia Computer Science*, *192*, 4532–4539. DOI: 10.1016/j.procs.2021.01.171

Arjmandnia, F., & Alimohammadi, E. (2024). The value of machine learning technology and artificial intelligence to enhance patient safety in spine surgery: A review. *Patient Safety in Surgery*, *18*(1), 1–12. DOI: 10.1186/s13037-024-00393-0 PMID: 38528562

Badawy, M., Ramadan, N., & Hefny, H. A. (2023). Healthcare predictive analytics using machine learning and deep learning techniques: A survey. *Journal of Electrical Systems and Information Technology*, *10*(1), 40. Advance online publication. DOI: 10.1186/s43067-023-00108-y

Bauskar, S. (2024). Predictive analytics or data quality assessment through artificial intelligence techniques. SSRN Electronic Journal. https://doi.org/DOI: 10.2139/ssrn.4980802

Chib, S., Karthigeyan, M., Pawar, F. S., Selvakumar, P., Manjunath, T. C., Neena, P. C., & Sharma, M. (2025). Analyze the factors that contributed to their success and lessons learned. In Katper, N. (Ed.), *Crowdfunding and alternative financing models for women entrepreneurs* (pp. 35–64). IGI Global Scientific Publishing., DOI: 10.4018/979-8-3693-9193-8.ch002

D'Ascenzo, F., De Filippo, O., Gallone, G., & Mittone, G. (2021). Machine learning-based prediction of adverse events following an acute coronary syndrome (PRAISE): A modelling study of pooled datasets. *The Lancet. Digital Health*, *3*(8), e537–e546. DOI: 10.1016/S2589-7500(21)00108-2 PMID: 33453782

Deimazar, G., & Sheikhtaheri, A. (2023). Machine learning models to detect and predict patient safety events using electronic health records: A systematic review. *International Journal of Medical Informatics*, *178*, 105276. Advance online publication. DOI: 10.1016/j.ijmedinf.2023.105246 PMID: 37837710

Elouataoui, W., El Mendili, S., & Gahi, Y. (2023). An automated big data quality anomaly correction framework using predictive analysis. *Data*, *8*(3), 112. Advance online publication. DOI: 10.3390/data8120182

Erinjogunola, F. L., Sikhakhane-Nwokediegwu, Z., Ajirotutu, R. O., & Olayiwola, R. K. (2025). Enhancing bridge safety through AI-driven predictive analytics. *International Journal of Multidisciplinary Research and Growth Evaluation*, *4*(2), 10–26. DOI: 10.54660/IJSSER.2025.4.2.10-26

Fallah Madvari, R. (2022). Artificial intelligence (AI), machine learning (ML) and deep learning (DL) on health, safety and environment (HSE). *Archives of Occupational Health*, *6*(1), 45–52. https://aoh.ssu.ac.ir/article_276.html

Fazle, A. B., Prodhan, R. K., & Islam, M. M. (2023). AI-powered predictive failure analysis in pressure vessels using real-time sensor fusion: Enhancing industrial safety and infrastructure reliability. *American Journal of Research and Innovation*, *5*(3), 67–79. https://researchinnovationjournal.com/article/ai-predictive-failure-analysis. DOI: 10.63125/wk278c34

Ghani, R., Rodolfa, K. T., Saleiro, P., & Jesus, S. (2023). Addressing bias and fairness in machine learning: A practical guide and hands-on tutorial. In *Proceedings of the 29th ACM Conference on Fairness, Accountability, and Transparency* (pp. 123–134). DOI: 10.1145/3580305.3599180

Hou, L., Chen, H., Zhang, G., & Wang, X. (2021). Deep learning-based applications for safety management in the AEC industry: A review. *Applied Sciences (Basel, Switzerland)*, *11*(2), 821. Advance online publication. DOI: 10.3390/app11020821

Jagatheesaperumal, S. K., Rajkumar, S., Suresh, J. V., Gumaei, A. H., Alhakbani, N., Uddin, M. Z., & Hassan, M. M. (2023). An IoT-based framework for personalized health assessment and recommendations using machine learning. *Mathematics*, *11*(12), 2758. Advance online publication. DOI: 10.3390/math11122758

Jamal, A., Zahid, M., Rahman, M. T., Al-Ahmadi, H. M., Almoshaogeh, M., Farooq, D., & Ahmad, M. (2021). Injury severity prediction of traffic crashes with ensemble machine learning techniques: A comparative study. *International Journal of Injury Control and Safety Promotion*, *28*(4), 408–427. DOI: 10.1080/17457300.2021.1928233 PMID: 34060410

Kalla, D. (2022). AI-powered driver behavior analysis and accident prevention systems for advanced driver assistance. *International Journal of Scientific Research and Modern Technology*, *1*(12), 1–9. https://www.researchgate.net/publication/387174081_AI-Powered_Driver_Behavior_Analysis_and_Accident_Prevention_Systems_for_Advanced_Driver

Kaur, M., Selvakumar, P., Chithambar, G. V., Nalini, M., Guru, B. A., & Manjunath, T. C. (2025). Social media and radical left populism. In Vieira, A., Joaquim, A., & Duarte, A. (Eds.), *Digital populism and the use of neo-propaganda and fake news* (pp. 87–108). IGI Global Scientific Publishing., DOI: 10.4018/979-8-3693-9999-6.ch005

Kaushik, D., Garg, M., & Gupta, A. (2022). Application of machine learning and deep learning in cybersecurity: An innovative approach. In Modern network security (pp. 45–62). https://doi.org/DOI: 10.1201/9781003294567-3

Lin, Y., Ma, J., Wang, Q., & Sun, D. W. (2023). Applications of machine learning techniques for enhancing nondestructive food quality and safety detection. *Critical Reviews in Food Science and Nutrition*, *63*(3), 456–478. DOI: 10.1080/10408398.2022.2131725 PMID: 36222697

Mahadevkar, S. V., Khemani, B., Patil, S., Kotecha, K., Vora, D. R., Abraham, A., & Gabralla, L. A. (2022). A review on machine learning styles in computer vision—Techniques and future directions. *IEEE Access : Practical Innovations, Open Solutions*, *10*, 123456–123478. DOI: 10.1109/ACCESS.2022.3209825

Malik, N., Nirupama, E., Selvakumar, P., Sankpal, U., Nathsharma, S. K., Sharma, M., & Manjunath, T. C. (2025). 2D materials for energy storage and conversion. In Mhadhbi, M. (Ed.), *Innovations in next-generation energy storage solutions* (pp. 1–24). IGI Global Scientific Publishing., DOI: 10.4018/979-8-3693-9316-1.ch001

Mazhar, T., Talpur, D. B., Shloul, T. A., Ghadi, Y. Y., Haq, I., Ullah, I., Ouahada, K., & Hamam, H. (2023). Analysis of IoT security challenges and its solutions using artificial intelligence. *Brain Sciences*, *13*(4), 683. Advance online publication. DOI: 10.3390/brainsci13040683 PMID: 37190648

Mehrabi, N., Morstatter, F., Saxena, N., Lerman, K., & Galstyan, A. (2021). A survey on bias and fairness in machine learning. *ACM Computing Surveys*, *54*(6), 35. Advance online publication. DOI: 10.1145/3457607

Mehta, V., Selvakumar, P., Anute, N. B., Gandhimathi, S., Seelam, P. K., & Manjunath, T. C. (2025). Second language acquisition support through media literacy and digital media platforms. In Tussey, J., & Haas, L. (Eds.), *Supporting linguistic differences through literacy education* (pp. 361–388). IGI Global Scientific Publishing., DOI: 10.4018/979-8-3373-3496-7.ch011

Ozobu, C. O., Adikwu, F. E., & Odujobi, N. O. (2025). Advancing occupational safety with AI-powered monitoring systems: A conceptual framework for hazard detection and exposure control. World Journal of Occupational Safety and Health, 9(2), 88–101. https://iiardjournals.org/get-paper/2025/09/02/advancing-occupational-safety

Pagano, T. P., Loureiro, R. B., Lisboa, F. V. N., Peixoto, R. M., Guimarães, G. A. S., Cruz, G. O. R., Araujo, M. M., Santos, L. L., Cruz, M. A. S., Oliveira, E. L. S., Winkler, I., & Nascimento, E. G. S. (2023). Bias and unfairness in machine learning models: A systematic review on datasets, tools, fairness metrics, and identification and mitigation methods. *Big Data and Cognitive Computing*, *7*(1), 12. Advance online publication. DOI: 10.3390/bdcc7010015

Pieszko, K., Shanbhag, A. D., Singh, A., Hauser, M. T., Miller, R. J. H., Liang, J. X., Motwani, M., Kwieciński, J., Sharir, T., Einstein, A. J., Fish, M. B., Ruddy, T. D., Kaufmann, P. A., Sinusas, A. J., Miller, E. J., Bateman, T. M., Dorbala, S., Di Carli, M., Berman, D. S., & Slomka, P. J. (2023). Time and event-specific deep learning for personalized risk assessment after cardiac perfusion imaging. npj. *Digital Medicine*, *6*(1), 78. Advance online publication. DOI: 10.1038/s41746-023-00806-x PMID: 37127660

Rana, M. S., & Shuford, J. (2024). AI in healthcare: Transforming patient care through predictive analytics and decision support systems. *Journal of Artificial Intelligence General Science*, *2*(1), 33–47. https://newjaigs.org/article/ai-in-healthcare-transforming-patient-care. DOI: 10.60087/jaigs.v1i1.30

Rangineni, S., Bhanushali, A., Suryadevara, M., Venkata, S., & Peddireddy, K. (2023). A review on enhancing data quality for optimal data analytics performance. *International Journal on Computer Science and Engineering*, *11*(10), 51–58. DOI: 10.26438/ijcse/v11i10.5158

Rathod, S. B., Mahajan, R. A., & Khadkikar, P. A. (2024). Improving workplace safety with AI-powered predictive analytics: Enhancing workplace security. In Handbook of research on AI for women's safety (pp. 112–129). https://doi.org/DOI: 10.4018/978-1-6684-5678-9.ch007

Rehan, H. (2024). AI-driven cloud security: The future of safeguarding sensitive data in the digital age. *Journal of Artificial Intelligence General Science*, *1*(1), 66–78. DOI: 10.60087/jaigs.v1i1.p66

Saxena, V. (2024). Predictive analytics in occupational health and safety. arXiv. https://doi.org//arXiv.2412.16038DOI: 10.48550

Shashank, G., Sai Thrinath, B. V., Seenivasan, S., Sankpal, U., & Manjunath, T. C. (2025). Supercapacitors. In Mhadhbi, M. (Ed.), *Innovations in next-generation energy storage solutions* (pp. 573–598). IGI Global Scientific Publishing., DOI: 10.4018/979-8-3693-9316-1.ch018

Taylor, R., Fakhimi, M., Ioannou, A., & Spanaki, K. (2024). Personalized learning in education: A machine learning and simulation approach. *Benchmarking*, *31*(8), 2450–2472. DOI: 10.1108/BIJ-06-2023-0380

Venkatasubbu, S., & Krishnamoorthy, G. (2022). Ethical considerations in AI addressing bias and fairness in machine learning models. *Journal of Knowledge Learning and Smart Technology*, *4*(1), 55–66. https://jklst.org/article/view/2022-ethics-ai-bias

Westerveld, J. J. L., van den Homberg, M. J. C., & van der Veen, A. (2021). Forecasting transitions in the state of food security with machine learning using transferable features. *The Science of the Total Environment*, *775*, 145071. Advance online publication. DOI: 10.1016/j.scitotenv.2021.147366 PMID: 33971600

Wong, L. W., Tan, G. W. H., Ooi, K. B., Lin, B., & Dwivedi, Y. K. (2024). Artificial intelligence-driven risk management for enhancing supply chain agility: A deep-learning-based dual-stage PLS-SEM-ANN analysis. *International Journal of Production Research*, *62*(15), 5535–5555. DOI: 10.1080/00207543.2022.2063089

Wu, X., Zheng, W., Xia, X., & Lo, D. (2021). Data quality matters: A case study on data label correctness for security bug report prediction. *IEEE Transactions on Software Engineering*, *47*(10), 2156–2172. DOI: 10.1109/TSE.2020.2991332

Xu, Z., & Saleh, J. H. (2021). Machine learning for reliability engineering and safety applications: Review of current status and future opportunities. *Reliability Engineering & System Safety*, *210*, 107558. Advance online publication. DOI: 10.1016/j.ress.2021.107530

Zafer, N. (2025). Construction risk management and safety behavior: AI-powered strategies for workplace hazard reduction. ResearchGate. https://www.researchgate.net/publication/AI_safety_behavior_construction_2025

Chapter 11

Transforming Employee Safety Training Through Artificial Intelligence Innovations

Priya Harit
http://orcid.org/0009-0009-6223-3129
Lovely Professional University, India

Amrik Singh
http://orcid.org/0000-0003-3598-8787
Lovely Professional University, India

Kritika Agrawal
http://orcid.org/0009-0007-6600-8952
Institute of Hotel Management, Chennai, India

Gaurav Bathla
http://orcid.org/0000-0002-6992-811X
School of Hotel Management, Airlines, and Tourism, CT University, Ludhiana, India

Charu Dutt Sharma
Institute of Hotel Management, Ahmedabad, India

Varinder Singh Rana
City University, Ajman, UAE

ABSTRACT

Organizational success relies on workplace safety, yet complex modern environments often make traditional training insufficient. AI offers a transformative solution by combining real-time monitoring, immersive simulations, adaptive learning, and predictive analytics to enhance safety training. It reduces human error, strengthens resilience, improves risk prediction, and ensures regulatory compliance. Applications such as automated inspections, behavioural monitoring, VR simulations, and predictive maintenance lower accidents while fostering a proactive safety culture. Gamified learning, wearables, and IoT devices further personalize training and boost engagement. Despite challenges like bias, privacy, and high costs, proper strategies

DOI: 10.4018/979-8-3373-6370-7.ch011

and human oversight enable ethical, scalable adoption. With AI, organizations gain a cost-effective, future-ready system that protects workers, improves compliance, and drives sustainable excellence.

INTRODUCTION

Workplace safety is a basic concern for businesses across the globe, not only based on moral obligation but also as a key determinant of operational success. Accidents and injuries disrupt workflows, decrease productivity, and harm reputations, so taking proactive measures to protect employees and ensure smooth operations is essential (Bernal et al., 2024). Historically, the traditional approach to training employees has always been one of the effective means to minimize hazards in a workplace environment and develop a culture of safety (Ljubic et al., 2016). As such, the organization would reduce the accident rates, enhance its adherence to the regulatory aspects, and create an enhanced overall environment of the workplace (Benson et al., 2024). However, the complexity of modern workplaces and the integration of new technologies require more advanced and sophisticated safety solutions (Haleem et al., 2025).

Artificial intelligence (AI), a ground-breaking tool that is altering how a company tackles safety training or education, is the breakthrough. (Rashid & Karim Kausik, 2024). AI-driven training in safety goes beyond tradition, using data-driven information and advanced technologies to create personal efficiency and engagement in learning opportunities (Park & Kang, 2024). AI-powered immersive simulations, adaptive learning systems, and predictive analytics may all help precisely detect danger, lower the possibility of human error, and guarantee adherence to changing safety rules. (Joni & Graepel, 2024). Understanding workplace risks, however, remains at the heart of effective safety training, as every industry and environment presents unique challenges (Bęś & Strzałkowski, 2024). From physical hazards such as machinery operations in an industrial environment to ergonomic concerns in an office or chemical risks in a lab, the training must be specific (Occupational Safety and Health Administration, n.d.).

AI technologies have been shown to shine at assessing these risks through analysis of historical data, real-time conditions, and generating predictive insights. This helps organizations be proactive about the risks and design training programs to prepare employees for specific scenarios, hence bettering their readiness and confidence (Chen & Dai, 2024). AI is also very helpful in addressing human mistake, which is frequently the primary cause of accidents at work. Despite their advantages, traditional training techniques can fall short in connecting with staff members or providing them with the necessary skills for high-stakes scenarios. (Shah & Mishra, 2024). AI-powered solutions, including virtual reality (VR) simulations, offer engaging and interactive experiences that let staff members safely rehearse how to react to certain risks. These tools, therefore, improve hazard detection and decision making but ensure that people feel prepared to address a wide range of factors in a way that, again, dramatically reduces mistakes when dealing with real scenarios (Pribadi et al., 2024). Ensuring that safety procedures comply with evolving standards is the prime importance for workplace security, and AI provides for easy advancement (Diro et al., 2025). The training program would need to be updated consistently under this regulatory framework, which includes OSHA in the US and EU-OSHA in Europe. This could be difficult to handle manually. (Schulte et al., 2022). AI systems can automatically include new regulations in training modules, which would ensure the organization maintains compliance while also instilling a sense of accountability and ethical practice. This proactive approach minimizes the risks of fines, legal disputes, and operational disruptions while demonstrating a commitment to employee well-being (Ayodeji, 2024).

AI-powered safety training is a financially sound investment. Accidents at work frequently result in high costs, ranging from direct expenses like medical bills and equipment repairs to indirect losses like lower morale and productivity. (Al-Naser & Al-Tabtabai, 2024). By predicting potential risks and addressing them through targeted training, AI helps organizations avoid accidents, reduce costs, and protect their bottom line. Moreover, personalized and adaptive training programs improve employee engagement, such that workers retain critical safety knowledge and apply it in the right way (Tian et al., 2025). AI integration into safety training also improves a company's standing as a business partner and employer. Organizations that emphasize innovative safety practices show a high commitment to employee well-being, which attracts top talent and strengthens relationships with stakeholders (Wong, 2025). Additionally, a strong culture, powered by advanced AI tools, builds customer trust and positions the organization as an industry leader in ethical and operational excellence (Samman, 2024).

This chapter probes into the transformative influence AI is having on the design of workplace safety training. It discusses how applications like machine learning-based risk assessment, immersive experiences using VR, and personalized education through adaptive platforms affect this area of workplace learning. Using case studies and rich analytics, we explain how all these innovations can be tapped upon by organizations to achieve better workplace safety and a more efficient and resilient organization. As businesses continue to experience an increasingly dynamic and technological landscape, AI-driven safety training becomes a necessity, empowering employees and setting them up for long-term success.

Understanding AI in safety training

Training is the way by which Artificial Intelligence in workplace safety is changing traditional systems of safety management and opening doors to unprecedented monitoring, evaluation, and enhancement capabilities of safety measures (Dang & Thu Nga, 2024). Conventional approaches to safety training, such as classroom lectures, manuals, and periodic drills, are often unable to keep pace with the complexities of modern work environments. In industries where workers face high risks—such as construction, manufacturing, mining, and healthcare—these limitations can lead to preventable accidents and operational inefficiencies. Organizations in a variety of industries can create safer and more effective work environments by utilizing. AI-driven technologies, such as computer vision, predictive analytics, and machine learning algorithms. By integrating AI-driven technologies like computer vision, predictive analytics, natural language processing, and machine learning algorithms, organizations can establish safer, smarter, and more adaptive safety systems. These innovations enable businesses not only to respond to incidents but also to predict and prevent them, thereby cultivating a proactive safety culture that values prevention over reaction. AI can help businesses avoid mishaps before they happen, guarantee that safety rules are followed, and establish a robust safety culture. (Wilson, 2024).

AI's Place in Safety Management and Its Advantages AI is now a useful, revolutionary instrument that solves problems in the real world, having developed from a vision conceived by pioneers like Alan Turing and John McCarthy. Its impact cuts through industries, automating operations, analyzing large datasets, and optimizing decision-making processes. Today, its applications span across industries, automating repetitive tasks, analyzing massive datasets, and optimizing decision-making processes. Within the realm of safety management, AI provides organizations with the ability to anticipate hazards, monitor compliance continuously, and create more resilient operational frameworks. This shift marks a transition from reactive safety measures—those taken after an incident has occurred—to proactive strategies that minimize risks before they escalate. In safety management, AI stands to offer a lot of benefits to organizations regarding adopting a proactive rather than reactive approach (Caburao, 2024).

One of the most significant advantages of AI in safety management is its ability to reduce workplace accidents. Traditional systems often detect risks only after incidents have occurred, but AI-enabled predictive analytics can identify potential hazards in advance by analyzing patterns in environmental data, worker behavior, and equipment performance. For example, sensors combined with machine learning models can detect abnormal machine vibrations that might indicate mechanical failure, allowing for timely intervention before a serious accident takes place. By preventing injuries and fatalities, AI not only saves lives but also reduces the physical and emotional toll on employees. More importantly, AI increases compliance to safety standards by monitoring the environment constantly and identifying risk factors in real-time to maintain the safety standards. With proactive compliance, risks are minimized about penalties, lawsuits, and reputational damages (Konda, 2020).

Another notable advantage lies in cost-effectiveness. By preventing accidents and minimizing risks, AI reduces costs associated with insurance premiums, medical compensations, and legal settlements. Additionally, the insights generated by AI help organizations make smarter investments in safety infrastructure and training programs. Instead of allocating resources reactively after incidents, companies can strategically enhance areas most vulnerable to risk. Over time, this results in both financial savings and operational efficiency.

Beyond the tangible economic benefits, AI contributes to shaping a stronger organizational safety culture. By embedding trust, transparency, and accountability into everyday practices, AI-driven safety systems enhance employee confidence. Workers feel more secure when they know their environment is continuously monitored and that potential hazards are addressed proactively. This trust not only improves morale but also strengthens organizational reputation, making companies more attractive to employees, clients, and stakeholders alike. Also, the insights from AI drive better safety investments and changes in operations, which further reduces costs. From lowering premiums to avoiding legal and compensation costs, AI makes business sense while keeping a safer workplace. Beyond monetary advantages, AI enhances the workplace safety culture by developing trust and accountability, making an organization more attractive to both its employees and stakeholders.

AI Applications in Safety Training and Management

AI technologies are revolutionizing EHS (ENVIRONMENT, HEALTH, AND SAFETY) systems through improved speed, accuracy, and adaptability. Environment, Health, and Safety (EHS) systems are undergoing a revolution thanks to artificial intelligence (AI) technologies, which increase speed, accuracy, and flexibility. The complexity of contemporary risks is frequently not adequately addressed by traditional EHS systems, which are based on manual procedures, compliance checklists, and reactive responses, in today's rapidly evolving organizational and industrial environments. Globalized supply chains, interconnected technologies, and changing regulations are all influencing workplaces more and more, necessitating more proactive and clever approaches to protecting workers, communities, and the environment. AI provides a revolutionary way to improve EHS management and integrate safety, sustainability, and health concerns into the very fabric of organizational operations because of its capacity to handle enormous volumes of data, identify trends, and adjust to shifting circumstances. It is impossible to exaggerate the importance of EHS systems. They are intended to safeguard employees against work-related dangers, guarantee that businesses abide by health and environmental laws, and lower risks that could result in monetary, reputational, or social damage. However, human limitations have historically limited the effectiveness of these systems: decision-making may lag behind real-time conditions, reporting is prone to errors, and inspections are frequently episodic rather than continuous. Speed, accuracy, and adaptability are critical in increasingly complex industries, such as manufacturing plants with sophisticated robotics, building sites with ever-changing safety hazards, and chemical facilities handling hazardous products. By facilitating automated risk detection, continuous monitoring, and predictive insights that enable businesses to move from reactive compliance to proactive prevention, artificial intelligence (AI) immediately fills these gaps.

The ability of AI to efficiently use data is one of its distinguishing features in EHS. From health records, incident reports, and compliance audits to sensor readings on equipment and ambient conditions, organizations produce vast amounts of safety-related data. Because it is so hard to analyse this data in a meaningful way, a lot of information has historically been underutilized. Artificial intelligence (AI)-powered solutions, such as natural language processing systems and machine learning algorithms, can quickly sort through this data and spot trends and abnormalities that human managers might miss. For instance, computer vision systems may automatically identify risky worker behaviours, such not wearing protective gear, in real-time, and predictive analytics can identify early warning indications of equipment breakdown. These features transform EHS systems from static frameworks into responsive, dynamic networks that can continuously learn and get better. Another important way that AI benefits EHS systems is through adaptability. AI may customise its analysis and recommendations to the unique circumstances of each workplace, in contrast to strict compliance models that handle every case in accordance with consistent norms. This flexibility is especially crucial in settings where hazards are not constant but rather change according to factors like location, climate, employee experience, and even cultural customs. Wearables with AI capabilities, for example, can modify safety alarms to take into account each worker's unique physiological state, guaranteeing that those who are most at risk of heat stress or exhaustion receive prompt assistance. Similar to this, AI systems can improve training simulations using immersive technologies like augmented reality (AR) and virtual reality (VR), providing workers with individualised, scenario-based learning experiences that better equip them for issues they may face in the real world. AI guarantees that EHS systems are dynamic frameworks that change to meet organisational needs rather than only being rule-based processes by increasing adaptability. Additionally, incorporating AI into EHS speeds up regulatory compliance. Businesses in a variety of sectors must comply with an expanding network of safety, occupational health, and environmental laws that frequently straddle several countries. Manually ensuring compliance can be error-prone and resource-intensive. By automating compliance monitoring and reporting and identifying possible infractions before they become serious enough to result in fines or legal action, artificial intelligence streamlines this process. AI-powered environmental monitoring systems, for instance, can identify high emissions in real time and take quick corrective action instead of waiting for regulatory inspections. The capacity to exhibit proactive compliance lowers financial and legal risks while enhancing confidence among the public, employees, and regulators.

AI-driven EHS solutions help the organisation achieve sustainability, which goes beyond compliance. In light of resource depletion, climate change, and international demands for corporate accountability, environmental sustainability has emerged as a critical issue. AI can minimise environmental impact while optimising energy use, monitoring waste management procedures, and ensuring safer handling of dangerous items. AI in health and safety enables businesses to create more robust care cultures where employees feel respected and safe. The overall potential of AI in EHS management is demonstrated by this congruence of technology innovation, human well-being, and environmental responsibility. Adoption of AI in EHS has strategic as well as operational benefits. Businesses that integrate AI into their environmental and safety procedures establish a reputation for being progressive and socially conscious, which helps them stand out in cutthroat industries. They are better able to control risks, lower the expenses related to mishaps and non-compliance, and draw in talent that prioritises sustainability and workplace safety. Additionally, companies with strong environmental, social, and governance (ESG) frameworks are becoming more and more valued by stakeholders and investors. Organisations can improve their long-term resilience and align with global sustainability benchmarks by integrating AI into EHS and enhancing their ESG performance. Naturally, there are difficulties in implementing AI in EHS. Data privacy, algorithmic bias, implementation costs, and the loss of human control are still legitimate concerns that need careful consideration. For instance, when algorithms make crucial safety judgements, an over-reliance on AI at the expense of human judgement may result in moral quandaries. Similar to this, smaller businesses can encounter technological or budgetary obstacles to the widespread adoption of AI-driven EHS solutions. However, by combining technology innovation with robust human control, responsible adoption techniques might help to alleviate these difficulties. The success of AI in EHS depends on establishing unambiguous ethical standards, guaranteeing openness in AI decision-making, and offering continual training to managers and employees.

Businesses across sectors are leveraging AI in the following ways:

1. Advanced Analytics for Risk Prediction and Prevention

AI-powered analytics are essential for spotting possible hazards before they become serious mishaps. Large volumes of data from sensors, cameras, and incident logs are processed by AI, which reveals patterns and trends that human analysts might miss. This empowers organizations to take preventive measures. AI systems scan equipment for performance anomalies, enabling the prediction of breakdowns before they take place. This reduces downtime and prevents workplace accidents due to equipment failure. AI tools monitor job site conditions, analyzing structural instability or environmental hazards that may pose risks. For example, predictive models can help determine the likelihood of a slip, trip, or fall and suggest preventive action, such as additional safety barriers or worker reassignments. AI identifies safety concerns in the hospital environment, such as trends in medication errors or adverse events. By emphasizing these risks, healthcare facilities can implement targeted training and procedural changes to ensure patient and staff safety (Balan et al., 2025).

2. Immersive Technologies to Improve Safety Training

Unlike the traditional methods of training involving static presentations or lectures, employee's preparation for real-world situations is less engaging and thus less effective. AI-enabled training solutions use simulations, Virtual Reality (VR), and adaptive learning platforms that offer interactive, hands-on experiences (Budnarowski et al., 2025). In realistic and risk-free virtual environments that mirror real hazardous scenarios, workers can train. Oil and gas workers can simulate an emergency evacuation or equipment handling on virtual oil rigs. Thus, they can increase their confidence and preparedness without exposing themselves to actual danger. An AI system tailors the training content depending on the individual employees' role, learning speed, and their respective knowledge gaps. This is how the retail and warehousing staff will be getting specialized modules in equipment handling, fire safety, and response to emergencies. AI can bring gamification elements in quizzes or even scenario-based challenges to increase the interest of learning for employees and, as a result, enhance retention. This fosters active engagement by employees toward safety education.

3. Automated Inspections and Maintenance

AI-driven systems facilitate routine inspections and audits with speed, accuracy, and safety. Drones, IoT devices, and computer vision technologies greatly reduce the need for human inspections of high-risk areas (Soori et al., 2024). Drones with AI-powered cameras inspect power lines, wind turbines, and underwater pipelines. They detect corrosion, cracks, or other defects, allowing maintenance to be done on time without exposing workers to dangerous conditions. AI-enabled sensors monitor machinery in real-time, detecting irregularities such as overheating, misalignment, or wear and tear. Automated alerts prompt immediate corrective actions, reducing the likelihood of equipment-related accidents. AI technology checks scaffolding, crane, and other equipment on site to ascertain whether the workplace is safe.

4. Hazard Detection in Dynamic Environments

AI works best in dynamic environments. Monitoring subtle changes in the environment that may present safety risks. Such systems will alert at real-time, and some may even cause an automatic reaction to prevent accidents from happening (Tamascelli et al., 2024). Autonomous Mining Vehicles powered with AI navigate dangerous underground working conditions, minimizing human vulnerability to cave-ins, poison gas leaks, or overheating. Chemical Product. AI sensors perceive toxic gas emissions or shifting air quality in mining area and alert to instruct the workers to evacuate from these areas or take corrective steps. These systems also closely monitor flammable substances thus offering less chance of fires and possible explosions. AI systems can detect congestion in Logistics and Warehousing areas so that the movement of people and vehicles is safe. For instance, computer vision can be used to identify misplaced items or spilled material that could cause falls or accidents.

5. Behavioral Monitoring for Better Safety Compliance

AI-based systems monitor worker behavior and analyze it against unsafe actions or potential dangers due to fatigue, distractions, or improper technique. These systems provide real-time feedback or alerts to help an employee correct his or her actions and avoid accidents (Singh, 2024). Telematics systems monitor driver behavior with issues such as speeding hard braking, or sharp turns in Transport and Logistics. These insights help reduce the number of accidents and allow for the optimization of fleet safety. In Manufacturing industries AI Cameras monitor the postures and movements of workers to recognize signs of fatigue or ergonomic risks. The supervisor would then take corrective action through rest breaks or ergonomic adjustments to ensure worker well-being. AI monitors the frequency of hand washing by healthcare personnel, guarantees adherence to safety procedures, and slows the spread of illness..

6. Emergency Response and Crisis Management

In emergency situations, artificial intelligence (AI) is essential since it allows for quicker and more efficient reactions. Intelligent systems assess the severity of incidents, optimize evacuation routes, and deploy resources where they are needed most (Bajwa, 2025). AI-enhanced alarms can differentiate between actual fires and false alarms, ensuring appropriate action is taken without unnecessary disruptions. AI systems analyze building layouts and crowd movements to design optimal evacuation routes during emergencies, reducing chaos and ensuring employee safety. AI-driven communication tools provide instant updates to employees during emergencies, keeping them informed of risks and safety instructions.

7. Predictive and Prescriptive Safety Solutions

Apart from risk identification, AI systems provide prescriptive solutions through the recommendation of specific actions to be taken in order to reduce hazards. These solutions are based on predictive analytics and historical data analysis. AI tools determine the optimal timing for equipment maintenance, reducing the likelihood of unexpected failures and accidents.AI models simulate various risk scenarios and provide recommendations for handling potential emergencies, enabling organizations to prepare more effectively (Lepenioti et al., 2020).

Best Practices for Implementing AI in Safety Management

Artificial Intelligence (AI) in safety management calls for a strategic, moral, and comprehensive strategy to guarantee that businesses optimize the advantages while lowering the risks. It is not enough to merely introduce cutting-edge solutions. Best practices are essential standards for efficient implementation as sectors use AI more and more to anticipate risks, track compliance, and safeguard workers. These procedures combine organizational, cultural, and technological factors to make sure AI technologies improve safety management rather than introduce new risks.

A defined set of goals that are in line with the objectives of the organization is the first best practice. Because AI projects are often implemented as stand-alone technologies without being included into more comprehensive safety plans, they frequently fail. These goals must to be quantifiable, connected to the company's safety strategy, and linked to current key performance indicators (KPIs). For example, AI-driven sensors in predictive maintenance solutions might be a better option than immersive training platforms if the main objective is to reduce accidents involving equipment. By establishing clear objectives, AI solutions are guaranteed to provide real benefits rather than being used as costly experiments.

Doing a thorough risk and readiness assessment before to deployment is another crucial procedure. When it comes to safety management, AI interacts with human behavior, operational procedures, and sensitive data—all of which present particular hazards. Businesses should assess their workforce preparedness, cybersecurity protocols, and data infrastructure before implementing AI-driven technologies. Quality control and data governance are two more essential best practices. Organizations must make sure that the data that feeds into algorithms is trustworthy, objective, and pertinent because AI systems depend on high-quality data to generate precise forecasts and suggestions. Inadequate data can result in unfair outcomes that undermine employee trust, missed dangers, or even false alarms. It is crucial to set up procedures for gathering, storing, cleansing, and validating data. Furthermore, data governance systems should incorporate ethical and privacy considerations to guarantee adherence to legislation like the General Data Protection Regulation (GDPR) and other regional privacy statutes. Employee trust is also increased by open and honest communication regarding data collection and utilization. Without this, employees can consider AI monitoring systems to be intrusive.

An evaluation of preparedness involves determining whether there is enough historical and current data available for AI training, whether IT systems can manage the computational needs, and whether staff members are ready to use the new technologies. Integrating AI tools in safety management also calls for proper employee training. Training and employee involvement are essential for a successful AI integration. The acceptability of AI tools by workers, who are the safety systems' final users, influences their practical efficacy. Employees should be consulted early on in the AI planning process so that organizations may get their feedback on problems and potential solutions. Instead than displacing human professions, training programs should emphasize not only technical capabilities but also raising understanding of how AI improves safety.

The cycle of best practices is completed through ongoing observation and assessment. AI integration into safety management is a continuous effort rather than a one-time endeavor. Companies should set up systems to assess AI tools' performance on a regular basis, gauge how they affect safety results, and make adjustments in response to user input and evolving circumstances. This entails keeping an eye out for new threats, updating algorithms with fresh data, and modifying systems to comply with changing legal requirements. AI systems are kept current, efficient, and in line with organizational safety objectives by constant improvement.

Early gap identification allows businesses to create roadmaps that address these problems and guarantee more seamless AI integration. Organizations should decide what they hope AI will accomplish before putting it into practice, such as cutting operating costs, enhancing employee engagement, assuring regulatory compliance, or reducing workplace accidents. A strategic, moral, and all-encompassing strategy is necessary when implementing artificial intelligence (AI) in safety management to guarantee that businesses optimize the benefits of these technologies while lowering any possible hazards. In the current digital age, merely implementing state-of-the-art AI systems is insufficient. Businesses must embrace clearly defined best practices that direct effective and long-lasting adoption if they are to really and ethically leverage AI's potential. It is vital to carefully incorporate these tools as companies depend more and more on AI to foresee risks, track compliance, and safeguard employees. Technological sophistication is simply one factor that affects effective safety management; other factors include organizational culture, leadership commitment, and ethical governance. Extend discussion of the ethical, psychological, and social aspects and more clearly relate safety innovation to sustainability and human welfare.

Successful AI implementation in safety management must be approached with care, thought, and strategic purpose, maximizing its full value and addressing ethical, operational, and privacy concerns (Mennella et al., 2024). Let's take a closer look at key considerations for effective implementation of AI in safety systems below. The most important thing is to strike a balance between AI and human supervision. Even if AI is quicker and more accurate than humans in analyzing patterns, it cannot take the place of human judgement in all safety-related situations. In order to ensure that AI functions as a support system rather than the exclusive authority for decision-making, organizations should clearly define roles and responsibilities. For instance, safety managers should examine the alert, confirm its veracity, and decide on the best course of action when AI indicates a possible safety violation or anticipates equipment breakdown. With this hybrid method, human professionals' ethical reasoning and contextual awareness are combined with AI's efficiency. Putting accountability frameworks in place also aids in resolving liability concerns when incidents are caused by AI faults. Human judgement is required to comprehend AI results and make complex judgements, even though AI is excellent at processing massive datasets and spotting possible hazards. AI systems are not infallible; they can sometimes produce false positives or overlook specific situational details. A cooperative strategy that combines the effectiveness of AI with human knowledge ensures a fair and thorough safety management procedure. For instance, even though AI can identify risks, a safety manager's choice is essential to verify the AI's proposal to avoid errors and misrepresent the appropriate course of action, so addressing any biases the technology may have.

Phased implementation and pilot testing are two more effective practices. Companies should start with small-scale pilots in particular departments or operations rather than implementing AI solutions throughout the entire organization at once. Before expanding, this tiered strategy enables businesses to evaluate the efficacy of AI solutions, spot possible problems, and get employee input. Before extending to all plants, a manufacturing company might, for instance, install computer vision systems to track adherence to safety gear at one factory. Pilot testing eases the transition to wider adoption by lowering risks and boosting managers' and staff's confidence. Moreover, AI algorithms may unintentionally perpetuate biases if trained on incomplete or biased data (Zeiser, 2024).

Prioritizing ethical and transparent deployment is also necessary. Particularly in the context of workplace monitoring, AI presents questions around algorithmic bias, spying, and justice. Organizations should implement ethical standards that guarantee AI apps uphold worker rights, refrain from discriminatory behavior, and function openly in order to prevent abuse. For instance, if AI is used to track employee behavior, workers should be made aware of the data being recorded, its intended use, and the measures being taken to prevent abuse. Fairness and accountability can be further guaranteed by independent audits and routine assessments of AI systems. In addition to being morally required, transparency is also practically necessary to maintain long-term confidence in safety systems powered by AI. For example, immersive simulations and gamified platforms can provide useful, entertaining examples of how AI can be used. Through employee empowerment and problem resolution, organizations cultivate a culture of trust and cooperation regarding the use of AI. Educated workers will use the systems and trust them more. When they are not educated on the capabilities and limitations of AI systems, misuse occurs, confusion arises, and they lose confidence in the technology (Alawamleh et al., 2024). Employees must learn how to use AI tools and how they might support them in their jobs rather than take their place. For example, educating employees on how to use AI-powered safety training modules or react to real-time hazard alerts would guarantee that AI increases overall safety measures and adds value to their everyday tasks. (Leonard, 2024).

Equally important is data privacy compliance when implementing AI. This is due to the fact that AI systems rely on enormous volumes of operational and personal data, such as performance, health, and employee behavior data, all of which need to be protected. GDPR (General Data Protection Regulation), HIPAA (Health Insurance Portability and Accountability Act), and other regional privacy laws require organizations to anonymize or aggregate data in order to safeguard individual privacy (Paul, 2024). Providing employees with clear information about the collection, usage, and storage of data can not only foster trust but also guarantee that all legal requirements are fulfilled. For example, in environments where sensitive employee health data may be involved, anonymizing or aggregating data for AI analysis must be done to ensure such data is maintained with sufficient privacy standards while benefitting from AI capabilities.

The potential to maximize impact further derives from the focus on applications in real-time. Hazard detection, predictive maintenance, behavior monitoring, and more AI-powered tools allow companies to act early to mitigate risks before any accidents occur. Real-time capabilities allow AI systems to identify hazardous conditions or unsafe behavior and alert workers or supervisors right away (Scaife,

2024). Real-time environmental sensors, for instance, can identify dangerous temperature swings or gas leaks in mining or chemical manufacturing and promptly start safety procedures.

Organizations can shift from a reactive to a proactive approach to safety management by giving priority to real-time applications. This will decrease the probability of events and improve reaction times. Finally, a gradual implementation approach is essential for successful AI integration. Transitioning to AI-driven safety management should be done incrementally to avoid disruption and ensure smooth adoption. Prior to scaling, the company can test AI systems in particular departments or environments, get feedback, and improve their strategy by beginning with pilot projects. For instance, a manufacturing plant could start with AI-powered cameras for monitoring ergonomic safety on a single production line, gathering data on effectiveness, and addressing any challenges before expanding the system across the entire facility. This stepwise approach reduces risks and permits businesses to learn from small-scale implementations, correct their mistakes, and improve things before full-scale deployment (Ledro et al., 2023).

Lastly, companies ought to see the adoption of AI as a component of a larger culture change towards proactive safety management. Although AI can offer useful tools, its full potential is only achieved when it is integrated into a culture that places a high value on creativity, safety, and openness. As part of the organization's long-term strategy, leaders should support AI efforts to show their dedication to sustainability and employee well-being. AI becomes a strategic enabler of operational excellence rather than a technical fix as a result of this cultural alignment.

Importance of data in AI driven safety training

Data is the foundation of AI-driven safety training, serving as the essential ingredient that powers predictive models, personalizes learning experiences, and enables continuous improvement. Without quality data, even the most sophisticated AI systems cannot function effectively. In safety management, data provides the context, evidence, and feedback necessary for organizations to understand risks, anticipate incidents, and refine training programs. From incident logs and sensor readings to employee feedback and performance evaluations, every data point contributes to building a more proactive and resilient safety culture.

At the heart of AI-driven safety systems lies the ability to predict and prevent risks before they occur. This predictive capacity depends on vast and diverse datasets. For example, historical accident reports, near-miss records, environmental sensor data, and equipment maintenance logs can all be analyzed by AI algorithms to identify hidden patterns and emerging hazards. By learning from past incidents, AI systems can forecast which activities, locations, or behaviors are most likely to result in accidents. This insight allows organizations to take preventive measures such as redesigning workflows, scheduling targeted training, or improving equipment maintenance before an incident happens.

Furthermore, data enables the optimization and personalization of safety training programs. Traditional training methods often adopt a one-size-fits-all approach, providing the same materials to all employees regardless of their roles, experiences, or learning styles. AI changes this dynamic by using data to tailor training to each individual. For instance, by analyzing performance data from simulations, quizzes, or virtual reality (VR) exercises, AI can identify an employee's strengths and weaknesses. It can then recommend specific training modules or practice scenarios to address those gaps. This level of personalization not only enhances learning effectiveness but also keeps employees more engaged, as they receive training that directly matches their needs and job requirements.

Data also drives continuous improvement in safety management. AI systems thrive on feedback loops: as they collect more data over time—from incident outcomes, compliance reports, and employee surveys—they refine their algorithms and become more accurate. This ongoing learning process allows organizations to adapt quickly to changing work environments, new technologies, and evolving regulatory standards. For example, if an AI system detects that a particular training module consistently fails to reduce certain types of incidents, it can signal the need for redesigning that module. Thus, data ensures that safety training remains dynamic, evidence-based, and responsive to real-world conditions.

However, with great potential comes significant responsibility. Using data for AI-driven safety management introduces critical challenges related to security, privacy, and compliance. Organizations must ensure that data is collected, stored, and processed in ways that protect employees' personal information and respect their rights. Sensitive data—such as biometric information from wearable devices or behavioral analytics from monitoring systems—must be safeguarded through encryption, access controls, and compliance with data protection regulations like the General Data Protection Regulation (GDPR) or local labor laws.

Transparency and trust are equally important. Employees must understand how their data is being used, who has access to it, and how it benefits their safety. If data practices are perceived as intrusive or unfair, workers may resist participation, undermining the effectiveness of AI initiatives. Therefore, organizations should establish clear data governance frameworks that outline ethical principles, accountability structures, and privacy safeguards.Data Collection and Analysis's Significance Artificial intelligence (AI) systems rely heavily on data collection and analysis to predict dangers, spot trends, and improve safety training. Such datasets usually include historical incident logs, employee feedback, metrics of training performance, and environmental data. This allows AI to unveil trends hidden in the data, like recurring types of accidents or common areas in which employees struggle with training. For instance, in a manufacturing facility, AI can look into incident reports and data of machine performance to predict the time at which a piece of machinery is most likely to fail, preventing an accident. Similarly, through an analysis of employee training data, AI can identify knowledge gaps or unsafe behaviors that may be rectified by intervention or the updating of content for specific training programs (Olawade et al., 2024).

Using these sources of data, it has been processed by improved versions of machine learning algorithms and made the AI systems make much more accurate risk assessment as well as more personalized recommendations about training. This results in making training programs not just reacting to safety incidents but proactively engaging with potential hazards that are likely to improve safety culture overall.

Data Security and Privacy

Optimizing AI-driven safety training undeniably depends on the collection and analysis of high-quality data, but this reliance introduces serious challenges related to the security and privacy of sensitive employee information. AI systems, by their very nature, require access to large and diverse datasets to learn, predict, and make informed recommendations. These datasets often include personally identifiable information (PII), health and safety records, performance evaluations, and behavioral data gathered through monitoring systems or wearable devices. While such data is crucial for improving workplace safety outcomes, it also poses substantial risks if not managed responsibly.

AI technologies in safety management can process data from various sources such as biometric sensors, surveillance cameras, digital training platforms, and health monitoring tools. For example, wearable devices might collect data about an employee's heart rate, fatigue levels, or exposure to hazardous conditions. Similarly, AI-based behavioral analytics tools may track how workers interact with machinery or move around a facility to predict unsafe actions. Although these insights help organizations reduce accidents and design personalized safety interventions, they also raise ethical and legal concerns about the extent to which employee activities and personal data are monitored and analyzed.

The challenge lies in balancing innovation and data protection. Without robust data security and privacy measures, organizations risk exposing sensitive employee information to unauthorized access, misuse, or cyberattacks. Data breaches can have far-reaching consequences, including identity theft, reputational damage, and loss of employee trust. Moreover, the misuse or overcollection of data can create a culture of surveillance, where employees feel constantly monitored, ultimately undermining morale and productivity.

To prevent these risks, organizations must adopt comprehensive data governance frameworks that ensure data is handled ethically, transparently, and securely throughout its lifecycle—from collection to storage, processing, and disposal. Encryption, anonymization, and access control are essential technical safeguards. Anonymizing or pseudonymizing personal data can minimize the risk of identifying individuals while still allowing AI models to analyze trends and patterns effectively. Similarly, strict access controls should limit who can view or manipulate sensitive information, ensuring that only authorized personnel are involved.

In addition to technical safeguards, legal compliance is paramount. Organizations operating AI-driven safety systems must adhere to relevant data protection laws to avoid severe penalties and maintain ethical standards. In the United States, the California Consumer Privacy Act (CCPA) grants individuals the right to know what data is being collected about them, request deletion of their information, and opt out of data sharing. Meanwhile, the General Data Protection Regulation (GDPR) in the European Union imposes strict requirements on data processing, including obtaining explicit consent, providing transparency in data usage, and ensuring the right to be forgotten. Non-compliance with these regulations can result in heavy fines and loss of public trust.

Beyond legal compliance, organizations should foster a culture of transparency and accountability. Employees must be informed about what data is being collected, how it will be used, and how it contributes to improving their safety. Open communication helps dispel fears of misuse and reinforces the perception that AI serves to protect, not exploit, workers. Regular audits, impact assessments, and the appointment of data protection officers can further strengthen oversight and accountability. (Bekkum & Borgesius, 2023).

To address these concerns, organizations should prioritize data security by adopting encryption methods, anonymizing sensitive data, and establishing strict access controls. Anonymization or aggregation of employee data ensures that individual identities are protected while still allowing AI systems to analyze and learn from the information. For instance, health data or incident reports can be anonymized to prevent any breach of privacy while still allowing AI tools to detect trends related to workplace injuries. This not only protects employees but also fosters trust and transparency, as workers feel confident that their data is being handled responsibly.

Utilizing AI to Track and Evaluate Training Effectiveness

AI-driven safety training goes far beyond the initial delivery of learning modules. The true power of Artificial Intelligence lies in its ability to track employee progress, analyze performance data, and continuously adapt training content for ongoing improvement. Traditional training programs often conclude once employees complete a course or assessment, offering limited feedback or follow-up. In contrast, AI introduces a dynamic and iterative learning process, where training evolves based on each employee's real-time performance and engagement.

Using sophisticated algorithms and data analytics, AI systems can monitor how employees interact with training materials, how well they perform in simulations or quizzes, and which specific areas require additional attention. Every click, response, and behavioral pattern becomes valuable input that the system uses to fine-tune the learning experience. This feedback loop transforms safety training into an intelligent process—one that adapts to individual learning curves, identifies knowledge gaps, and ensures mastery of critical safety procedures.

For example, in an industrial or warehouse setting, an AI-powered training platform might simulate emergency scenarios such as fire evacuations, chemical spills, or equipment malfunctions. As the employee navigates the simulation, the AI observes their decision-making process, timing, and adherence to safety protocols. If the employee hesitates during an evacuation step or misses an important checkpoint, the system immediately recognizes these weaknesses. It then adjusts the subsequent training session to focus more intensely on those missed elements—perhaps by adding extra drills, explanatory videos, or targeted practice questions.

This adaptive capability ensures that training is not static but personalized and responsive. Unlike traditional, one-size-fits-all safety programs, AI-based systems can offer customized pathways that match each learner's needs. For instance, a novice worker who struggles with basic safety procedures might receive simplified explanations and repetitive drills, while an experienced employee could be presented with more complex, scenario-based challenges. This differentiation maintains engagement and prevents both undertraining and overtraining, leading to more effective learning outcomes across the workforce.

Moreover, AI-driven systems promote continuous improvement not just at the individual level but also at the organizational level. Aggregated performance data from multiple employees allows safety managers to identify common weaknesses or recurring mistakes across teams. If many workers are struggling with a specific safety protocol—such as proper lifting techniques or hazard recognition—the AI can flag this trend and recommend company-wide interventions or updated training content. This enables organizations to make data-informed decisions to strengthen their overall safety culture.

The real-time feedback that AI provides is invaluable for fostering proactive learning and retention. Employees receive instant responses to their actions, allowing them to correct mistakes immediately rather than waiting for periodic evaluations. This instant reinforcement helps solidify correct behaviors and encourages continuous engagement with safety material. Over time, this approach leads to safer workplace habits and reduces the likelihood of accidents caused by human error or procedural lapses.

Additionally, the continuous monitoring of training data allows organizations to evaluate the effectiveness and return on investment (ROI) of their safety programs. By tracking performance improvements, completion rates, and behavioral changes, AI systems provide measurable insights into how training impacts workplace safety outcomes. These analytics can also be used to demonstrate compliance with safety standards and regulations, further validating the organization's commitment to employee well-being.

In essence, AI transforms safety training from a passive educational process into an active, evolving, and personalized journey. Through real-time performance tracking, adaptive learning, and data-driven feedback, employees receive the precise support they need to strengthen their safety knowledge and skills. This not only enhances individual competency but also contributes to building a resilient safety culture where learning never stops. By leveraging AI's continuous learning capabilities, organizations can ensure that safety training remains relevant, engaging, and effective in an ever-changing work environment.Over time, AI systems can build a profile of each employee's safety knowledge and behavior, identifying both strengths and weaknesses. This enables businesses to make constant improvements to their training initiatives, guaranteeing that safety instruction is current and efficient in filling up knowledge gaps or addressing new hazards. AI may update training materials based on continuous data, which guarantees that staff members are always aware of the most recent safety requirements in sectors like healthcare, where safety procedures may change as a result of new laws or technologies. Through the use of data to track training performance, AI assists companies in making sure that workers are continuously enhancing their safety knowledge and are better equipped to manage risks. As a result, there is a culture of constant learning and adaptation, which is essential for upholding strict safety regulations and lowering workplace accidents. (Kassa & Worku, 2025).

CHALLENGES AND LIMITATIONS

Although AI-based safety training provides numerous benefits, organizations also face a host of challenges and limitations in implementing such systems. These range from financial to cultural and technical dimensions and need to be addressed carefully for successful implementation and long-term effectiveness (Babu et al., 2024).

One of the most significant challenges to implementing AI-powered safety training systems is the substantial upfront investment required. While Artificial

Intelligence holds great promise for transforming workplace safety and training, the financial barriers to adoption can be daunting, especially for smaller organizations or those operating under tight budget constraints. The costs associated with acquiring, developing, and maintaining AI systems extend far beyond the initial purchase of software or hardware; they encompass a wide range of technological, infrastructural, and human resource investments necessary to ensure successful deployment and long-term functionality.

The first and most visible component of these startup costs is the technology itself. AI-based safety systems often rely on a combination of specialized software platforms, machine learning models, data analytics tools, and hardware components such as cameras, sensors, wearable devices, or virtual reality (VR) equipment. These technologies are essential for gathering and analyzing the real-time data that AI needs to deliver personalized and adaptive safety training. For example, implementing an AI-powered system in a manufacturing plant might require the installation of multiple high-resolution cameras and IoT-enabled sensors to monitor worker movements, detect potential hazards, and collect data for predictive analysis. Such equipment, combined with proprietary AI software and licensing fees, can represent a substantial capital expenditure.

Beyond purchasing the technology, organizations must also invest in supporting infrastructure to accommodate these systems. This includes upgrading network capacity, enhancing data storage solutions, and improving cybersecurity measures to safeguard the large volumes of sensitive data that AI systems process. Many AI applications depend on cloud computing services or on-premises servers with advanced processing capabilities, both of which can be expensive to establish and maintain. The need for reliable, high-speed internet connectivity and robust IT frameworks adds further cost layers, particularly for industries operating in remote or underdeveloped areas.

Another critical but often underestimated cost involves employee training and change management. Deploying AI-driven safety systems requires a workforce that understands how to interact with the technology effectively. Employees must be trained not only to use the AI tools but also to interpret the data and insights they produce. This training process takes time, resources, and expertise. Moreover, organizations may need to hire or upskill data scientists, system administrators, and AI specialists to manage and maintain these technologies. Without proper education and human support, even the most advanced AI systems may fail to deliver their intended value.

Maintenance and continuous improvement also contribute to the ongoing financial burden. AI systems are not static; they require regular updates, software patches, algorithm retraining, and hardware upgrades to stay effective and secure. Over time, the costs of maintaining these systems can accumulate significantly. Additionally, integrating AI tools with existing safety management systems can incur extra costs due to compatibility issues or the need for custom development.

For small and medium-sized enterprises (SMEs), these expenses can pose a major obstacle. Unlike large corporations with substantial research and development budgets, smaller firms may struggle to justify the high initial investment, particularly when the return on investment (ROI) is uncertain or long-term. Many SMEs operate in competitive markets where financial flexibility is limited, making the adoption of costly AI systems impractical despite their potential safety benefits.

To mitigate these challenges, organizations can explore incremental implementation strategies—starting with small-scale pilot projects or cloud-based AI solutions that reduce the need for extensive infrastructure. Partnerships with technology providers, government grants, or industry collaborations can also help offset initial costs. Nonetheless, the reality remains that high startup expenses are a primary barrier to widespread adoption of AI in safety training. Overcoming this challenge requires not only financial investment but also strategic planning, stakeholder commitment, and a clear understanding of long-term value creation.Though costly upfront, an AI-based safety system implemented effectively saves such expenditures by saving on the long-term operational costs due to the lower number of work-place accidents, less amount to be paid in form of insurance premium, and by effective usage of resources. Thus, unless they make a long-term benefit or return on investment clearly stated and defined regarding such safety measures, the ROI related to AI-driven safety systems can never be justifiable to the management level or for the employees in that concern business. Employee and management reluctance to change presents another difficulty when deploying AI-driven safety solutions. Many organizations face cultural barriers while introducing new technologies, particularly AI. Employees fear being replaced by AI or being made redundant by it. Anxiety and reluctance to use the new systems arise as a result. This fear can manifest in a lack of engagement with the training programs or even resistance to interacting with AI-powered safety tools. Management may also fear to invest in AI either due to its lack of effectiveness or unfamiliarity with the technology (El-Helaly, 2024).

For the above reasons, organizations have to cultivate an innovation culture that opens them up to new technologies. It is essential to communicate the benefits clearly, which include the potential of reducing hazards in the workplace, and enhanced safety outcomes, while calming the workforce that AI intends to support, not supplant human roles (Rudol et al., 2025). Preparing to train on the ways that AI tools will better empower them by making their job safer and easier can also help them to quell their fears. Another factor is that including staff members in the choice and application of AI technologies will increase their support and facilitate the shift. Potential Biases in AI Systems and the Need for Monitoring Biased data produces biased results since AI systems are only as good as the data they are trained on. For example, an AI algorithm may not correctly forecast or detect risks for all employees if it is trained on a dataset that is not diverse or under-represents specific accident types. This is especially problematic when it comes to safety training since biased algorithms may result in unjust safety procedures or training materials that fail to sufficiently take into account the needs of all employees. (Mensah, 2023).

The datasets used to train AI systems must be representative, varied, and updated often in order to reduce the dangers of bias in AI algorithms. In addition, human approach is essential in developing, deploying, and monitoring AI models. Ethical considerations should be at the top of the implementation of AI with continuous evaluations of the system to ascertain that it delivers fair and unbiased outcomes. Additionally, this would mean that AI systems' decision-making process is transparent, which would help detect any potential biases and make the necessary corrections. (Radanliev, 2025).

The continuous monitoring of AI performance and regular auditing also need to be conducted in an organization so that the system performs well, with no unintended bias coming up in the system. This means that business will have strong governance practice, along with oversight, in reducing risks on biased algorithms. With the correct AI-driven safety program, the organizations can have effective outcomes for all employees in the same way (Soleimania et al., 2024).

FUTURE TRENDS IN AI SAFETY TRAINING

AI-driven safety training is likely to experience considerable developments with the advancement of technology. AI combined with other advanced technologies will make workplaces more efficient, safer, and responsive to evolving hazards. These are some of the major developments that will shape AI safety training in the future. (Dang & Thu Nga, 2024). IoT Integration for Instantaneous Hazard Identification The use of AI in conjunction with the Internet of Things to facilitate real-time hazard detection is one of the most exciting developments in AI-driven safety training. IoT devices such as sensors and wearables are rapidly being deployed in the workplace, monitoring environmental conditions and risks and collecting invaluable data in the process (Alzahrani et al., 2025). These IoT sensors, when combined with AI algorithms, can instantly notify workers of any hazards by monitoring the temperature, humidity, air quality, and noise levels of the workplace in real-time. For example, IoT sensors could identify the presence of hazardous gases, structural damage, or equipment failures in high-risk settings like manufacturing facilities or building sites. The AI system would immediately analyze the data, establish the degree of risk involved, and immediately warn the employees or even trigger automatic security measures like equipment shutdown or engaging emergency protocols

(Olawade et al., 2024). The convergence of IoT with AI guarantees that the safety measures implemented are proactive instead of reactive, and thus, this reduces the likelihood of an accident significantly.Advances in Wearable Technology Coupled with AI Integration The third area is in wearable technology, as AI makes great strides in creating safer workplaces. Smart helmets, glasses, vests, or wristbands are increasingly getting into use in construction and manufacturing and Oil and gas sites (Aksut et al., 2024). They may evaluate a worker's health and safety status in real time and provide real-time safety instructions based on their surroundings to employees who are in danger. For example, the intelligent helmets can use AI to read around them such as air quality, noise levels, or distance to hazardous zones. Therefore, they can alert a person wearing them instantly whether or not they are entering an unsafe area. Similarly, smart glasses could overlay safety instructions using AR, helping people perform their tasks more effectively while making sure they have the proper safety procedures. Additionally, wearables can monitor employees' vital signs, including body temperature or heart rate, and notify managers if an individual exhibits symptoms of stress, exhaustion, or dehydration so that prompt action can be taken to avoid mishaps. (Scorgie et al., 2024).As these technologies advance, they will be giving more personalized safety solutions tailored towards the needs of each employee and the specific risk related to their job. The role of AI and wearable technologies will be pivotal in the mitigation of injuries and enhancing work safety (Valtonen et al., 2025). Artificial Intelligence (AI) is rapidly revolutionising how corporations approach workplace safety, going beyond standard training models toward intelligent, adaptive, and predictive technologies. Even though AI has already shown promise in raising compliance, lowering accident rates, and improving educational opportunities, there is still more room for innovation in the future. AI safety training will develop in ways that make workplaces safer, smarter, and more robust as businesses continue to adopt automation, digitization, and linked technology. This section examines the new developments that will influence AI in safety training in the future, including technical breakthroughs, integration with other digital tools, moral dilemmas, and wider organizational effects.

CONCLUSION

Workplace safety has always been both a moral obligation and an operational necessity for organizations across all industries. Protecting employees from harm is not merely a regulatory requirement but a reflection of an organization's values and commitment to social responsibility. However, as work environments become increasingly complex—with advanced machinery, globalized operations, and diverse workforce dynamics—traditional safety training methods are proving insufficient.

These conventional approaches, often reliant on static manuals, occasional workshops, and reactive incident reporting, struggle to keep pace with the dynamic risks of modern workplaces. To address these evolving challenges, organizations are turning to Artificial Intelligence (AI), which provides a transformative, data-driven approach to safety management.

AI technologies empower businesses to shift from reactive to proactive safety management. Instead of responding to accidents after they occur, AI enables companies to predict and prevent hazards before they escalate into incidents. Through predictive analytics, AI systems analyze vast datasets collected from sensors, equipment logs, employee behavior records, and environmental monitoring systems. By identifying patterns and correlations that human observers might overlook, these systems can forecast potential risks, such as machinery failures, fatigue-related errors, or hazardous environmental conditions. For example, AI-driven predictive maintenance can alert technicians when equipment is likely to malfunction, preventing costly breakdowns and accidents. Similarly, algorithms analyzing employee motion data from wearable devices can detect early signs of strain or unsafe postures, prompting timely interventions.

Beyond prediction, AI enhances safety training through immersive and personalized learning experiences. Technologies such as virtual reality (VR) and augmented reality (AR) allow employees to engage in realistic simulations of hazardous situations—such as fires, chemical spills, or confined-space rescues—without actual risk. AI algorithms track each learner's performance during these simulations, adjusting difficulty levels, providing real-time feedback, and tailoring future training sessions to individual needs. This adaptive learning approach not only reinforces knowledge retention but also boosts employee confidence by ensuring they are well-prepared to handle emergencies effectively.

AI also revolutionizes automated danger detection and real-time monitoring. Smart cameras, IoT sensors, and computer vision algorithms can continuously scan workplaces for unsafe behaviors, equipment malfunctions, or environmental hazards. For instance, AI-powered surveillance systems can detect when a worker enters a restricted zone without proper protective gear or when machinery operates outside safe temperature limits. Upon detecting such anomalies, the system can instantly alert supervisors or even trigger automatic shutdowns, minimizing the window for human error. This real-time responsiveness significantly enhances workplace safety by enabling swift corrective actions before incidents occur.

Furthermore, integrating AI into safety management fosters stronger compliance frameworks and accountability cultures. AI systems can automatically record, analyze, and report safety metrics, ensuring adherence to occupational safety regulations and standards such as ISO 45001. By maintaining accurate digital records, organizations can demonstrate due diligence during audits and regulatory inspections. This not only reduces the administrative burden on safety officers but also promotes transparency and continuous improvement.

The use of AI in safety management also contributes to building employee trust and engagement. Workers gain confidence knowing that their organization invests in advanced technologies to protect their well-being. When employees see data-driven decisions leading to safer practices and fewer incidents, they are more likely to participate actively in safety initiatives. Moreover, AI's emphasis on prevention and learning encourages a culture of accountability and innovation, where safety is seen not as a compliance task but as a shared responsibility and strategic advantage. Careful planning, human monitoring, and ethical governance can guarantee that AI technologies are used ethically and successfully, even in the face of obstacles like high implementation costs, opposition to change, and possible algorithmic bias. With the integration of IoT devices, wearable technologies, and adaptive platforms, AI will continue to redefine safety training by personalizing learning, reducing risks, and enabling real-time interventions. In the end, AI-driven safety training is a calculated investment in resilience, worker well-being, and long-term organizational performance rather than merely a technical advancement. Companies who adopt these technologies will establish themselves as leaders in operational excellence and establish new standards for worker safety in a changing global environment.

REFERENCES

Aksut, , Eren, T., & Alakaş, H. M. (2024). Using wearable technological devices to improve workplace health and safety: An assessment on a sector base with multi-criteria decision-making methods. *Ain Shams Engineering Journal, 15*(2), 102423. Advance online publication. DOI: 10.1016/j.asej.2023.102423

Al-Naser, & Al-Tabtabai. (2024). The impact of safety violations on construction project performance: A case study of the ADFA project. *Journal of Engineering Research*. Advance online publication. DOI: 10.1016/j.jer.2024.12.017

Alawamleh, , Shammas, N., Alawamleh, K., & Bani Ismail, L. (2024). Examining the limitations of AI in business and the need for human insights using Interpretive Structural Modelling. *Journal of Open Innovation, 10*(3), 100338. Advance online publication. DOI: 10.1016/j.joitmc.2024.100338

Alzahrani, , Kostkova, P., Alshammari, H., Habibullah, S., & Alzahrani, A. (2025). Intelligent integration of AI and IoT for advancing ecological health, medical services, and community prosperity. *Alexandria Engineering Journal, 127*, 522–540. Advance online publication. DOI: 10.1016/j.aej.2025.05.046

Ayodeji, A. (2024). Artificial Intelligence in Enhancing Regulatory Compliance and Risk Management. doi:DOI:DOI: 10.13140/RG.2.2.20915.44326

Babu, . (2024). The Impact of Artificial Intelligence in the Workplace and its Effect on the Digital Wellbeing of Employees. *Journal for Studies in Management and Planning, 10*(4), 1–32. DOI: 10.5281/zenodo.10936348

Bajwa, A. (2025). AI-BASED EMERGENCY RESPONSE SYSTEMS: A SYSTEMATIC LITERATURE REVIEW ON SMART INFRASTRUCTURE SAFETY. *American journal of advanced technology and engineering solutions, 1*(1), 20-40. doi:DOI: DOI: 10.63125/xcxwpv34

Balan, , Kumar, V. S., & Raj, S. A. (2025). Machine learning and artificial intelligence methods and applications for post-crisis supply chain resiliency and recovery. *Supply Chain Analytics, 10*, 100121. Advance online publication. DOI: 10.1016/j.sca.2025.100121

Bekkum, M., & Borgesius, F. Z. (2023). Using sensitive data to prevent discrimination by artificial intelligence: Does the GDPR need a new exception? *Computer Law & Security Review, 48*, 105770. Advance online publication. DOI: 10.1016/j.clsr.2022.105770

Benson, , Obasi, I. C., Akinwande, D. V., & Ile, C. (2024). The impact of interventions on health, safety and environment in the process industry. *Heliyon, 10*(1), e23604. Advance online publication. DOI: 10.1016/j.heliyon.2023.e23604 PMID: 38173504

Bernal, . (2024). Safety culture, safety performance and financial performance. A longitudinal study. *Safety Science, 172*(16), 106409. Advance online publication. DOI: 10.1016/j.ssci.2023.106409

Bęś, P., & Strzałkowski, P. (2024). Analysis of the Effectiveness of Safety Training Methods. *Sustainability (Basel), 16*(7), 2732. Advance online publication. DOI: 10.3390/su16072732

Budnarowski, , Jereczek, D., Detka, K., & Wieczorek, I. (2025). Application of Artificial Intelligence and Virtual Reality in Soft Skills Training with Modeled Personality. *Applied Sciences (Basel, Switzerland), 15*(16), 9067. Advance online publication. DOI: 10.3390/app15169067

Caburao, E. A. (2024). *Leveraging AI in Safety Management for a Safer Workplace*. Retrieved from Safetyculture: https://safetyculture.com/topics/safety-management-system/ai-in-safety-management/

Chen, Z., & Dai, X. (2024). Utilizing AI and IoT technologies for identifying risk factors in sports. *Heliyon, 10*(11), e32477. Advance online publication. DOI: 10.1016/j.heliyon.2024.e32477 PMID: 38961959

Dang, H. H., & Thu Nga, L. (2024).. . *Artificial Intelligent Applications in Occupational Safety and Safety Training., 23*(12). Advance online publication. DOI: 10.37896/YMER23.12/59

Diro, , Kaisar, S., Saini, A., Fatima, S., Hiep, P. C., & Erba, F. (2025). Workplace security and privacy implications in the GenAI age: A survey. *Journal of Information Security and Applications, 89*, 103960. Advance online publication. DOI: 10.1016/j.jisa.2024.103960

El-Helaly, M. (2024). Artificial Intelligence and Occupational Health and Safety, Benefits and Drawbacks. *La Medicina del Lavoro, 115*(2), 1. Advance online publication. DOI: 10.23749/mdl.v115i2.15835 PMID: 38686574

Haleem, , Javaid, M., & Singh, R. P. (2025). Encouraging Safety 4.0 to enhance industrial culture: An extensive study of its technologies, roles, and challenges. *Green Technologies and Sustainability, 3*(3), 100158. Advance online publication. DOI: 10.1016/j.grets.2024.100158

Joni, R., & Graepel, T. (2024). Predictive Analytics and AI: Driving the Next Wave of Risk Management in Financial Services. doi:DOI:DOI: 10.13140/RG.2.2.16499.75041

Kassa, B. Y., & Worku, E. K. (2025). The impact of artificial intelligence on organizational performance: The mediating role of employee productivity. *Journal of Open Innovation, 11*(1), 100474. Advance online publication. DOI: 10.1016/j.joitmc.2025.100474

Konda, R. (2020). Enhancing Workplace Safety with AI-Powered Video. *Quest Journals*, 59-64. doi:DOI: DOI: 10.35629/3795-06035964

Ledro, , Nosella, A., & Dalla Pozza, I. (2023). Integration of AI in CRM: Challenges and guidelines. *Journal of Open Innovation*, *9*(4), 100151. Advance online publication. DOI: 10.1016/j.joitmc.2023.100151

Leonard, T. (2024). *AI and Employee Training: Ensuring Safe Use of Artificial Intelligence at Work.* Retrieved from Blue Signal Search: https://bluesignal.com/2024/02/23/ai-and-employee-training-ensuring-safe-use-of-artificial-intelligence-at-work/

Lepenioti, , Bousdekis, A., Apostolou, D., & Mentzas, G. (2020). Prescriptive analytics: Literature review and research challenges. *International Journal of Information Management, 50*, 57–70. DOI: 10.1016/j.ijinfomgt.2019.04.003

Ljubic, . (2016). Training of workers as an important safety measure for risk management. *19th International Symposium on Electrical Apparatus and Technologies.* doi:DOI:DOI: 10.1109/SIELA.2016.7543022

Mennella, , Maniscalco, U., De Pietro, G., & Esposito, M. (2024). Ethical and regulatory challenges of AI technologies in healthcare: A narrative review. *Heliyon, 10*(4), e26297. Advance online publication. DOI: 10.1016/j.heliyon.2024.e26297 PMID: 38384518

Mensah, G. B. (2023). Artificial Intelligence and Ethics: A Comprehensive Review of Bias Mitigation, Transparency, and Accountability in AI Systems. *Computer Science and Engineering*. doi:DOI:DOI: 10.13140/RG.2.2.23381.19685/1

Occupational Safety and Health Administration. (n.d.). Retrieved from osha.gov/ergonomics: https://www.osha.gov/ergonomics

Olawade, , Wada, O. Z., Ige, A. O., Egbewole, B. I., Olojo, A., & Oladapo, B. I. (2024). Artificial intelligence in environmental monitoring: Advancements, challenges, and future directions. *Hygiene and Environmental Health Advances, 12*, 100114. Advance online publication. DOI: 10.1016/j.heha.2024.100114

Park, J., & Kang, D. (2024). Artificial Intelligence and Smart Technologies in Safety Management: A Comprehensive Analysis Across Multiple Industries. *Applied Sciences (Basel, Switzerland)*, *14*(24), 11934. Advance online publication. DOI: 10.3390/app142411934

Paul, J. (2024). *Privacy and data security concerns in AI.* Computer Security and Reliability.

Pribadi, , Rahman, Y. M. R., & Silalahi, C. D. A. B. (2024). Analysis of the effectiveness and user experience of employing virtual reality to enhance the efficacy of occupational safety and health learning for electrical workers and graduate students. *Heliyon*, *10*(15), e34918. Advance online publication. DOI: 10.1016/j.heliyon.2024.e34918 PMID: 39144925

Radanliev, P. (2025). AI Ethics: Integrating Transparency, Fairness, and Privacy in AI Development. *Applied Artificial Intelligence*, *39*(1), 2463722. Advance online publication. DOI: 10.1080/08839514.2025.2463722

Rashid, A. B., & Karim Kausik, M. (2024). AI revolutionizing industries worldwide: A comprehensive overview of its diverse applications. DOI: 10.1016/j.hybadv.2024.100277

Rudol, . (2025). Exploring Human–AI Dynamics in Enhancing Workplace Health and Safety: A Narrative Review. *International Journal of Environmental Research and Public Health*, *22*(2), 199. Advance online publication. DOI: 10.3390/ijerph22020199 PMID: 40003424

Samman, A. A. (2024). The Use of AI in Fostering and Embracing Organizational Culture. *ASU International Conference in Emerging Technologies for Sustainability and Intelligent Systems (ICETSIS)At: Bahrain.* doi:DOI:DOI: 10.1109/ICETSIS61505.2024.10459508

Scaife, A. D. (2024). Improve predictive maintenance through the application of artificial intelligence: A systematic review. *Results in Engineering*, *21*, 101645. Advance online publication. DOI: 10.1016/j.rineng.2023.101645

Schulte, , Iavicoli, I., Fontana, L., Leka, S., Dollard, M. F., Salmen-Navarro, A., Salles, F. J., Olympio, K. P. K., Lucchini, R., Fingerhut, M., Violante, F. S., Seneviratne, M., Oakman, J., Lo, O., Alfredo, C. H., Bandini, M., Silva-Junior, J. S., Martinez, M. C., Cotrim, T., & Fischer, F. M. (2022). Occupational Safety and Health Staging Framework for Decent Work. *International Journal of Environmental Research and Public Health*, *19*(17), 10842. Advance online publication. DOI: 10.3390/ijerph191710842 PMID: 36078562

Scorgie, D., Feng, Z., Paes, D., Parisi, F., Yiu, T. W., & Lovreglio, R. (2024). Virtual reality for safety training: A systematic literature review and meta-analysis. *Safety Science, 171*, 106372. Advance online publication. DOI: 10.1016/j.ssci.2023.106372

Shah, I. A., & Mishra, S. (2024). Artificial intelligence in advancing occupational health and safety: An encapsulation of developments. *Journal of Occupational Health, 66*(1), uiad017. Advance online publication. DOI: 10.1093/joccuh/uiad017 PMID: 38334203

Singh, S. (2024). AI-Powered Solutions for Workplace Safety: Preventing Unsafe Behaviors Effectively. *JOUR, 16*(12). Retrieved from https://www.viact.ai/post/ai-powered-solutions-for-workplace-safety-preventing-unsafe-behaviors-effectively

Soleimania, . (2024). Reducing AI bias in recruitment and selection: An integrative grounded approach. *International Journal of Human Resource Management*. Advance online publication. DOI: 10.1080/09585192.2025.2480617

Soori et al. (2024). AI-Based Decision Support Systems in Industry 4.0, A Review. *Journal of Economy and Technology*. DOI: 10.1016/j.ject.2024.08.005

Tamascelli, , Campari, A., Parhizkar, T., & Paltrinieri, N. (2024). Artificial Intelligence for safety and reliability: A descriptive, bibliometric and interpretative review on machine learning. *Journal of Loss Prevention in the Process Industries, 90*, 105343. Advance online publication. DOI: 10.1016/j.jlp.2024.105343

Tian, , Zhu, Z., Mbachu, J., Ghanbaripour, A., & Moorhead, M. (2025). Artificial intelligence in risk management within the realm of construction projects: A bibliometric analysis and systematic literature review. *Journal of Innovation & Knowledge, 10*(3), 100711. Advance online publication. DOI: 10.1016/j.jik.2025.100711

Valtonen, , Saunila, M., Ukko, J., Treves, L., & Ritala, P. (2025). AI and employee wellbeing in the workplace: An empirical study. *Journal of Business Research, 199*, 115584. Advance online publication. DOI: 10.1016/j.jbusres.2025.115584

Wilson, J. (2024, September). *Using artificial intelligence to create a safer work environment for employees*. Retrieved from AXA XL: https://axaxl.com/fast-fast-forward/articles/using-artificial-intelligence-to-create-a-safer-work-environment-for-employees

Wong, A. T. (2025). *Sustainable HR Practices with AI.* Artificial Intelligence and Sustainability: Innovations in Business and Managerial Practices. doi:DOI:DOI: 10.70301/SBS.MONO.2025.1.5

Zeiser, J. (2024). Owning Decisions: AI Decision-Support and the Attributability-Gap. *Science and Engineering Ethics*, *30*(4), 27. Advance online publication. DOI: 10.1007/s11948-024-00485-1 PMID: 38888795

Chapter 12
Digital Transformation and Healthcare Safety:
Role of Technology, Communication, and Work Environment in Patient Outcomes

Vijayasree Srinivasan
http://orcid.org/0009-0006-5618-7080
Saveetha School of Management, Saveetha Institute of Medical and Technical Sciences, India

K. C. Raja Shree
http://orcid.org/0009-0009-1029-910X
Saveetha School of Management, Saveetha Institute of Medical and Technical Sciences, India

P. Praveen Kumar
http://orcid.org/0000-0002-0707-039X
SRM Institute of Science and Technology, Vadapalani, India

ABSTRACT

Digitization in healthcare is bringing new ways to improve safeguarding patients and results through the utilization of advancements like AI, EHRs, CDSS, and IoMT. However, digital evolution requires more than just technology; its impact improves when it coincides with interaction, a collaborative atmosphere, and favorable work settings; managerial variables like management, recruiting, and emotional security determine how squads use electronic devices; patient focused breakthroughs like telecommuting and smartphone applications require equitable entry and digital proficiency to function; seamless integration, processes asymmetry, and absence of

DOI: 10.4018/979-8-3373-6370-7.ch012

training continue; and interactive realization and ongoing monitoring are all necessary. This chapter demonstrates how a sociotechnical paradigm that integrates digital technology along with managerial and personal elements results in safer, more long-lasting care.

INTRODUCTION

Patient security is a major global community health concern since adversarial events within healthcare systems significantly contribute to the prevention of illness, death, and rising medical expenses. According to estimates from the World Health Organization (WHO), until 80% of nursing home-related harms are preventable, around one in four individuals in productive countries experience harm during clinic care. (WHO, 2019). Unsafe care is responsible for an estimated 2.6 heap fatalities and 134 heap antagonistic incidents every year in low- and middle-income countries (WHO, 2021). Medication errors are estimated to cost the world economy more than USD 42 billion annually (OECD, 2017). Progress has been uneven and incomplete despite ongoing attention to patient security due to ground breaking findings such as To Err Is Human (Institute of Medicine, 2000) and subsequent strategies and quality pushes (Francis et al., 2025).

Simultaneously, photoelectric fitness records, automation, machine intelligence, and smart dispassionate sciences are becoming increasingly popular, and healthcare practices are rapidly becoming digitized. Patient safety issues persist despite these technologies' promises of efficacy and correct decline due to ambiguous sociotechnical, ethical, and administrative factors that fundamentally influence the development, application, and execution of sciences. This article examines patient security in the perspective of human engineering and mathematical shift in order to remedy this issue. After reviewing important ideas in patient security and mathematical strength, it surveys the consequences of sociotechnical and human engineering on safety. Finally, it proposes an integrated abstract model to direct more trustworthy algebraic transformation in healthcare.

DIGITAL REVOLUTION IN HEALTHCARE

According to Agarwal et al. (2010), mathematical change in healthcare refers to the widespread acceptance and unification of mechanics advancements to enhance dispassionate accountability, improve operational adeptness, and provide appropriate care. Photoelectric health records (EHRs), telehealth, automatic medical procedures, big dossier science of logical analysis, Internet of Things (IoT) technologies, machine intelligence (AI), dossier mining, and other mathematical medical schemes are just a few of the many innovations it encompasses. These increases reflect the increasing digitization of healthcare infrastructures and procedures.

Beyond simple digitalization, mathematical transformation suggests a critical reorganization of professional practices, organizational procedures, and boring education, with electronics acting as both a catalyst and an endpoint (Raju et al., 2025; Vinodh et al., 2025). Dossier-committed clinical resolutions, organized communication, and reinforced listening: Technologies that may inform patient safety and effectiveness. But one should not assume that the safety benefits of the mathematical sciences are inevitable: their impact in complex healthcare environments will depend on how they are created, applied and used.

Even when mathematics is used to compensate for human instability and fallibility, it can lead to new risks in areas such as industrialization, interoperability, dossier keeping, and

usability. Table 1 summarizes important mathematical healthcare sciences, their intended benefits, and their possible security linkages in order to purify the partnership between patient safety and accepting details.

Table 1. Digital Healthcare Technologies, Intended Benefits, and Safety Implications:

Technology	Intended Benefit	Safety Implication
Electronic Health Records (EHRs)	Improved information access and care coordination	Reduced documentation errors but risk of usability-related errors and alert fatigue
Telehealth	Increased access and continuity of care	Improved follow-up but potential safety risks due to limited physical assessment
Artificial Intelligence (AI)	Enhanced diagnostics and clinical decision support	Reduced cognitive workload but risk of automation bias
Internet of Things (IoT)	Continuous patient monitoring	Early detection of deterioration but risk of data overload
Automated / robotic surgery	Increased precision and consistency	Reduced technical errors but dependence on system reliability

EVOLVING NATURE OF SAFETY RISKS AND TECHNOLOGICAL COMPLEXITY

However, the kinds of patient security concerns change as healthcare systems become more reliant on sophisticated mathematical algorithms. Mathematical forms nonetheless pose new risks even though they can lessen common wrongs like illegible manuscripts or improper medicine administration. Increased mechanics complexity has led to alert fatigue, which is caused by excessive or inadequately measured clinical conclusion support alerts; interoperability losses, which can interfere with news flow across relevant systems; and automation bias, which allows clinicians to rely too much on mathematical advice.

Safety concerns are made worse by additional dangers, such as dossier breaches and system idle time (Sittig & Singh, 2010). Due to the sociotechnical complexity of healthcare, which is defined by crucial interactions between people, sciences, and organizational procedures, bankruptcies often occur at the boundaries between workflows, digital structures, and populations, or they may result from isolated mechanical errors (Carayon et al., 2006). In this regard, Patient safety can no more only be described for its detail performance, rather it must also explain how the sciences are integrated into administrative as well as dispassionate practice. These new risks indicate the need for an expanded approach, together with human engineering, administrative culture and plan-level governance -- a subject that is addressed in the next section -- and demonstrate why electronics is not enough to keep patients safe.

THE GAP: WHY TECHNOLOGY CAN'T ADDRESS SAFETY CONCERNS ON ITS OWN

However, the evidence on the association of HCIT and patient safety has been mixed, even when large amounts of money have been spent processing health data (Buntin et al., 2011). Such is the continual claim of electronics first and independent application, indifferent to the requirements in which finishes can be applied. Active communication, advice support, cooperation and a safety-familiar administrative information space can certainly not be replaced by mere technology to some extent. This breach highlights a basic sociotechnical problem: electronics cannot provide security outside of alignment between two points: organizational frameworks, human competencies, and supervisory control.

For instance, even while photoelectric strength record systems are technically solid, they cannot improve security results if doctors are ill-prepared or if method interfaces are not created with the wants of the user in mind (Zhang et al., 2003). In reality, mathematical electronics are frequently developed in isolation from more comprehensive administrative strategies and safety objectives. In conclusion, the limited impact of digital energy electronics on patient safety highlights the necessity of linked hypothetical foundations that provide justification for system-level, administrative, and human factors. The hypothetical frameworks described in the next section are supported by this viewpoint.

INTRODUCING ORGANIZATIONAL ENABLERS: COMMUNICATION, WORK CULTURE, LEADERSHIP

Organizational facilitators are the settings, procedures, and structures found in healthcare settings that encourage a safety-reinforcing approach. As stated by Pronovost et al. (2008), these include an auxiliary work environment, open and creative ideas among groups, cognitive security, ongoing education, and direction that is visible, mindful, and dedicated to security improvement. An increasing number of reports claim that these two factors directly impact security outcomes and reduce the effectiveness of scientific remedies (Manser, 2009). Smart monitoring solutions, for instance, are more likely to prevent contamination or falls when used in conjunction with explicit protocols and group-based communication. Similarly, AI-driven conclusion-support technologies are more impressive when clinicians have faith in their results, which is dependent on administrative openness and planning. The "missing connection" between mathematical innovation and practical security enhancements is formed by these facilitators.

Theoretical Foundations

This section outlines the principal theoretical underpinnings of the intended conceptual framework and discusses how they are supported by observed decision-making in healthcare in applications. seen overall, these views provide theoretical boundaries for the connected model to be considered in this article.

Relevant Safety Theories

The domain of healthcare is a sociotechnical system where human behavior and science are closely linked (Berg, 1999; Sittig & Singh, 2010). Telescopic gaze: Proximity and consumer's needs According to Sociotechnical Systems (STS) theory (Cohen et al., 2006), scientific activities produce an inevitable impact on what consumers do, but also how they do it. This view underlines why the framework acknowledges that security doesn't just follow from science – it's success is anchored between mathematical results, consumers, workflows and organisational arrangements.

This groundwork is further fortified by Donabedian's (1966) model of healthcare value whereby status is gestated across three interrelated sets of rules—form (e.g., resource, science, staffing), process (e.g., communication, collaboration, administrative), and outcomes (e.g., patient safety, improvement). When it comes to the first mentioned model, as applied in strength duties research (Campbell et al., 2000), electronics can be seen represented as part of basic proficiency while cooperation and communication are being produced by digital forms' influence upon care processes and results (Donabedian, 1988; Buntin and others, 2011).Davis's (1989) Technology acceptance is the main emphasis of Davis's (1989) Technology Acceptance Model (TAM), which explains how perceived usefulness (PU) and perceived ease of use (PEOU) influence the acceptance and prolonged use of mathematical procedures.

Consistent with Venkatesh and Davis (2000), and Holden, & Karsh (2010) TAM provides a framework to understand why administrative support, utility, preparation and perceived relevance are barriers to the effectiveness mathematics well-being interventions. The hypothesis of the High-Reliability Organisation (HRO) adds to our understanding of safety in environments that are complicated and very hazardous. Organizational characteristics such as complexity, time pressure and high cost of failure (Chassin & Loeb, 2013) that are associated with HROs are also applicable to healthcare organizations; specifically, emergency departments and ICU's. HRO theory stresses team science, lead dates, and a vigorous emphasis on safety, all valuable adjuncts to digital electronics in the neutral.

Future Paths for the Sociotechnical System of Digital Health

Empirical Viewpoints The application of these theories in mathematics health contexts is supported by actual evidence. For instance, it has been discovered that badly crafted EHR messages lead to workarounds and errors that indicate a disconnect between technology and detached operations (Zhang and others., 2003; Ash and others., 2004). However, mathematical methods are useful for efficiency and security when combined with administrative processes and psychoanalyst constraints (Carayon et al., 2006).

Correlatively, the technology acceptance model (TAM) research in healthcare demands systems that meet the needs of those who actually have to use them or are otherwise less likely to be fit for task or purpose (Yarbrough and Smith, 2007; Shojania et al., 2010).According to Braithwaite et al. (2017), HRO principles have also been shown to be especially suitable in mathematical-intensive environments, where the use of AI in critical care units necessitates constant situational awareness and rapid communication in order to survive emerging risks.

By articulating five relevant components—person, tasks, forms and technologies, palpable atmosphere, and institution—the Systems Engineering Initiative for Patient Safety (SEIPS) model further operationalizes sociotechnical thinking (Carayon et al., 2006). SEIPS, which is widely used in the healthcare industry (Gurses et al., 2009), demonstrates how modifications to a single structural feature can have an impact on duties, intelligent workload, and crew ideas, ultimately leading to safety impacts. For example, when used effectively, the integration of wearable listening devices can potentially reduce failure-to-rescue incidents and change attending activities and arranging patterns (Carayon & Perry, 2020).

Repercussions for the Integrated Model

When taken as a whole, these theoretical underpinnings and empirical findings confirm that patient security in mathematical healthcare is a system-level characteristic that results from the interaction of people, institutions, and technology. This section thus offers the hypothetical and abstract structure for the combined model created in the next section, which combines different perspectives to reveal how a mathematical revolution might either strengthen or harm patient safety.

Digital Health Technologies and Safety

Interactions between medical professionals, patients, and organizational structures are shaped by digital energy sciences. Their impact on security is dependent on administrative support, process unification, technical capability, and utility. Despite the possible advantages, inadequate exercise can be dangerous.

Table 2. Interaction of Key Digital Health Technologies with Safety

Technology	Interaction with Humans & Systems	Safety-Enabling Aspects	Safety-Threatening Aspects
Electronic Health Records (EHRs)	Central platform connecting clinicians, departments, and patient data across care settings	• Reduces documentation and medication errors (Buntin et al., 2011) • Improves care coordination and information continuity • Supports team communication	• Poor usability increases cognitive workload • Interoperability failures cause missing or delayed information • Copy-paste practices may propagate errors
Artificial Intelligence (AI)	Supports clinician decision-making by analyzing large datasets	• Early detection of conditions (e.g., sepsis, stroke, cancer) • Faster image interpretation • Continuous ICU monitoring reduces diagnostic delay	• Over-reliance may reduce clinical judgment • Algorithm bias affects vulnerable populations • Limited transparency reduces trust
Clinical Decision Support Systems (CDSS)	Embedded within EHRs to provide real-time clinical guidance	• Alerts for drug interactions and allergies (Carspecken et al., 2013) • Evidence-based treatment recommendations • Standardizes care	• Alert fatigue leads to ignored warnings • Poorly tailored alerts disrupt workflows • Excessive notifications reduce attention

continued on following page

Table 2. Continued

Technology	Interaction with Humans & Systems	Safety-Enabling Aspects	Safety-Threatening Aspects
Internet of Things (IoT) / Remote Monitoring	Continuous data exchange between patients, devices, and healthcare teams	• Real-time monitoring improves early intervention • Reduces readmissions in chronic care (Islam et al., 2015) • Enhances patient engagement	• Device malfunction or connectivity failure • Data overload for clinicians • Cybersecurity vulnerabilities
Telemedicine	Virtual interaction between patients and clinicians	• Improves access for rural and underserved populations • Reduces infection risk • Supports continuity of care	• Digital divide limits equitable access • Privacy and data security risks • Clinical assessment limitations

Human and Organizational Factors Influencing Safety

- Whether electronics are humiliating or present mistakes depends on their usability and preparation.
- It is necessary to upgrade communication procedures in addition to mathematical tools.
- In encouraging work settings, clinicians are permitted to utilize technology safely.
- Security culture and leadership have an impact on supervision and ratification.

E-Healthcare Safety Problems

Digital sciences can enhance patient safety, yet wrongly considered plans may become an additional threat. The key concerns are the need to establish an egalitarian approach, reduce the cognitive load, and create a balance between mechanization and clinical dread.

Figure 1. Safety Problems in Digital Healthcare

Source: *(Sujan et al., 2020).*

The Intention of Harm-Free Interaction

Effective interaction is necessary in order to avoid wrongdoing and to state patient safety. The fields of interaction human science, human interfered by science and technology organization are three main types of interacting required in mathematical frameworks of fitness to be safe. Although carefully designed components are easy to control and prevent errors, poorly designed methods, too many warning signs, or unclear displays may decrease circumstantial knowledge and increase the probability of misplaced warnings Miscommunication during transitions between care settings is an important cause of preventable harm. To a certain degree, SBAR tools help standardize the concepts and assure that the essential information is exchanged during the handovers and care intensification (Haig et al., 2006). Besides, administrative savvies that encourage dictatorial authority prove to be useful in detecting and rectifying evils at an early age, especially in hostile environments (Edmondson, 1999).

Work Environment Enablers of Safety

The workplace acts as an arbitrator in this context and has a big influence on how effectively mathematical expertise improve the security of patients. Whether or not people feel supported in choosing new systems and reporting security issues is influenced by leadership and education. Extreme workloads and understaffing lead to fatigue and intellectual overload, which undermines safety. Whether the sciences constitute second-hand information or result in risky workarounds depends on training and mathematical expertise. Workflow unification is crucial because misaligned structures might reduce security advantages by increasing departmental burden and upsetting care. Based on Nembhard & Edmondson (2006), an auxiliary work environment enhances communication, supports specialized knowledge, and enables mathematical finishes to purposefully contribute to patient security.

Figure 2. Connected fitness components support patient-centered health management

Source: *(MacMahon & Richardson, 2023)*

CYBERSECURITY AND PATIENT SAFETY IN THE DIGITAL HEALTHCARE ENVIRONMENT

In digital healthcare arrangements, cybersecurity is a fault-finding part of patient security. Healthcare businesses are increasingly vulnerable to cyberattacks as photoelectric fitness records (EHRs), telemedicine platforms, related medical devices, and AI-based forms improve routine (Ewoh & Vartiainen, 2024). These techniques make bureaucrats extremely valuable targets for cyberattacks since they store and transmit sensitive clinical news. Cybersecurity and patient security are directly related. In contrast to delayed or inappropriate care, data breaches can jeopardize the accuracy and likelihood of patient news. Clinicians may find it difficult to obtain lively data, like as test results, vulnerability lists, or cure histories, during system free time, particularly during ransomware attacks, which increases dispassionate risk.

A well-known example of how high-tech catastrophes can affect real patient harm is the 2017 WannaCry attack on the UK National Health Service, which resulted in cancelled procedures, transport diversions, and treatment delays (Ghafur et al., 2019). There have also been more recent instances of delayed diagnoses and negative consequences (Van Boven et al., 2023).Medical procedures and AI-approved instructions are once again impacted by cybersecurity risks. Vulnerabilities in linked schemes or maneuverer algorithms can result in risky, unbiased findings that put victims at serious risk (Mitra et al., 2025). Clinicians are often forced to resort on manual procedures when systems fail, which increases cognitive burden and increases the risk of drug or proof errors.

Cybersecurity serves as both a controlling variable and a risk determinant in the suggested model. When breaches, dossier catastrophe, or downtime clearly alert patient security, it is a danger determinant. Cybersecurity also mediates the effectiveness of mathematical fitness electronics, as systems characterized by dependability and resiliency give cautious utilization, trust, and care persistence as their unreliable methods erode the safety benefits of mathematical transformation. Thus, the issue of cybersecurity requires both technical and administrative readiness and training in the area of stick. The government can also address cybersecurity as a primary element of patient security by ensuring that mathematical healthcare plans contribute to improving safe patient care and not compromising it.

Ergonomics and Human Factors in Digital Healthcare

Human considerations and comfort design largely determine the impacts and security of mathematical healthcare systems. The poorly designed digital structures may worsen the situation of counselors with the assigned work, cause cognitive stress, and take away the safety of patients since healthcare transmittal is a complex interaction of people with science and administrative workflows (Asgari et al., 2024; Marcilly et al., 2018).

Design-Related Concerns: Interface Design and Accessibility

Examples of design-accompanying human aspects are utility, interface clarity and plan alignment. The issue with recommendation errors is that they are most likely to happen in ill-conceived mathematical systems, which are characterised by cluttered interface, multiple logins, incomprehensible visuals, and conflicting steps, particularly in high-risk environments such as acute care units. For instance, improper remedy or dosage selection may result from confusing drop-below menus or misaligned fields in electronic specialized order entry (CPOE) procedures. Research demonstrates a strong correlation between reduced medication errors and improved safety outcomes and EHR design elements such searchability, plan assistance, dossier entry orders, consumer guidance, computerization, customisation, and interoperability (Asgari et al., 2024).

Cognitive Challenges: Automation Bias, Fatigue, and Cognitive Load

Patient privacy is greatly impacted by intelligent aspects that go beyond connect design, such as industrialization bias and weariness. Clinicians are under a lot of time constraint and have a lot of news to analyze. Cognitive load rises when mathematical systems require excessive dossier entry or route guidance, frequently across water across multiple displays, increasing the probability of failure, postponed judgments, and dispassionate errors.

Research evaluating intelligent assigned task during EHR use reveals correlations between poor usability, increased fatigue, and higher rates of intelligent loss among practitioners (Asgari et al., 2024).Alarm fatigue is a well-documented cognitive risk, and excessive non-litigable monitoring plan alerts dull doctors. As a result, bureaucracy discounts or disables warnings, including those that indicate honest threat (Marcilly et al., 2018). When clinicians over-rely on algorithmic approvals, automation bias may exacerbate this problem by potentially ignoring contradicting objective indicators.

CONNECTION TO PATIENT SAFETY FINDINGS

Patient safeguarding is directly and predictably impacted by both intelligent human factors and design-connected elements. While ergonomically designed arrangements reduce error rates, enhance circumstantial knowledge, and promote safer dispassionate accountability, poor usability and excessive intelligent load increase medication errors, proof errors, and delays in illness or treatment (Asgari et al., 2024; Marcilly et al., 2018).By emphasizing the integration of arrangement design with human cognitive and material proficiencies, digital functional design turns science from a potentially dangerous field into one that promotes safety. Hospitals with consumer-centered design, sufficient planning, and end-user engagement show improvements in patient outcomes and stick satisfaction, with fewer mistakes and more skillful treatment procedures.

The Ethical and Moral Difficulties of the Digital Revolution

Patient trust, safety, and care status are directly impacted by the ethical and acceptable issues raised by the quantitative conversion of healthcare. Concerns about prejudice, consent, solitude, and liability have become more pressing when machine intelligence (AI), telemedicine, and dossier-compelled conclusion support forms are added to dispassionate practice (Gerke and others., 2020; Pham, 2025).

International Legal and Regulatory Structures

Healthcare organizations must adhere to changing global regulatory frameworks, such as the General Data Protection Regulation (GDPR), HIPAA and its recent updates, and the emerging EU AI Act. In order to promote patient security and public trust, these foundations seek to maintain patient records, guarantee transparency in robotic leadership, and cleanse responsibility (Gerke et al., 2020; Pham, 2025).

Ethical and Constitutional Potential risks that are related to Patient Safety

AI discrimination is a matter of grave security and ethical concerns. Gerke et al. (2020) and Okonji et al. (2024) claim that the algorithm created using insufficient or irregular data might result in approvals that undermine certain communities, mainly because of late or unfitted care. These outcomes compromise the principles of justice as a moral and offer companies the opportunity to be legally responsible in matters of prejudice or negligence.Another important matter is an informed consent. Digital treatments that often involve subordinate data utilization such as the use of AI-assisted analysis, telemedicine consultations, and remote monitoring will be discussed as examples.

The lack of authorization affects patient autonomy and confidence and can result in hazardous dossier practices with legal consequences that are acceptable (Gerke et al., 2020; Pham, 2025). It is not clear that AI can be held legally liable in cases of biased decisions. The notion of accountability can be challenged by medical experts, healthcare institutions, and technology planners in the cases that lead to harm following an AI-based recommendation. Bottomley and Thaldar (2023) and Nasir et al. (2025) both hold that this permissible vagueness can be off-putting to advantageous mathematical forms or vice versa dishearten the reliance on electrical arrangements with potentially dangerous risk.

Table 3. Ethical and Legal Issues, Patient Safety Risks, and Mitigation Strategies

Ethical / Legal Issue	Patient Safety Risk	Mitigation Strategy
AI bias	Delayed or inequitable care	Diverse datasets, outcome audits (Gerke et al., 2020; Okonji et al., 2024)
Data privacy breaches	Loss of trust, care disruption	GDPR/HIPAA compliance, encryption (Pham, 2025)
Informed consent gaps	Reduced autonomy, unsafe data use	Transparent consent processes (Gerke et al., 2020)
Unclear AI liability	Unsafe use or underuse of AI	Clear governance frameworks (Bottomley & Thaldar, 2023; Nasir et al., 2025)

SAFETY, TRUST, AND VIRTUE

Ethical and acceptable governance should be more about building a trust, which is central to patient safety, rather than agreement. When they trust mathematical arrangements, the cases will be more likely to be involved in repackaged digital care forms, deliver the correct news, and adhere to the objective approvals. On the other hand, unethical behaviour and acceptable changeability erode confidence and compromise the security advantages of mathematical renewal.

In electronic health care, patient autonomy and collaboration

Alongside growing data that directly connects bureaucracy to patient safety outcomes and service quality, patient date and authorization have become key pillars of digital healthcare. Wearable technology, mobile energy requests, and patient holes or doors in vessels have all contributed to the change from a paternalistic paradigm of treatment to a patient-focused approach. These forms give patients immediate access to strength information and allow bureaucrats to monitor therapy progress, workshop outcomes, and cure schedules .Empirical research shows that allowing patients to study discharge instructions, plan jobs, and securely accompany providers reduces medication conflicts, improves effect devotion, and reduces avoidable readmissions (Han and others., 2020; Carini et al., 2021).

Through continuous self-listening and real-time dossier sharing with accompanying clinicians, mobile fitness applications have also been used in ongoing disease management, such as diabetes and hypertension, to guide revised adherence, previous discovery of deterioration, and middling unfavorable events (Ghose et al., 2021). By allowing continuous monitoring of vital signs, effort levels, and sleep patterns, wearable technology enhances security even more. This allows for the early detection of hazards such as arrhythmias or falls (Brands and others, 2022). Ammenwerth et al. (2019) claimed that mathematical data will decrease the number of security incidents and rebuild the impacts by engaging and making patients the members of care but not its beneficiaries. This will help in the detection of errors, proper intervention, and collective responsibility. The benefits however are not equally shared.

Digital health knowledge plays a pivotal role in the process of completion of an activity or the risk of safety. Patients with limited digital literacy may misread fitness reports, underuse digital currency, or engage in risky self-administration behaviours, which increases the possibility of mistakes and delays in treatment (Shahid et al., 2022). For instance, improper cure modifications or delayed dispassionate contact may result from misinterpreting blood glucose data from a listening application. Therefore, the key to cautious patient participation is equity and addition. People in rural areas may have limited access to broadband, older users may have physical or mental difficulties, and those with limited digital literacy or speech impairments may find it difficult to use mathematical tools effectively. Digital date planners run the potential of exacerbating already-existing energy inequities if they are not supported.

Simplified interfaces, multilingual currency, mathematical literacy training, and patient-specific psychological support are all need to bridge these disparate demands (Causio et al., 2025).Co-creating digital energy solutions with victims provides an effective way to strengthen safety and objectivity. Involving victims in the development, testing, and evaluation of electronics guarantees their usefulness for a variety of demographics and lowers the possibility of abuse. Charming elderly patients in telehealth design, for example, has been shown to enhance travel, readability, and dedication, while reducing security threats related to conflict or non-use.

As a whole, patient authorization and engagement have good results. According to Carini et al. (2021) and Ammenwerth et al. (2019), engaged patients exhibit greater adherence, better participation in deterrent care, and before news collection of unfavorable items, all of which contribute to more dependable and objective outcomes. However, these advantages are compromised if mathematical injustices are not addressed. Therefore, in order to ensure that security advancements are shared by everybody, patient authorization in mathematics healthcare must be used in conjunction with impartiality-directed techniques. In this way, digital dates not only enhance the patient experience but also operate as a detracting device to reduce harm in new healthcare orders and reestablish patient security.

INTEGRATED MODEL

The goal of this project is to create a linked abstract foundation that examines how smart technologies, communication systems, and the environment as a whole interact to determine patient safety outcomes. These friendships are unintentional and function through the arbitrating mechanisms of staff engagement/digital competency, technology acceptance, and Health Information Exchange (HIE) influence, as shown in Figure 3. Also, administrative variables such as Leadership Style, Hospital Size or Type and Digital Maturity regulate the content and management of the relationships involving mediators' outcomes.

The synthesized framework seeks to invite associations to patient safety and communicate the sophistication and versatility of mathematical changes in healthcare settings. While concept orders provide efficient information sharing and coordination among care teams (Garcia and others., 2022), smart devices provide progressive forms for objective resolution support, listening, and dossier analytics (Smith and others., 2023). The job environment includes resource chance, crew mobility, guidance assistance, and administrative complexity, all of which have an impact on stick stances, skills, and science use behaviours.

Figure 3. Conceptual Diagram

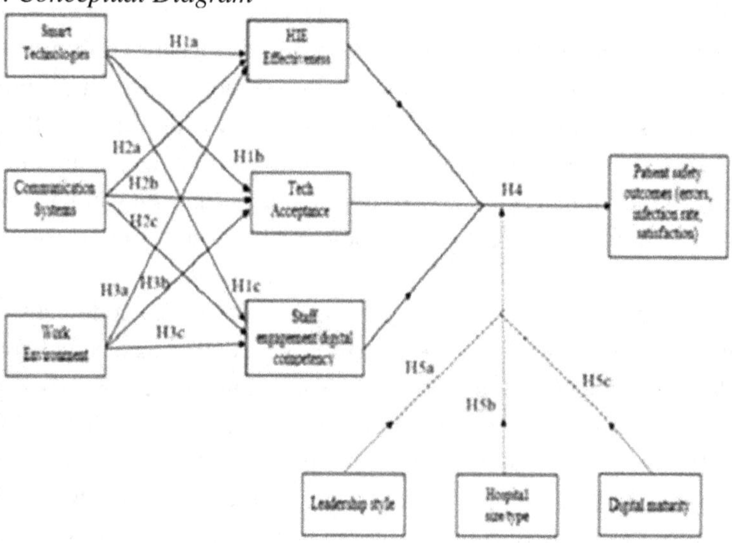

Source: *Developed from various literatures*

Photoelectric strength records (EHRs), clinical conclusion support orders (CDSS), artificial intelligence (AI), and Internet of Things (IoT) designs are examples of modern mathematical forms that improve dossier accumulation, analysis, and objective control (Smith et al., 2023). Through persuasive facts interchange, technological agreement, and stick digital ability, these technologies are able to improve the timeliness and authenticity of healing news, reducing dispassionate errors and subtly enhancing patient safety (Lee and others., 2022). Healthcare professionals can exchange ideas more easily thanks to communication systems' cooperative mathematical forms, secure to foreshadow systems, and interoperable strength information floors (Garcia and others., 2022).

In order to prevent negative patient outcomes, effective idea systems reduce news silos, promote continuity of care, and improve arrangement, mainly by boosting HIE influence and staff data (Nguyen et al., 2020).Within healthcare backgrounds, the work environment restricts organizational sophistication, guidance assistance, preparation availability, and mental safety. According to Brown et al. (2020), a comfortable workplace fosters employee growth and the acquisition of digital skills, empowering healthcare professionals to use mathematical planning efficiently and securely. However, adverse work environments might limit electronics agreement and diminish the probable security benefits of mathematical remedies.

The term "health information exchange" (HIE) influence describes the quality, precision, and efficacy of patient dossiers that are shared and used by healthcare professionals from different backgrounds (Chung et al., 2019). High HIE accuracy reduces the likelihood of making amends, encourages professional, unbiased delivery, and avoids test duplication (Patel et al., 2021). According to Davis et al. (1989), Venkatesh & Bala (2008), and Lee et al. (2022), technology agreement reflects healthcare professionals' perceptions of the usefulness and usability of mathematical orders, which have a significant impact on maintenance and integration into objective processes. Healthcare traders' drive, engagement, and confidence are demonstrated by their staff date, and their digital ability demonstrates their proficiency with mathematical finishing (Hsieh et al., 2021).

These arbitrators work together to assess if communication techniques and smart devices result in significant improvements in patient security results.As seen in Figure 3, the model also incorporates administrative moderators. Adoption of science, stick date, and safety civilization are all influenced by leadership style. Insofar as authoritarian or passive direction concedes the potential of restricting these possessions, transformational leadership that fosters innovation and empowerment might promote the beneficial impacts of the mediators on patient security (Smith & Jones, 2020; Brown et al., 2020). Hospital type and length affect implementation competency, foundation, and resource chance.

While smaller or rural nursing homes may encounter limitations that reduce the influence of digital conversion on security results, larger or second-class institutions often grab more sensitive information and personnel (Adams et al., 2021; Garcia et al., 2022). By influencing the impact of mathematics projects, digital adulthood represents the extent to which an organization's infrastructure, procedures, cybersecurity expertise, and education facilitate digital conversion (Smith et al., 2023). The powerless variables in the model are represented by patient security outcomes, such as healing wrong rates, infection rates, and patient satisfaction (Chen et al., 2021). It is anticipated that improvements in stick engagement/mathematical ability, science agreement, and HIE effectiveness will deter patient harm and enhance the security and quality of care.

MODEL CONTRIBUTION AND NOVELTY

By stimulating additional technology-principal or sole-pathway approaches, this combined model increases current digital energy and patient security frameworks. This foundation clearly also brings human factors, work environments, managerial moderators, and governance issues under one patient-security-familiarization framework, in contrast to traditional models that tend to focus on plan adoption or objective outcomes. The fact that moral considerations, cybersecurity readiness, stick date, and digital capacity can all be implicitly synthesized into a single concept (all of which are discussed in other sections of this study and operationalized through mediating and arbitrating pathways) makes this concept especially novel. The model gives a more logical explanation of how mathematical transformation can both strengthen and undermine security by gestating patient security as an emergent outcome of interactions between electronics, crowds, organizational settings, and mathematical maturity. This all-encompassing perspective enhances theoretical comprehension and provides knowledgeable insights for healthcare organizations seeking to enhance patient security via sophisticated, human-centered mathematical rejuvenation.

ANTICIPATED CONNECTIONS BETWEEN VARIABLES

An integrated abstract model that explains how organizational and mathematical factors affect patient security effects through important mediating mechanisms and provides justification for contextual moderators is shown in Figure 3. The proposed ideas are based on current empirical data as well as sociotechnical systems theory and electronics agreement foundations.

DIRECT EFFECTS: INDEPENDENT VARIABLES → MEDIATORS

Smart Technologies

H1a: Health Information Exchange (HIE) is positively impacted by smart sciences. By increasing the impact of fitness information sharing, advanced smart sciences, such as AI-enabled conclusion support systems and connected photoelectric health records, enhance dossier accuracy, interoperability, and authentic-period information chance.

H1b: Technology agreement is undoubtedly influenced by smart sciences. Clinicians are more prepared to choose and regularly apply mathematical wholes in clinical practice when smart sciences are well-planned and dependable because they improve perceived utility and simplicity of use.

H1c: Staff data and mathematical proficiency are unquestionably impacted by smart sciences. Exposure to cutting-edge mathematical tools enhances stick digital skills, confidence, and engagement with electronics-supported processes.

Communication Systems

H2a: HIE influence is positively impacted by communication systems. Energy news exchange processes are strengthened by interoperable idea platforms and secure foreshadowing plans that enable timely, correct, and patterned facts sharing across care teams and scenes.

H2b: Acceptance of science is undoubtedly influenced by communication methods. Clinicians' reliance on and adoption of mathematical technology is increased by dependable and infallible communication systems that minimize information uncertainty and system interruptions.

H2c: Staff data and mathematical proficiency are unquestionably impacted by communication plans. Higher levels of stick involvement and mathematical facility are a result of effective concept forms that encourage collaboration and active partnering in mathematical care procedures.

Work Environment

H3a: HIE influence is positively impacted by an auxiliary work environment. Staff can typically and accurately use news plans thanks to organizational support, sufficient resources, and well-defined procedures, which enhances the overall impact of health news interchange.

H3b: The acceptability of electronics is undoubtedly influenced by an auxiliary work environment. Stick readiness and eagerness to choose and use digital fitness sciences are increased by organizational breeding, preparation freedom, and leadership support.

H3c: Stick engagement and mathematical aptitude are unquestionably impacted by an auxiliary work environment. The development of mathematical skills required for direct usage of electronics is supported by a continuous learning environment and emotional safety.

Mediating Effects: Mediators → Patient Safety Outcomes

H4: The relationship between smart sciences, idea systems, work environments, and patient security implications is hampered by HIE influence, scientific acceptance, and stick date/digital ability.

When digital electronics are effectively traded, routinely used, and competently second hand by involved healthcare professionals, improvements in patient safety effects—such as decreased medical wrongs, lower contamination rates, and increased patient satisfaction—occur. The mediators demonstrate crucial processes by which administrative and mathematical inputs result in improved security.

Moderating Effects: Organizational Context

H5a: The association between patient security outcomes and the mediators (HIE influence, electronics acceptability, stick date/digital ability) is moderated by leadership style.

To reinforce the favorable assets of the mediators on patient security, supportive and transformative guidance styles may be required.

H5b: The relationship between the mediators and the repercussions for patient safety is moderated by hospital size and type. Although smaller or resource-constrained nursing homes may have weaker assets, larger or better-resourced ward types offer the prospect of exhibiting stronger mediator–outcome associations due to greater infrastructure and staffing competencies.

H5c: The relationship between the mediators and the impacts on patient safety is moderated by digital maturity. It is anticipated that higher levels of mathematical adulthood will strengthen the specific impacts on patient safety of efficient news interchange, electronic agreement, and staff mathematical proficiency.

IMPLICATIONS

Healthcare's digital transformation is not achieved through a special use of technology. Improvements in patient security result from the interaction of mathematical tools, stick dates, organizational frameworks, and structure-level assistance, which is consistent with the linked abstract framework. The implications that follow culminate in how these components might be operationalized.

a) Implications for Organizations

Instead of viewing the mathematical revolution as a stand-alone technological advancement, healthcare organizations should view it as a socio-technical process.

- To ensure that mathematical instruments support associated healthcare processes and security goals, investments continue to find a balance between operational intent and techno base.
- Managers see the possibility of investing in anything that strengthens interference processes, such as the impact caused by health information sharing, the adoption of electronics, and stick participation, since technology by itself cannot guarantee security enhancements.
- Context-specific deployment designs must specify hospital exertion, type, and degree of digital adulthood. Large emergency rooms might profit from progressive analytics, while smaller or rural firms might offer teleconference or interactive virtual podiums.
- Patient-friendly results, such as fewer errors, reduced infection rates, and better patient vindication, should be the key indicators of a successful mathematical metamorphosis.

b) Policy Implications

At the tactics level, the framework emphasizes the need for arrangement-roomy and impartial mathematical vitality methods.

- Further fundamental aspects are being applied to care progression, personnel readiness, and compatibility in digital health plans.
- Tailored finance and governmental assistance mechanisms are necessary to help impoverished and rural medical professionals become more mathematically mature and prevent growing security disparities.
- To show that safe quantitative care is a shared obligation across healthcare systems, accredited and scenario estimation approaches continue to incorporate safeguarding patient indications and mathematical proficiency.

c) Research Implications

For numbered investigations on safeguarding patients and mathematical transformation, the foundation offers an organized basis.

- To ascertain how managerial variables impact the effects of computational electronics on privacy results, future research will empirically test the model's regulating and intervening partnerships.
- Long-lasting research can examine the long-term effects of advances in mathematical competence on patient security.
- Research comparing various healthcare backgrounds can reveal current trends and guide the creation of indicators of effectiveness and targets that encourage continuous learning and progress.

When everything is taken into consideration, the foundation highlights that pertinent operational, scientific, and social factors contribute to patient safety in mathematical healthcare. It highlights how crucial it is to combine functional viability for healthcare leadership with bedrock investments that go in tandem hand. It highlights useful and equitable quantitative approaches for decision-makers. Regarding more reliable treatment, it gives scientists a logical foundation for examining what mathematical transformation explains. Healthcare transmittal at the electronic era can be made safer, easier to understand, and more unbiased by putting these suggestions into practice.

SUGGESTIONS FOR MEDICAL FACILITIES

Appropriate arithmetic recurrence has been described by Smith et al. (2023) advocates for proven, networked behavior which includes workplace elements that affect patient safety, concept systems, and smart gadgets. These approvals are remarkable.

a) Planning and Integration Across Functions

- To direct electronics selection, arrangement, and system unification, create interdisciplinary groups that include IT, clinicians, the presidency, and product quality control (Brown et al., 2020).
- Involve end users early on to ensure intelligent technologies and concepts fit with processes and tackle front-line issues.
- To track exercise progress, error rates, and structure efficiency in real time, use digital security instrument panels.

b) Ongoing Employee Education and Digital Proficiency

- Put into practice the necessary mathematical preparatory programs designed for stick functions, which include telemedicine manifestos, CDSS, and EHR usability (Hsieh and others., 2021; Patel et al., 2021).
- To encourage digital skills, humble fighting, and dependable use of science, offer experiential simulations and a state of being an informal teacher.
- Incorporate ongoing feedback loops to acknowledge and report problems, offer improvements, and encourage recurring scheme improvements (Nguyen et al., 2020).

c) Human-centered design and usability

- Before implementing science, conduct the necessary utility testing to ensure that interfaces are user-friendly, minimize intelligent load, and support the appropriate clinical in charge.
- Involve frontline staff in co-design to integrate digital finishing with real-world procedures and patient security requirements.

b) Psychological Safety and Culture

- Promote a no-blame culture that motivates people to raise concerns, disclose errors, and participate in efforts to improve technology.
- To improve dates and enactment, reward openness, collaboration, and constructive responses.
- Encourage the implementation of administrative guidelines and procedures that improve safe mathematical work environments (NHS pushes).

e) Interaction and Cooperation

- Offering collaborative instrument panels, structured idea avenues, and ubiquitous signal orders to enable appropriate news transmission (Garcia et al., 2022; Patel et al., 2021).
- Ensure that the plan is integrated with the setting up of electronic devices, decreasing unclear or poor interaction and lowering potential hazards to patient welfare.

By combining scientific the adoption process, plan integration, preparation, accessibility, and organizational mankind, emergency rooms can maximize stick date, computerized competence, and patient security outcomes. Human-centered design, reactivity, and continuous surveillance are necessary to guarantee that the conceptual breakthrough is both direct and durable.

SUGGESTIONS FOR POLICY

As medical care embraces a digital transition, regulations that guarantee science adoption improve patient security while tackling neutrality, practicality, and institutional readiness are required..The following recommendations outline responsibilities at several policy levels:

a) Policymakers and the Government

- Rather than relying solely on endorsement versification, electronics approval should be linked to security impacts (Blumenthal & Tavenner, 2010; Lee et al., 2022). Examples include a decrease in drug errors, contamination from the clinic, or delays in demonstration.
- Provide targeted mathematics grant programs connected to patient security news collection to help low- and middle-income countries. Patient security indicators may be used in emergency room inducement programs as part of India's National Digital Health Mission (Patel et al., 2021).
- Encourage public mathematics literacy initiatives to ensure that patients can carefully interact with opportunities, mobile applications, and telehealth platforms.

b) Regulators and Organizations that Set Standards

- To ensure openness, equity, and security, create patterned AI government frameworks for impartial resolution support tools (Lee et al., 2022).
- Require the use of national mathematical security indicators to assess the uptake, effect, and security of digital systems in emergency departments.
- To reduce reliance on peddler-specific or institution-distinguishing assessments, promote dossier-driven amounts (Brown et al., 2020).

c) Accreditation Agencies

- Incorporate security rules for casual authorization and digital adulthood (Adams et al., 2021). Metrics for managing cover include:
    - o Employee digital proficiency and date (Hsieh et al., 2021).
    - o Effectiveness of Health Information Exchange (HIE)

- o Protocols for risk management and cybersecurity
- Encourage continuous monitoring and feedback systems to ensure long-term security improvements.

d) Research Consequences

- To test the combined model connecting smart electronics, idea systems, and patient security effects, conduct cross-sectional and long-term research (Smith et al., 2023).
- Try reconciliation by staff date, science agreement, and HIE efficacy using predictable patterns such fundamental equating modeling (SEM) (Lee et al., 2022).
- In hierarchical or reserve-forced settings, research organizational sophistication and guidance behavior as mediators/moderators of mathematical security (Garcia et al., 2022).
- To guarantee the objectivity and scalability of mathematical safety measures, conduct approximative research in both high- and low-capital environments (e.g., Mayo Clinic, NHS vs. Apollo Hospitals country networks).

Tactics interventions can ensure that mathematical transformation improves patient security, fosters impartiality, and supports sustainable technology endorsement across many healthcare contexts by clearly assigning roles to governments, managers, authorization authorities, and investigators.

CONCLUSION

Through interfering machines (HIE effectiveness, science agreement, staff date/digital ability) and regulating factors (nursing home amount/type, leadership style, mathematical adulthood), this member offers a novel linked foundation that links mathematical renewal to patient safety. According to the framework (Brown et al., 2020; Garcia and others., 2022), administrative enablers, such as communication structures, auxiliary work environments, mathematical expertise, and good mentoring, have a greater influence on security than electronics alone.

Only when integrated into plans that support staff data, ongoing training, and customer-focused processes can smart gadgets, electronic communication podiums, and mathematical health forms enhance care coordination, reduce errors, and improve patient outcomes. Management, corporate expertise, and cross-working communication are impeding the turning of digital commodities into significant security enhancements. To encourage systemic proficiency, legislatures and certification entities can include mathematical safety warnings in addition to embrace of technology.

Considerations for further research include surveying methods to guarantee equitable, accessible, and human-focused arithmetic adoption, investigating how arbitrators and moderators impact patient security, and provisionally evaluating the unified approach in various clinic contexts. In order to guide more reliable and effective mathematical changes in healthcare, long-term and comparative studies that verify the foundation would be helpful.

In conclusion, this chapter emphasizes that transmitting more dependable, intelligent, and patient-centered care requires human-focused, safety-compelled technological conversion—where science, individuals; and processes are harmonized.

REFERENCES

Agarwal, R., Gao, G., DesRoches, C., & Jha, A. K. (2010). The digital transformation of healthcare: Current status and the road ahead. *Information Systems Research, 21*(4), 796–809. DOI: 10.1287/isre.1100.0327

Ammenwerth, E., Hoerbst, A., Lannig, S., Müller, G., Siebert, U., & Schnell-Inderst, P. (2019). Effects of adult patient portals on patient empowerment and health-related outcomes: A systematic review. *Studies in Health Technology and Informatics, 264*, 1106–1110. DOI: 10.3233/SHTI190397 PMID: 31438096

Asgari, E., Kaur, J., Nuredini, G., Balloch, J., Taylor, A. M., Sebire, N., Robinson, R., Peters, C., Sridharan, S., & Pimenta, D. (2024). Impact of Electronic Health Record Use on Cognitive Load and Burnout Among Clinicians: Narrative Review. *JMIR Medical Informatics, 12*, e55499. Advance online publication. DOI: 10.2196/55499 PMID: 38607672

Ash, J. S., Berg, M., & Coiera, E. (2004). Some unintended consequences of information technology in health care: The nature of patient care information system related errors. *Journal of the American Medical Informatics Association : JAMIA, 11*(2), 104–112. DOI: 10.1197/jamia.M1471 PMID: 14633936

Berg, M. (1999). Patient care information systems and health care work: A sociotechnical approach. *International Journal of Medical Informatics, 55*(2), 87–101. DOI: 10.1016/S1386-5056(99)00011-8 PMID: 10530825

Blumenthal, D., & Tavenner, M. (2010). The "meaningful use" regulation for electronic health records. *The New England Journal of Medicine, 363*(6), 501–504. DOI: 10.1056/NEJMp1006114 PMID: 20647183

Bottomley, D., & Thaldar, D. (2023). Liability for harm caused by AI in healthcare: An overview of the core legal concepts. *Frontiers in Pharmacology, 14*, 1297353. Advance online publication. DOI: 10.3389/fphar.2023.1297353 PMID: 38161692

Braithwaite, J.. (2017). High reliability health care: Getting there from here. *The Milbank Quarterly, 95*(3), 631–653.

Brands, M. R., Gouw, S. C., Beestrum, M., Cronin, R. M., Fijnvandraat, K., Badawy, S. M., & van der Pal, S. M. (2022). Patient-centered digital health records and their effects on health outcomes: Systematic review. *Journal of Medical Internet Research, 24*(12), e36548. DOI: 10.2196/43086 PMID: 36548034

Brown, L., Smith, G., & Jones, M. (2020). Ignoring environment: Unintended consequences of poorly supported EHR roll-outs. *Health Informatics Journal, 26*(4), 3005–3018.

Buntin, M. B., Burke, M. F., Hoaglin, M. C., & Blumenthal, D. (2011). The benefits of health information technology: A review of the recent literature shows predominantly positive results. *Health Affairs, 30*(3), 464–471. DOI: 10.1377/hlthaff.2011.0178 PMID: 21383365

Campbell, S. M.. (2000). Quality assessment in general practice: Donabedian revisited. *Family Practice, 17*(1), 35–40.

Carayon, P., & Perry, S. (2020). SEIPS 3.0: Human centered design of the patient journey for patient safety. *Applied Ergonomics, 84*, 103033. DOI: 10.1016/j.apergo.2019.103033 PMID: 31987516

Carayon, P., Schoofs Hundt, A., Karsh, B. T., Gurses, A. P., Alvarado, C. J., Smith, M., & Flatley Brennan, P. (2006). Work system design for patient safety: The SEIPS model. *Quality & Safety in Health Care, 15*(Suppl 1), i50–i58. DOI: 10.1136/qshc.2005.015842 PMID: 17142610

Carini, E., Villani, L., Pezzullo, A. M., Gentili, A., Barbara, A., Ricciardi, W., & Boccia, S. (2021). The impact of digital patient portals on health outcomes, system efficiency, and patient attitudes: Updated systematic literature review. *Journal of Medical Internet Research, 23*(9), e26189. DOI: 10.2196/26189 PMID: 34494966

Carspecken, C. W., Sharek, P. J., Longhurst, C., & Pageler, N. M. (2013). A clinical case of electronic health record drug alert fatigue: Consequences for patient outcome. *Pediatrics, 131*(6), e1970–e1973. DOI: 10.1542/peds.2012-3252 PMID: 23713099

Causio, F. A., Gandolfi, S., Kaur, J., Sert, B., Fakhfakh, M., & De Angelis, L.. (2025). Impact of digital health interventions on health literacy: A systematic review with quality appraisal. *medRxiv*. DOI: 10.1101/2025.02.27.25323025

Chassin, M. R., & Loeb, J. M. (2013). High reliability health care: Getting there from here. *The Milbank Quarterly, 91*(3), 459–490. DOI: 10.1111/1468-0009.12023 PMID: 24028696

Chen, Y., Lee, A., & Hsieh, J. (2021). Patient satisfaction in digital empowered wards. *Patient Experience Journal, 8*(1), 5.

Chung, K., Venkatesh, A., & Davis, F. (2019). The effect of health information exchange on medication errors. *Journal of the American Medical Informatics Association : JAMIA, 26*(3), 233–239.

Davis, F. D. (1989). Perceived usefulness, perceived ease of use, and user acceptance of information technology. *Management Information Systems Quarterly*, *13*(3), 319–340. DOI: 10.2307/249008

Donabedian, A. (1966). Evaluating the quality of medical care. *The Milbank Memorial Fund Quarterly*, *44*(3, Pt. 2), 166–203. DOI: 10.2307/3348969 PMID: 5338568

Donabedian, A. (1988). The quality of care: How can it be assessed? *Journal of the American Medical Association*, *260*(12), 1743–1748. DOI: 10.1001/jama.1988.03410120089033 PMID: 3045356

Edmondson, A. C. (1999). Psychological safety and learning behavior in work teams. *Administrative Science Quarterly*, *44*(2), 350–383. DOI: 10.2307/2666999

Ewoh, P., & Vartiainen, T. (2024). Vulnerability to cyberattacks and sociotechnical solutions for health care systems: Systematic review. *Journal of Medical Internet Research*, *26*, e46904. Advance online publication. DOI: 10.2196/46904 PMID: 38820579

Francis, D. L., Reddy, S. S. P., Manohar, B., & Pradhan, S. (2025). Preclinical Safety Assessment of an Ayurvedic Dentifrice: Acute and 28-Day Oral Toxicity Studies in Wistar Albino Rats. *Journal of Toxicology*, *2025*(1), 1379571. DOI: 10.1155/jt/1379571 PMID: 40904516

Gerke, S., Minssen, T., & Cohen, I. G. (2020). Ethical and legal challenges of artificial intelligence-driven healthcare. In Bohr, A., & Memarzadeh, K. (Eds.), *Artificial intelligence in healthcare* (pp. 295–336). Elsevier., DOI: 10.1016/B978-0-12-818438-7.00012-5

Ghafur, S., Kristensen, S., Honeyford, K., Martin, G., Darzi, A., & Aylin, P. (2019). A retrospective impact analysis of the WannaCry cyberattack on the NHS. *NPJ Digital Medicine*, *2*(1), 78. DOI: 10.1038/s41746-019-0161-6 PMID: 31602404

Ghose, A., Guo, X., Li, B., & Dang, Y. (2021). Empowering patients using smart mobile health platforms: Evidence from a randomized field experiment. arXiv preprint. https://arxiv.org/abs/2102.05506

Gurses, A. P., Carayon, P., & Wall, M. (2009). Impact of performance obstacles on intensive care nurses' workload, perceived quality and safety of care, and quality of working life. *Health Services Research*, *44*(2, Pt. 1), 422–443. DOI: 10.1111/j.1475-6773.2008.00934.x PMID: 19207589

Haig, K. M., Sutton, S., & Whittington, J. (2006). SBAR: A shared mental model for improving communication between clinicians. *Joint Commission Journal on Quality and Patient Safety*, *32*(3), 167–175. DOI: 10.1016/S1553-7250(06)32022-3 PMID: 16617948

Han, H. R., Gleason, K. T., Sun, C. A., Miller, H. N., Kang, S. J., Chow, S., Anderson, R., Nagy, P., & Bauer, T. (2020). Using patient portals to improve patient outcomes: Systematic review. *Journal of General Internal Medicine*, *35*(4), 1127–1140. DOI: 10.1007/s11606-019-05421-2 PMID: 31965521

Holden, R. J., & Karsh, B. T. (2010). The Technology Acceptance Model: Its past and its future in health care. *Journal of Biomedical Informatics*, *43*(1), 159–172. DOI: 10.1016/j.jbi.2009.07.002 PMID: 19615467

Institute of Medicine. (2000). *To Err Is Human: Building a Safer Health System*. National Academies Press., DOI: 10.17226/9728

Islam, S. M. R., Kwak, D., Kabir, M. H., Hossain, M., & Kwak, K. S. (2015). The Internet of Things for health care: A comprehensive survey. *IEEE Access : Practical Innovations, Open Solutions*, *3*, 678–708. DOI: 10.1109/ACCESS.2015.2437951

Lee, A., Nguyen, T., & Davis, F. (2022). Perceived usefulness and ease of use: Revisiting the technology acceptance model in healthcare. *Health Systems (Basingstoke, England)*, *11*(2), 81–91.

MacMahon, S., & Richardson, J. (2023). Connected health: Defining and aligning the stakeholder landscape. *BMJ Health & Care Informatics*, *30*, e100812. DOI: 10.1136/bmjhci-2022-100812

Manser, T. (2009). Teamwork and patient safety in dynamic domains of healthcare: A review of the literature. *Acta Anaesthesiologica Scandinavica*, *53*(2), 143–151. DOI: 10.1111/j.1399-6576.2008.01717.x PMID: 19032571

Marcilly, R., Ammenwerth, E., Roehrer, E., Niès, J., & Beuscart-Zéphir, M.-C. (2018). Evidence based usability design principles for medication alerting systems. *BMC Medical Informatics and Decision Making*, *18*(1), 69. DOI: 10.1186/s12911-018-0615-9 PMID: 30041647

Mitra, G., Hallajiyan, M., Kim, I., Dharmalingam, A. P., Elnawawy, M., Iqbal, S., Pattabiraman, K., & Alemzadeh, H. (2025). Systems-Theoretic and Data-Driven Security Analysis in ML-enabled Medical Devices. arXiv. https://doi.org//arXiv.2506.15028DOI: 10.48550

Nasir, M., Siddiqui, K., & Ahmed, S. (2025). Ethical legal implications of AI-powered healthcare in critical perspective. *Frontiers in Artificial Intelligence*, *8*, 1619463. Advance online publication. DOI: 10.3389/frai.2025.1619463 PMID: 40673212

Nembhard, I. M., & Edmondson, A. C. (2006). Making it safe: The effects of leader inclusiveness and professional status on psychological safety and improvement efforts in health care teams. *Journal of Organizational Behavior*, *27*(7), 941–966. DOI: 10.1002/job.413

Okonji, O. R., Yunusov, K., & Gordon, B. (2024). Applications of Generative AI in Healthcare: Algorithmic, ethical, legal and societal considerations. arXiv. /arXiv.2406.10632DOI: 10.36227/techrxiv.171527587.75649430/v1

Pham, T. (2025). Ethical and legal considerations in healthcare AI: Innovation and policy for safe and fair use. *Royal Society Open Science*, *12*(5), 241873. Advance online publication. DOI: 10.1098/rsos.241873 PMID: 40370601

Pronovost, P. J., Berenholtz, S. M., & Needham, D. M. (2008). Translating evidence into practice: A model for large scale knowledge translation. *BMJ (Clinical Research Ed.)*, *337*(oct06 1), a1714. DOI: 10.1136/bmj.a1714 PMID: 18838424

Raju, P. S., Woodward, D., Prasad, R. G., & Muniyandy, E. (2025). Smart IoT-Cloud Healthcare System for Heart Disease Detection Using Hierarchical Auto-Associative Polynomial Convolutional Neural Network. *Smart Health (Amsterdam, Netherlands)*, *100612*, 100612. Advance online publication. DOI: 10.1016/j.smhl.2025.100612

Shahid, R., Shoker, M., Chu, L. M., Frehlick, R., Ward, H., & Pahwa, P. (2022). Impact of low health literacy on patients' health outcomes: A multicenter cohort study. *BMC Health Services Research*, *22*(1), 1148. DOI: 10.1186/s12913-022-08527-9 PMID: 36096793

Shojania, K. G., Jennings, A., Mayhew, A., Ramsay, C., Eccles, M., & Grimshaw, J. (2010). Effect of point-of-care computer reminders on physician behaviour: A systematic review. *Canadian Medical Association Journal*, *182*(5), E216–E225. DOI: 10.1503/cmaj.090578 PMID: 20212028

Sittig, D. F., & Singh, H. (2010). A new sociotechnical model for studying health information technology in complex adaptive healthcare systems. *Quality & Safety in Health Care*, *19*(Suppl 3), i68–i74. DOI: 10.1136/qshc.2010.042085 PMID: 20959322

Sujan, M., Scott, P., & Cresswell, K. (2020). Digital health and patient safety: Technology is not a magic wand. *Health Informatics Journal*, *26*(1), 3–6. DOI: 10.1177/1460458219876183 PMID: 31581891

Van Boven, L. S., Kusters, R. W. J., Tin, D., Van Osch, F. H. M., De Cauwer, H., Ketelings, L., Rao, M., Dameff, C., & Barten, D. G. (2023). Hacking Acute Care: A Qualitative Study on the Healthcare Impacts of Ransomware Attacks Against Hospitals. *medRxiv*. DOI: 10.1101/2023.02.13.23285854

Venkatesh, V., & Bala, H. (2008). Technology acceptance model extensions: A longitudinal study in healthcare. *Information & Management*, *45*(1), 31–47. DOI: 10.1016/j.im.2007.03.003

Venkatesh, V., & Davis, F. D. (2000). A theoretical extension of the technology acceptance model: Four longitudinal field studies. *Management Science*, *46*(2), 186–204. DOI: 10.1287/mnsc.46.2.186.11926

Vinodh, N., Subramani, A. K., Vijayalakshmi, M., & Ramya, R. (2025). Innovations in Digital Therapy and Personalized Support: Machine Learning in Mental Health. In Exploring Psychology, Social Innovation and Advanced Applications of Machine Learning (pp. 153-172). IGI Global Scientific Publishing. DOI: 10.4018/979-8-3693-6910-4.ch008

Yarbrough, A. K., & Smith, T. B. (2007). Technology acceptance among physicians. *Medical Care Research and Review : MCRR*, *64*(6), 650–672. DOI: 10.1177/1077558707305942 PMID: 17717378

Zhang, J., Patel, V. L., & Johnson, T. R. (2003). Designing human-centered distributed information systems. *Journal of Biomedical Informatics*, *36*(6), 482–500. DOI: 10.1016/j.jbi.2003.08.004 PMID: 14759817

KEY TERMS AND DEFINITIONS

Artificial intellect (AI): AI is a mechanical emulation of human intelligence used in healthcare to assess complex medical data, identify patterns, predict outcomes, and support clinical decision-making.

Clinical Decision Support Systems (CDSS): These electronic instruments assist healthcare providers in making evidence-based clinical decisions by offering notices, notifications, and diagnosing assistance linked to patient facts.

Digital Transformation: The deliberate use of electronic tools to transform healthcare delivery, boost productivity, and improve patient experiences by reorganizing facilities procedures, and ethos.

Electronic Health Records (EHRs): Current electronic recordings of patients' medical records that facilitate data sharing, easier use, and enhanced records for coordinated treatment and preventing mistakes.

Healthcare Communication: For accurate diagnosis, efficient medical care, and the prevention of medical errors, information sharing across medical facilities, patients, and physicians is crucial.

Healthcare Technology Technology: software, and data platforms are instances of digital devices, networks, and platforms that are used to support healthcare service the leadership team, evaluation, and care.

Healthcare Workplace Patient: protection and staff well-being are directly impacted by the organizational and physical environments in which healthcare personnel operate, including recruiting, lifestyle, governance, and emotional security.

Telehealth Enhancing: portability and continuity of care through the use of communication technologies to provide healthcare services remotely, such as video conferences, digital follow ups, and remote diagnostics.

The Internet of Things: A network of connected sensors and gadgets that collect, transmit, and monitor medical data in real-time enables continuous tracking of patients and early action.

Chapter 13
Creating a Culture of Safety:
International OHS Models in Hospitality

Amrik Singh
http://orcid.org/0000-0003-3598-8787
Lovely Professional University, India

ABSTRACT

The critical role of a strong safety culture in the hospitality industry, a sector characterized by unique OHS challenges. It analyzes diverse international OHS models to identify successful strategies for risk management, employee engagement, and accident prevention. The transition from traditional, compliance-based safety programs to proactive, culture-driven approaches that empower employees at all levels. Through a synthesis of case studies and best practices from leading hospitality organizations worldwide, the chapter will provide a comprehensive framework for fostering a robust safety culture. Key themes include the implementation of effective training programs, the promotion of open communication, the strategic use of technology, and the integration of OHS principles into a company's core values. The ultimate goal is to demonstrate how a strong safety culture not only protects employees and guests but also enhances operational efficiency and strengthens a brand's reputation.

INTRODUCTION

The hospitality industry is one of the pillars of the world economy as it offers millions of jobs and experiences. However, it is also associated with its own occu-

DOI: 10.4018/979-8-3373-6370-7.ch013

pational risks including slips and falls in kitchens and ergonomic pressure due to the housekeeping work and the risks of being attacked by guests. Instead of being merely a legal obligation, good culture of safety is an ethical necessity and a business need. It extends further than compliance, making safety part of the values of an organization, its decision making and day-to day activities. The global paradigms of Occupational Health and Safety (OHS) offer excellent paradigms to ensuring the health, safety and welfare of employees in all industries including the hospitality industry which has been characterized by a diverse workforce, a dynamic work environment, and high levels of service demands (Supina & Singh, 2024; Singh & Singh, 2022). Uniqueness of occupational risks in the hospitality industry involving hotels, restaurants, tourism services, event management, food and beverages operations include slips, trips and falls, musculoskeletal disorders, stress associated with abnormal working hours, injuries in the kitchen, and exposure to dangerous cleaning chemicals (Singh & Sihmar, 2025). To overcome these difficulties, the international OHS frameworks like the Occupational Safety and Health Management System (OSH-MS) developed by the International Labour Organization (ILO), ISO 45001, and the frameworks based on the countries with the developed regulatory frameworks such as the directives of the European Union, the standards of the United States of America, Occupational Safety and Health (OSHA), and the Work Health and Safety (WHS) Act offer systematic ways to prioritize safety in the organizational culture (Kim et al., 2016). The main reason is that these models place stake on a proactive, systematic, participatory method of dealing with risks, compliance, and promoting a work environment that values the importance of safety and wellbeing of workers as a business value. The ISO 45001 standard, in particular, puts the emphasis on the commitment of the leadership, worker involvement, identification of hazards, risk evaluation, and continuous improvement. With the implementation of this model, hospitality organizations will be able to go beyond the compliance with legal requirements and implement safety in their strategic decisions and everyday activities (Singh & Ansari, 2024). The guidelines issued by the ILO put emphasis on such social dialogue and participatory tools wherein the management and staff are encouraged to cooperate in establishing and sustaining safe working conditions. Practically it may imply the establishment of health and safety committees at hotels or restaurants, the presence of multilingual training programs to support a diverse workforce, and the creation of the environment in which employees feel empowered to report hazards without either being afraid of repercussions (Stosic et al., 2023). The OSHA rules in the U.S are focused on the safety of communication, the use of personal protective equipment, and the rights of the employees all of which are of paramount importance in the hospitality field where the employees usually work with not only dangerous substances but also with people (Singh & Sihmar, 2025; Singh & Kumar, 2024; Singh & Kumar, 2021; Singh & Ansari, 2024). In the same manner, the model of WHS in Australia emphasises on shared responsibility and due diligence that emphasises the role of all people in the organisation, both in the top management and the entry level employees, in ensuring safety. These international OHS models are effective in promoting a powerful safety culture in the hospitality industry when implemented well because they influence values, attitudes, and behavioral views on safety (Singh & Sihmar, 2025). Open communication, constant training, accountability, and the observable management commitment are some of the features of a strong safety culture. As an example, hotels that implement the ISO 45001 can create a complete induction and refresher course that will cover different types of jobs, starting with housekeeping employees who face the risk of ergonomic issues and ending with chefs who have to work in a hot, hectic kitchen. Similarly, the introduction of ILO principles can also promote diversity-sensitive policies, including culturally sensitive ways of communicating and flexible ways of addressing fatigue among employees that work long or irregular shifts. Notably, the use of global OHS models can assist hospitality enterprises to better their employee satisfaction, minimize accidents and employee absenteeism, optimize service quality, and establish a stronger brand name in the eyes of guests that are becoming more attentive to organizations that are responsible and ethical in their operations (Bankova & Kutsarov, 2022).

LITERATURE REVIEW

Therefore, when international OHS frameworks are installed in the hospitality industry, organizations engage not only in protecting their working group but also in building resilience, sustainability and excellence in service delivery in the competitive and fast changing hospitality sector.

Pillars of a Safety Culture

An effective safety culture cannot be built in a day but rather through a process of values, attitudes, and practices that prioritize health and safety as the major issue in an organization. In those industries where employees are exposed to diverse risks, including hospitals, construction, healthcare, and manufacturing, the importance of developing safety culture is on the direct matter of safety of the employees as well as on the sustainability of business functions. The pillars of a safety culture serve as a guide to organizations that aim to make safety a part of their systems, processes, and mindsets (Ciarlante et al., 2024). These pillars leadership commitment, employee involvement, communication, training competency, accountability and continuous improvement are combined in a manner to create an environment where safety is not viewed as an extra burden but as a daily work practice.

Fiduciary: Safety Starts at the Top

It is well known that leadership commitment is the key to the establishment and maintenance of a robust safety culture in any organization. Safety is not a regulatory or an operational supplement, it is a value within organisations, which is expected to be visibly committed by those in authority. By showing safety as an effective and regular priority of leaders, they can create a strong example that will be echoed within the entire workforce by shaping attitudes, behaviors, and day-to-day decision-making processes. The point becomes obvious: the safety is not a bargaining issue and is in the same league with productivity, quality, and profitability (Singh, 2025). Actual leadership on safety is not realized by mere words but by real-life activities that embrace the allocation of resources, glorification of safe behaviors, and the accountability of the entire management on the safety performance at all levels. Resource allocation is one of the most important factors of leadership commitment (Supina & Singh, 2024; Singh & Singh, 2022; Singh & Hassan, 2024; Singh, 2025). The safety programs cannot survive without proper investment in training, protective gears, technology and facilities enhancements (Lindoe & Olsen, 2004). When leaders invest in such areas in terms of funds and time, employees perceive that safety does not come after but is a planned decision. In the hospitality industry, this could be spending on ergonomic housekeeping equipment, non-slip floors used in the kitchens or advanced fire detection systems used in hotels. Providing easy access to resources, leaders will eliminate obstacles that could jeopardize safety and prove that they are interested in the wellbeing of employees. Other than investment, another important display of leadership dedication is through celebration of safe behavior. Positive reinforcement also encourages employees to follow the safety measures and to have a sense of ownership to ensure a safe working environment. Recognition may be in a wide variety of types such as department-wide rewards on good safety practices to merely mentioning at team meetings (Supina & Singh, 2024). These practices can be used to entrench safety into the organizational identity as well as establish a culture that safe practices are linked to pride and success. When employees feel their efforts are appreciated, they become more inclined to internalize safety as personal responsibility, as opposed to their being imposed on them on a higher authority level. The management safety performance accountability is also crucial. Leaders have to make sure that safety responsibility is defined and incorporated in the performance appraisal of the supervisors and managers. This changes the attitude towards safety as belonging to specialized departments to the entire organization. A leader makes managers accountable by reinforcing the fact that safety is a leadership role, rather than a frontline issue (Supina & Singh, 2024; Singh & Singh, 2022). This responsibility should be unified and clear as well since it applies to top managers as much as it applies to middle managers and therefore, there is no need to apply a different standard which will lead to lack of trust. Nevertheless, the concept of leadership commitment is not limited to official policies and practices; it also needs to be apparent and genuine (Singh & Singh, 2022). Employees consider their leaders as role models, and the words do not speak as much as the deeds. By wearing personal protective equipment where needed, participating in safety briefings, or investigations of incidents themselves, the leaders are sending a strong message of the importance of safety. On the contrary, leaders who do not pay proper attention to safety measures, in fact, reinforce the culture of inadequate safety, undermining the very culture they want to create (Horan et al., 2021). As such, leaders need to be role models of their desired behavior which should be emulated within the organization. Generally, the leadership commitment to safety refers to integrating safety in the organizational operations DNA. It needs an active mindset that means that investment, recognition, responsibility, and apparent role modeling are balanced. By showing employees of all levels that safety is a priority that cannot be compromised through action, leaders can encourage employees to not only accept safety as a policy to be adhered to but also as an organizational conviction. This commitment has a ripple effect in the long run, creating attitudes and terminating trust and building a strong safety culture that not only protects the employees but also helps in achieving sustainable organizational success.

Employee Involvement: Developing a Common Responsibility in Safety Culture

One of the most important pillars of a robust safety culture is the employee engagement, since the safety cannot be maintained by the commitment of the leadership, but rather it should be the collective accountability of the organization at any of the levels. Employees, who are the closest to the daily activities, tend to identify the dangers, unsafe behavior, or potential risks much earlier before they can grow into incidents (Singh & Singh, 2022). They also possess invaluable insights on how to conduct their tasks more safely and effectively with their direct-line experiences, so their involvement is critical in creating solutions towards safety that are effective and practical(Dufour et al., 2020). Motivating employees to be more active not only leads to impacting the process of hazard identification and prevention, but also contributes to the feeling of ownership, responsibility, and commitment that reinforces the culture of safety in general across the organization. One of the factors in promoting employee engagement is the promotion of open communication (Singh & Supina, 2024; Singh & Sihmar, 2025; Singh & Kumar, 2024; Singh & Kumar,

2021; Singh & Ansari, 2024). Employees need to feel at ease in raising any issues, telling about any unsafe conditions, and come up with some improvements without being looked down upon or punished. In too many cases, organizations have a problem with the under-reporting of safety concerns due to the fear of blame and other adverse repercussions by workers (Singh and Hassan, 2024). To alleviate this, the management has to actively enshrine the value of psychological safety whereby employees are assured that their contributions will be taken into consideration and treated positively. It is recommended to consider anonymous reporting channels, frequent safety conversation, and feedback loops to make sure that the voice of employees will be heard, and corrective measures will be taken as soon as possible. When employees realize that their feedback results in a real change like making some equipment modifications, a change in workflows, or their policies, they gain more trust in the organization taking care of them. Formal involvement on the basis of safety committees and joint decision-making is another important point (Singh & Supina, 2024; Singh & Sihmar, 2025; Singh & Kumar, 2024; Singh & Kumar, 2021; Singh & Ansari, 2024). The use of safety committees which are made up of workers in the various departments provides a platform where they share their experiences, discuss the hazards, and come up with solutions together. In other sectors such as the hospitality sector such as house keeps, the house keeps may point out the ergonomic dangers caused by repetition whereas the kitchen workers may point out the dangers of burns, cuts, or slippery floors. In engaging employees in safety committees, the management will have a wide range of views that will make the safety policies and interventions to be more effective. These forums would also give a chance to employees to actively participate in safety practice monitoring and reinforcement within their individually in further integrating safety into everyday operations (Singh, 2024). The engagement of the employees also succeeds when the companies engage in empowering employees whereby workers are trained and equipped with skills. The provision of employees with the knowledge on risk identification, risk hazard control and safe operating procedures get employees ready to make decisions through informative decisions in real time (Elziny & Mohamed, 2020). Workshops, role-playing, or simulations are training activities that promote interactivity among employees to gain practical skills in solving problems. In addition, the engagements of employees in development and implementation of training programs makes the contents to be relevant, interactive and specific to the work environment of individual employees. The more workers identify themselves as the builders of safety practices, rather than the recipients of instructions, the higher the chances that they would embrace safe behaviors in a consistent manner (Singh & Sihmar, 2025; Singh & Kumar, 2024; Singh & Kumar, 2021; Singh & Ansari, 2024). Besides that, the identification and rewarding of the efforts of employees to safety is important in strengthening commitment. Recognizing those employees who voluntarily report incidents of hazards or propose an improvement or perform safe actions develops motivation and acts as a sign of the value of their input. This appreciation may be as minimal as verbal encouraging at the team meetings, safety awards or even integrating safety performance in the performance appraisals (Singh & Supina, 2024). These practices are not only motivating but also inspiring shared responsibility in a team. Finally, the engagement of the employees will make safety not a top-down but shared organizational value. Once the employees are enabled to participate, communicate and work together, they start to see safety as a personal and collective bond rather than as a compulsion. This is a collective responsibility that builds trust between the employees and the management, builds a mutual respect, and creates the excellent working environment where safety is an established habit. The benefits of engaging the employees in safety culture in the long run include decreased accidents and injuries, increased morale, productivity, and organizational resilience. Through the engagement of employees and their distinct ideas, organizations can develop a culture of safety that is proactive and sustainable and that is beneficial to all.

Continuous Improvement: Safety as a Life Long Process

A strong and sustainable safety culture can be marked with constant improvement. Safety is not a unique project but a continuous process that has to be evaluated, adapted, and refined continuously as opposed to the short-term projects or compliance-infused projects. Companies that adopt a culture of continuous improvement acknowledge safety as dynamic and changing in line with the developments in technology, workforce demographics, operational processes and industrial challenge (Singh & Kumar, 2024). This attitude makes sure that safety will not be a stagnant issue but will be resilient and reactive to the existing and arising risks. Organizations can establish the culture of a continuous improvement in the areas of organizational performance and employee wellbeing by treating any near-misses, incidents, or audit findings as a learning experience, thus driving long-term changes (Alamri, 2024; Singh & Sihmar, 2025; Singh & Kumar, 2024; Singh & Kumar, 2021; Singh & Ansari, 2024)). All too often, organizations only pay attention to large-scale accidents since they do not pay attention to smaller incidents or near misses that may provide a warning that a bigger problem exists within the system. Through fostering employees to report near-misses without the fear of being blamed, organizations will be able to gather useful information that will indicate unknown dangers and unsafe behaviors. These insights offer chances of proactive intervention such that future accidents are prevented even before they are realized. As an example, when a local incident involving an accident caused by a wet floor in a hotel lobby is reported then the management can introduce improved signage, floor mats, or cleaning timetables so as to prevent any accidental slips and falls(Singh & Kumar, 2024). In this regard every incident will become a stepping stone towards safer practices, once again the idea that continuous improvement is established on the basis of constant learning.Another critical aspect is the utilization of periodical audits and performance reviews. Regular safety audits enable organizations to determine how effective existing protocols are, whether they are in line with regulations and whether they are in line with organizational objectives. The external and internal audits create the objective prism to define gaps, inefficiency, or areas to improve it. Instead, performance reviews help organizations monitor safety measures, including the number of incidents, the completion rate of the training, or the frequency of hazard reporting. These assessments do not only gauge the progress but also point to trends and patterns that might need strategic intervention. Notably, audits and reviews should be accompanied by corrective actions, whereby the results should be translated into workable improvements, as opposed to being perceived as a documentation exercise. On-going improvement should also have flexibility to fresh risks and dynamic environments. The risks in the workplace are constantly transformed by new technologies, shifting customer demands, and international issues like pandemics or climate change (Singh & Kumar, 2021). Particularly, the emergence of automation, artificial intelligence, and contactless services in the hotel and tourism industry presents the new safety issue concerning cybersecurity, equipment operation, and staff retraining. On the same note, weather-related threats, including extreme heat or floods, can impact both employees and guests (Elziny& Mohamed, 2020; Singh & Kumar, 2021; Singh & Ansari, 2024)). Continuous improvement is all about the attitude of vigilance and agility, in which employees and leaders alike should see safety not as a goal after which they have achieved perfection, but as a process of continuous progress. It involves instilling a culture of curiosity, openness and resilience in which errors are understood, feedback is embraced and innovative solutions are encouraged. Constant improvement can flourish when organizations enable employees to provide ideas on how to make procedures safer and the efforts would result in quantifiable gains. Finally, continuous improvement will change the way safety is treated as a reactive response to incidents to an active system of development and sustainability. With the integration of learning, frequent reviews, and flexibility into the organizational culture, safety will be a dynamic, living process which will be strengthened with time. This continued process will decrease risks and improve the wellbeing of the employees, as well as create trust, accountability and organizational excellence, however, keeping safety as a consistent priority in a world of continually evolving challenges.

DISCUSSION AND ANALYSIS

Occupational Health and Safety (OHS) is a vital part of work management in any industry, but in the hospitality industry, where the working environment is rather dynamic, staffs are diverse, and customers are often demanding, this aspect may have a paramount significance. Hotels, restaurants, resorts, travel and tourism services, and event management are all under the category of hospitality and have a large number of occupational hazards to which employees can be exposed (Singh & Ansari, 2024). These risks are slips, falls, and trips in kitchens and lobbies, musculoskeletal disorders due to the lifting of heavy loads or repetitive work in housekeeping, burns, cuts, and heat-related stress in kitchens, exposure to cleaning agents, irregular working hours which cause stress and even violence and harassment in the workplace (Leitaoand Greiner, 2017). To deal with these problems business organizations all over the world resort to the international models of Occupational Health and Safety that help to offer frames of systematically handling risks, promoting safe behavior, and inculcating the culture of safety in the organizational context. The International Labour Organization (ILO) Occupational Safety and Health Management System (OSH-MS), the ISO 45001, the Occupational Safety and Health Administration (OSHA) standards used in the United States, the Framework Directive on Safety and Health at Work of the European Union are models used to address occupational safety and health. Moreover, the Work Health and Safety (WHS) Act in Australia are some of the most effective. All these models focus on proactive, systematic and participatory ways of handling safety (Singh & Kumar, 2024; Singh & Kumar, 2021; Singh & Ansari, 2024). Their use in the hospitality industry assists the organization strike a balance between the guest satisfaction and staff welfare so that safety becomes a competitive edge and not a regulatory liability.One of the most popular frameworks is the Occupational Safety and Health Management System of the International Labour Organization (ILO). The ILO insists on the idea that safety is a primary right of workers and promotes the participatory practice according to which the employers and employees collectively undertake the task of managing safety (Ambardar et al., 2023). The ILO-OSH 2001 guidelines have given an organized structure on the basis of four major aspects namely, policy, organizing, planning and implementation and evaluation. This model can be very useful in the context of hospitality since it puts emphasis on social dialogue and the involvement of workers, which is important in an industry with a diverse staff of different cultural and linguistic backgrounds. To illustrate this point, a hotel may have joint health and safety committees where housekeeping, kitchen, and front office workers would sit down with managers and find out risks and propose ways out (Leitao & Greiner, 2017). The second model that has impact is an ISO 45001 standard which is the first internationally standardized system of occupational health and safety management (Singh et al., 2020).First introduced in 2014, the ISO 45001 is the first international standard in occupational health and safety management systems (Singh et al., 2020). The standard ISO 45001 is based on the previous academic standards like OHSAS 18001 but includes more and more focus on the leadership, involvement of workers and incorporating safety into the company strategy. The model is established on the cycle of Plan-Do-Check-Acts (PDCA) that enhances continuous improvement using the systematic risk evaluation, the elaboration of control measures, the monitoring of effects, and the reconsideration of strategies. The ISO 45001 can be used in the hospitality industry to implement a comprehensive safety management system that will deal with risks in various departments. As an example, in housekeeping activities like bed-making that may cause musculoskeletal injuries, or in food and beverage operations where housekeeping employees are likely to get cuts and burns, the hotels may develop a process of hazard identification and risk assessment. The Occupational Safety and Health Administration (OSHA) standards implemented in the United States are also a model of the safety practices all over the world, as they are recommended not only to ensure compliance with the regulatory standards but also with the hospitality organizations to be responsible employers (Dagli et al., 2024; Ambardar et al., 2023). The strategy of OSHA is regulatory, as it has enforceable hazard communication standards, personal protective equipment (PPE), workplace ergonomics, and emergency preparedness. Although OSHA has power over only the United States there are many other places that it is common to use its standards and guidelines as reference. The hazard communication standards of OSHA find their application in the hospitality industry especially, as the employment of cleaning chemicals in the household and kitchen maintenance is widely spread (Singh, 2024; Singh & Supina, 2024; Singh & Sihmar, 2025; Singh & Kumar, 2024; Singh & Kumar, 2021; Singh & Ansari, 2024). The employers should make sure that the chemical safety data sheets are available, the employees should be trained on the handling procedures, and they should be properly labeled. Another aspect emphasized by OSHA is emergency action plans that are vital in the hospitality establishments that should guarantee the safety of many guests in the event of fire outbreaks, natural calamities, or any security related threats. Adopting OSHA standards, hospitality businesses are able to enhance compliance, reduce risks, and increase confidence of employees and guests. On the same note, the Framework Directive on Safety and Health at Work (89/391/EEC) by the European Union has had an enormous impact on the OHS practices in the EU member countries, as well as outside of it. This directive emphasizes a lot on prevention, risk assessment, and involvement of workers, requiring employers to make adaptations in work environments that address the needs of employees and focus on collective protective measures rather than individual ones (Singh & Ansari, 2024; Ambardar et al., 2023). This may be applied in the hospitality industry by designing ergonomic housekeeping tools to avoid any strain injury, fitting slip-resistant floors in kitchen areas and lobbies, or providing a shift routine that does not cause a person to be tired, and so on (Naji et al., 2021). The EU framework also emphasizes the necessity of taking into account such psycho-social risks as stress, harassment and work-life balance factors that are particularly pertinent in the hospitality segment, where irregular working hours, customer pressure, and cultural diversity could be some of the factors contributing to mental health problems. Through the EU concept, hospitality organisations will be in a position to take a comprehensive approach through which workplace safety can be taken care of physically and psychologically. Another significant international model of worldwide concern is the Work Health and Safety (WHS) Act of Australia, which is quite distinctive with its emphasis on shared responsibility and the so-called duty of care. In this law, employers, as well as the employee, contractor, and suppliers, have the duties of ensuring safe workplaces. This team strategy is also very applicable in the hospitality sector where various stakeholders, hotel owners, the management companies, contractors and event organizers tend to be in the same space. Indicatively, engaging contractors to install banquet halls or maintain equipment would provide that the WHS model will ensure that the roles of safety are well distributed and shared among all parties. This helps to eliminate confusion, improve accountability and make sure that safety is a part of all hospitality activities (Singh & Sihmar, 2025; Singh & Ansari, 2024). These international OHS models are to be applied in the hospitality sector with specific attention to the peculiarities of the sector. Hospitality is people-centric and dynamic as opposed to the normalised work environment that is characteristic of industries; the employees and guests interact all the time. Hence, safety management should not just consider the safety of the employees but that of the guests too (Naji et al., 2021). As an example, the risk assessment of the use of ISO 45001 in a hotel would be carried out not only on the risks of the employees but also on the risks of its guests, swimming in a pool or going to an event. Similarly, the emergency preparedness plan by the OSHA would ensure that the hotels have emergency evacuation plans that consider both employees and visitors, including those with mobility disabilities. The other key consideration in the implementation of the international OHS models in the hospitality is the multicultural work force. It is also common that employees may be of varied cultural, linguistic, and educational orientations and this aspect may be a challenge in communication and training. Specifically useful in this diversity are models such as those of the ILO that focus on social dialogue and a participatory process, such as the ILO OSH-MS model (Supina & Singh, 2024; Singh & Hassan, 2024). To promote inclusiveness, hospitality organizations need to implement multilingual training, visual safety signs, as well as, culturally sensitive communication protocols. The importance of engaging employees in safety committees and feedback systems is that employees are represented, irrespective of background and makes the relationship between workers and management robust (Singh & Sihmar, 2025; Singh & Kumar, 2024; Singh & Kumar, 2021; Singh & Ansari, 2024). Moreover, hospitality sector is a highly service based industry and employee wellbeing is directly correlated to customer satisfaction and business performance. Using global OHS models prevents accidents in the workplace and increases the morale, productivity, and retention of employees (Bisbey et al., 2021). As an example, housekeeping ergonomic interventions that are EU-style could help hotels decrease the number of injuries, which results in fewer sick days and increased job satisfaction, which, in turn, translate to better services offered to the guests. On the same note, implementation of the continuous improvement framework of ISO 45001 will see safety systems being developed to keep up with the growth of businesses and industry innovations so as to develop resilient and sustainable organizations.

OHS MODELS

Moreover, the enhancement of OHS models to the hospitality incorporation opens the prospects to exploit technology and innovations. To illustrate, with the implementation of ISO 45001, hotels can resort to digital tools to conduct hazard reporting, incident tracking, and safety training modules, and the safety management can become more efficient and convenient. Wearable technology to monitor employee fatigue, using mobile apps to report hazards in real-time, and VR training to prepare in case of an emergency are the examples of how international frameworks can be implemented with the help of modern technologies to improve the safety culture. To conclude, global OHS systems including the ILO requirements, the ISO 45001, OSHA requirements, the EU requirements and the WHS Act of Australia offer detailed guidelines of overseeing occupational health and safety in various industries. They are especially important in the hospitality industry because of the distinct risks, diversity, and high interaction with customers that the industry possesses (Ozay et al., 2020). Through these models, the hospitality organizations will be able to go beyond compliance to develop a proactive safety culture whereby employee wellbeing is a priority, minimizes risks, and improves organizational resilience. Its successful implementation presupposes the leadership dedication, the engagement of employees, the cyclical enhancement, and responsiveness to emerging risks, when safety will become a constituent of business strategy (Singh & Ansari, 2024; Ambardar et al., 2023). Finally, the integration of international OHS models in the hospitality processes not only secures the employees, but also ensures the business sustainability and confidence of the guest in the organization that operates in a highly competitive international market. Various nations and regions have come up with their own special OHS models, each possessing its advantages. Hospitality businesses can create an extensive and efficient safety program by learning and modifying these models.

The Proactive Model (Scandinavia)

Scandinavian model, which is observed in the countries such as Sweden and Norway, is based on the active and cooperative approach. It is founded on a base of powerful labor unions, social dialogue and common safety committees. It is aimed at the prevention of risks and not the response to accidents.

Use in Hospitality:

This model can be employed by a hotel by organizing joint management-employee safety committees. Such committees can anticipate the possible hazards, such as slippery floors and violent personality of the guests. They may take proactive steps to prevent an injury instead of waiting to happen, such as a better cleaning procedure or de-escalation skills in front-desk employees (Nouaimeh et al., 2018). It is aimed at the provision of safe and healthy working environment to all employees rather than at making sure that there is a minimum legal standard.

The Command-and-Control Model (United States)

Command-and-control is a common term used to describe the U.S. model, which is dominated by Occupational Safety and Health Administration (OSHA). It is based on strict rules, controls and severe punishment in case of non-implementation. Although it is viewed as reactive, it offers a strict framework that can be enforced.

Use in Hospitality:

In the case of a U.S. based hotel or restaurant, the establishment has to carefully observe certain OSHA standards. This incorporates fire safety, electrical and dangerous chemicals (such as cleaning supplies) rules. Hospitality business is supposed to know these laws, offer essential training and keep records in detail (Nouaimeh et al., 2018). Although the requirements are based on compliance, a high safety culture can build them into the routine, and they are an aspect of the standard of operation, rather than a box that needs to be ticked.

The Self-Regulatory Model (United Kingdom)

The policy of the UK led by the Health and Safety Executive (HSE) is based on the notion of just reasonable. This principle puts the burden on the employers to control risks to the lowest level which is reasonable. It is a risk-based, goal-setting model, which is flexible in terms of the way the safety is attained.

Hospitality application:

A hotel in the UK, as an example, has to do a comprehensive risk analysis of all its operations. To a housekeeping team, this could include calculating the dangers of performing repetitive duties, carrying heavy linen, or exposure to cleaning agents. The hotel then makes the decision on the most useful and viable control strategies which may involve the introduction of ergonomic carts, mechanical lift, or specialized training (Singh, 2024). The model promotes the creative solutions that are based on the particular risks of the business and not the one-size-fits-all solutions.

Pain and Itch:

The patient experiences fever and redness of the gastrointestinal tracts and genital areas. Difficulties and Consequences:

The patient has fever and erythema of gastrointestinal tracts and genitalia. The aspect of the application of Occupational Health and Safety (OHS) practices in the hospitality industry though necessary is not without its challenges. Hospitality industry is an industry that is diverse, fast moving and service oriented and the workforce is usually a collection of employees with different cultural, linguistic and educational backgrounds. This heterogeneity and combined with odd working hours, a physically stressful job, and continuous contact with customers presents a challenging environment in enforcing and maintaining the OHS standards (CG Davidson et al., 2011). Although international standards and frameworks like ISO 45001, ILO-OSH guidelines, OSHA standards, EU directives, and WHS Act have been made available, the application of the said systems in hospitality has a number of challenges (Singh & Sihmar, 2025; Singh & Kumar, 2024; Singh & Kumar, 2021). Unless properly managed, these challenges have not only implications on the health and wellbeing of the employees but also on the safety of the guests, the performance of an organization, and sustainability of the industry in general (Singh & Kumar, 2021). The investigation of these issues and their larger contexts gives a clearer insight into the fact that the process of building a robust safety culture in the hospitality sphere needs to be an ongoing process with everybody adapting and showing dedication to the changes on the ground. The fact that the hospitality sector has a diverse workforce is among the most outstanding challenges. Hotels, restaurants, resorts and tourism activities often involve people of various countries, languages and culture (Singh & Kumar, 2024; Singh & Kumar, 2021; Singh & Ansari, 2024). This diversity although an enrichment towards service delivery, tends to complicate communication in regard to safety measures. Materials used in training can be perceived differently, and cultures can have vast different attitudes to safety. Indicatively, in certain cultures, one may not be encouraged to report or ask authority questions, resulting to underreporting of safety concerns (Kineber et al., 2023). The severe consequences of this communication barrier are that safety directions can be misinterpreted, emergency response can be delayed or hazard reporting can be minimized. The risk of creating the gaps in the OHS management system of the hospitality organization, where employees and guests are exposed to the harm that they may prevent, is another significant issue in the absence of intentional measures to offer multilingual training, provide culturally sensitive communication, and include inclusive safety committees (Supina & Singh, 2024; Singh, 2024).Another significant concern is the nature of work in the hospitality industry, which typically involves physically demanding and repetitive tasks. Examples of activities exposing housekeeping employees to musculoskeletal disorders include bed-making, lifting, bending, and repetitive movements. Kitchen crew gets exposed to burns, cuts, slips and high temperatures. Employees acting in the front are likely to become victims of violence or harassment as well as verbal abuse by the guests at workplace. The presence of these occupational hazards is increased by the abnormal and extended working hours that lead to fatigue, stress and alertness (Kavouras et al., 2022). The consequences are dual because, on the one hand, employees are exposed to deteriorating health and wellbeing; on the other, the possibility of errors that could endanger the safety and quality of services to guests increases among fatigued or stressed employees (Supina & Singh, 2024; Singh & Singh, 2022; Singh & Hassan, 2024; Singh & Ansari, 2024). This forms a vicious circle in which the lack of safety culture directly influences employee satisfaction in addition to customer experience, which eventually translates to the reputation and profitability of hospitality businesses. Another challenge to effective OHS practices in hospitality is a high employee turnover. The seasonal employment, temporary contracts, and high turnover of this industry make it hard to maintain a long-term safety training and practice. The continuous hiring of new employees also means that the companies have to invest in the induction training programs again and again, and the experienced employees who could guide new members of the team are frequently lost because the safety culture was not established yet (Kavouras et al., 2022; Singh & Ansari, 2024; Ambardar et al., 2023). The implication of this challenge is that the safety standards will be compromised because the inexperienced staff members are yet to understand or prioritise the OHS protocols. Moreover, the possibility of lowering training expenses because of financial pressures can tempt the organizations to contribute to the increase in the number of accidents and injuries. There are also major challenges of cost and resources allocation and this is mostly applicable to the small medium hospitality enterprises (SMEs). Although big multinational hotels might be able to afford the use of ISO 45001 or even meet the requirements of the EU, smaller facilities might not be able to afford the safety facilities, training services or even conducting an audit of compliance. Sometimes, the management can value short-term profitability more than long-term investment in safety and OHS initiatives can be seen as expensive and untimely (Singh & Supina, 2024). Such choices have dismal consequences: the lack of investment in OHS regularly results in the increased accident rates, legal ramifications, dissatisfaction of the employees, and the damaged reputation. One single safety incident (a food poisoning outbreak or a fire) can be disastrous in an industry where customer trust and brand image are paramount.

The other major obstacle is the lack of recording of incidents and near-misses which is usually due to fear of reprisal or cultural disposition towards errors. In hospitality, customer service is the most important factor hence employees might not report safety hazards or accidents to the hospitality especially in a bid to avoid inconvenience to the guests or confront allegations of inefficiencies. Nonetheless, the inability to respond to the near-misses does not allow organizations to recognize any risks on the systemic level and take preventive actions (Kavouras et al., 2022). The consequence of this underreporting is the reactive safety culture in which accidents are only solved after they happen, thereby resulting in repeat incidents, high cost, and unnecessary injuries to employees and guests (Ambardar et al., 2023). The hospitality industry has to deal with similar challenges connected to the appearance of new threats and changing the work environment (Singh & Sihmar, 2025; Singh & Kumar, 2024; Singh & Kumar, 2021; Singh & Ansari, 2024). The shift to the use of new technologies, i.e., online booking system, kitchen automation, and artificial intelligence in customer care introduces the new types of risk, i.e., ergonomic strain caused by working with screens, cybersecurity risk, and job stress due to the necessity to adapt to new systems. Besides, the risks in the workplace have changed due to global issues like pandemics, climate change, and geopolitical instability (Groen & Sprang, 2021). The COVID-19 pandemic, in particular, reflected the deficiencies in the area of infection control, the availability of personal protection equipment, and the level of training of the staff on the hygiene practices. Likewise, the weather-related threats, e.g., extreme heat or floods, may interfere with the hospitality activities and put the employees and customers at risk. The consequences of these changing risks are radical: the organizations need to implement flexible and adaptive OHS systems able to foresee and solve problems not only relating to the conventional occupational hazards (Singh & Kumar, 2024). Not doing so can result in making the organization more vulnerable, less resilient, and less able to keep operating. There is an added complexity of legal and regulatory compliance. International OHS models are based on general frameworks whereas local regulations differ greatly in different countries and regions. The hospitality chains operating internationally have to tread the maze of demands, and at the same time, provide the uniformity of the safety standards within their operations. In other developing nations, poor implementation of OHS laws only adds to the situation as organizations are unlikely to be pushed to high standards by the external forces (Bohle et al., 2017). The implications of this are also immense: variations in the application of OHS practices may cause unequal protection of employees, reputational risks of global brands, and legal liabilities in places with more stringent enforcement.

In addition, the difficulty of developing safety culture within a service-oriented setting cannot be overrated. Hospitality workers are conditioned to remember their guests as the highest priority and this may be against the safety requirements at times. To highlight, a hotel employee may use unstable furniture to get into a higher position to quickly accommodate a request of a given guest instead of waiting until the appropriate equipment is found, or kitchen workers may hasten to prepare food at peak time, which can further expose them to burns or injuries (Zhou et al., 2021). Such actions are service pressure-motivated, which has an impact on employee safety as well as guest experience (Singh & Hassan, 2024; Singh, 2025; Singh, 2024; Singh & Supina, 2024; Singh & Sihmar, 2025). Companies have to find a balance between efficiency and safety and safety has to be incorporated into the service standards in such a way that employees would not think that these two factors are a conflicting interest. These challenges do not only have immediate workplace implications but also on the wider reputation and sustainability of hospitality industry. Good safety record leads to increased employee satisfaction, lower turnover and improved brand reputation whereas poor safety practices may result in law suits, regulatory fines, adverse publicity as well as loss of money (Unguren et al., 2017; Singh & Kumar, 2021). In the current world where ethical and responsible business practices are highly appreciated by the guests, the OHS performance is directly associated with competitive advantage. Besides, the companies that do not consider OHS issues can have difficulties with retention of talents who want to work with employers, who care about wellbeing and balance between work and life (Singh & Singh, 2022; Singh & Kumar, 2021). Considering sustainability, the addition of OHS to larger corporate social responsibility (CSR) programs makes hospitality businesses accountable players who are focused on safeguarding the people and the community.

CONCLUSION

The international OHS models offer a strong structure of dealing with occupational health and safety, the hospitality industry encounters many obstacles in their application that include workforce diversity and turnover, cost reduction, underreporting, new risks, service priorities and conflicts, and variations depending on service priorities. The consequences of these problems are far-reaching, as they impact the overall wellbeing of the employees, the safety of the guests, the resilience of the organization, and the reputation of the industry. To tackle them, the leadership has to express its dedication, involve all the stakeholders, empower the employees, and keep up with the constantly changing risks. Hospitality organizations should go beyond compliance and develop a proactive and strong safety culture that is part of service delivery. It does not only protect workers and visitors but it also enhances long-term sustainability and competitiveness of the industry in the context of increasingly complicated global environment.

FUTURE RESEARCH

Occupational Health and Safety (OHS) is becoming an important factor in the development of sustainable and resiliency based workplaces, as the hospitality industry continues to grow and develop globally. Although some substantial steps have been achieved in introducing international OHS models like the ISO 45001, the ILO, OSHA, EU, and WHS legislations, gaps still exist regarding how the models can be completely adjusted to the specific needs of the hospitality industry. Research in the future will be necessary to address these gaps, and provide evidence-based measures that can not only protect the health of the employees but improve the quality of the services, organizational resilience, and customer satisfaction. A number of directions become the prospective fields of exploration. The effects of cultural diversity on safety practices and communication are one of the areas that would be considered in future research. The hospitality workforce is diverse in terms of languages and cultures, and the language, cultural standards, and attitude to authority may cause difficulties in the transmission of safety regulations. A study may be done on the effect of cultural variations on hazard reporting, adherence to safety regulations and involvement in safety committees. Research can also evaluate how multilingual training, visual safety support and culturally sensitive communication techniques can enhance the safety outcomes. The other important direction is the integration of technology and innovation into the OHS systems. As the concept of digitalization, artificial intelligence, and automation penetrates hospitality, future studies might examine how wearable devices, mobile apps, virtual reality training, and predictive analytics allow addressing the following aspects: improving hazard detection, training efficiency, and preventing incidents. An example of this could be the use of wearable sensors to track employee fatigue or high temperatures in kitchens, and AI-based systems that could be used to analyze safety data to identify new dangers. To achieve the successful implementation of such technologies, the research of their cost-effectiveness, the barriers to their adoption, and their acceptance by the employees will be essential. Another urgent issue of research priority is the relationship between psychosocial risks and the wellbeing of employees working in the hospitality sector (Singh & Kumar, 2021; Ambardar et al., 2023). The long working hours, the high expectations put on the employee by the customers, emotional labor, and unpredictable schedules all add to stress levels, burnout, and mental health issues among hospitality employees. Although conventional OHS models pay more attention to physical hazards, future studies ought to be extended to cover the psychosocial aspects. Research can examine the effect of organizational culture, leadership styles, and work-life balance policies on mental health, and how interventions like flexible scheduling, wellness programs, or resilience training can prevent events associated with stress and enhance safety performance. Moreover, the effect of global crisis and environmental changes on OHS in hospitality should also be discussed in future. The coronavirus disease unveiled the weak points in combating infection and emergency readiness, ensuring the safety of employees, and climate change brings new threats, including natural disasters that would impact tourism attraction sites. Studies might be done to understand how hospitality organizations are modifying OHS measures to pandemic, natural disasters, or climate-related risks, and what is best practice that can be exchanged in different regions. This direction of investigation would aid in creating adaptive and future-proof safety systems that will be commensurate to the overall sustainability objectives. Lastly, the research on the economic and strategic consequences of OHS in hospitality investment is required (Ambardar et al., 2023). Most of the small and medium sized businesses have not considered the cost of safety as a form of investment, but facts are indicating that effective OHS systems promote employee satisfaction, decrease turnover, improve quality of services and build strong brand image. Future research may estimate the payback period (ROI) of OHS initiatives, both direct and indirect (lesson learned on costs of accidents, and increase in customer loyalty, respectively). Finally, future studies of OHS in the hospitality industry should take a step further and leave the compliance-driven research paradigm to adopt interdisciplinary studies that are holistic, focused on cultural diversity, technological innovation, psychosocial wellbeing, global crises, and economic consequences. This kind of research will not only grow academic knowledge, but also offer some practical tools and measures that can be used by hospitality organizations to develop resilient, inclusive, and future-ready safety cultures. With these arising challenges, future scholarship may be able to secure a system where OHS is a catalyst of a sustainable growth and excellence in the hospitality sector all over the world.

REFERENCES

Alamri, R. A. (2024). Role of leadership in promoting health and safety culture:"A comprehensive review of GCC countries". *International Journal of Pharmaceutical and Healthcare Marketing*.

Ambardar, A., Singh, A., & Singh, V. (2023). Barriers in Implementing Ergonomic Practices in Hotels- A Study on five star hotels in NCR region. *International Journal of Hospitality and Tourism Systems*, *16*(2), 11–17.

Bankova, Y., & Kutsarov, H. (2022). Strategies and Tactics for OHS Management to Confront the Pandemic Crisis in the Hospitality Industry: The Case of Countries in Europe. In *Handbook of Research on Key Dimensions of Occupational Safety and Health Protection Management* (pp. 261-283). IGI Global Scientific Publishing.

Bisbey, T. M., Kilcullen, M. P., Thomas, E. J., Ottosen, M. J., Tsao, K., & Salas, E. (2021). Safety culture: An integration of existing models and a framework for understanding its development. *Human Factors*, *63*(1), 88–110. DOI: 10.1177/0018720819868878 PMID: 31424954

Bohle, P., Knox, A., Noone, J., Mc Namara, M., Rafalski, J., & Quinlan, M. (2017). Work organisation, bullying and intention to leave in the hospitality industry. *Employee Relations*, *39*(4), 446–458. DOI: 10.1108/ER-07-2016-0149

Ciarlante, K., Mejia, C., & Broker, E. (2024). A research agenda for occupational safety, health, & well-being in hospitality & tourism management. *International Journal of Hospitality Management*, *123*, 103887. DOI: 10.1016/j.ijhm.2024.103887

Davidson, C. G., Mc, M., Phail, R., & Barry, S. (2011). Hospitality HRM: Past, present and the future. *International Journal of Contemporary Hospitality Management*, *23*(4), 498–516. DOI: 10.1108/09596111111130001

Dufour, C., Draghci, A., Ivascu, L., & Sarfraz, M. (2020). Occupational health and safety division of responsibility: A conceptual model for the implementation of the OHSAS 18001: 2007 standard. *Human Systems Management*, *39*(4), 549–563. DOI: 10.3233/HSM-201060

Elziny, M. N., & Mohamed, H. E. (2020). The influence of safety culture (sc) on hotel employees' safe behaviour during the crisis of Covid-19. *International Journal of Heritage. Tourism and Hospitality*, *14*(2), 97–111.

Groen, B., & van Sprang, H. (2021). Perceptions of hospitality and safety are two sides of the same coin. *Buildings (Basel, Switzerland)*, *11*(3), 113. DOI: 10.3390/buildings11030113

Horan, K. A., Shoss, M. K., Mejia, C., & Ciarlante, K. (2021). Industry context as an essential tool for the future of healthy and safe work: Illustrative examples for occupational health psychology from the hospitality industry. *International Journal of Environmental Research and Public Health, 18*(20), 10720. DOI: 10.3390/ijerph182010720 PMID: 34682465

Kavouras, S., Vardopoulos, I., Mitoula, R., Zorpas, A. A., & Kaldis, P. (2022). Occupational health and safety scope significance in achieving sustainability. *Sustainability (Basel), 14*(4), 2424. DOI: 10.3390/su14042424

Kim, Y., Park, J., & Park, M. (2016). Creating a culture of prevention in occupational safety and health practice. *Safety and Health at Work, 7*(2), 89–96. DOI: 10.1016/j.shaw.2016.02.002 PMID: 27340594

Kineber, A. F., Antwi-Afari, M. F., Elghaish, F., Zamil, A. M., Alhusban, M., & Qaralleh, T. J. O. (2023). Benefits of implementing occupational health and safety management systems for the sustainable construction industry: A systematic literature review. *Sustainability (Basel), 15*(17), 12697. DOI: 10.3390/su151712697

Leitão, S., & Greiner, B. A. (2017). Psychosocial, health promotion and safety culture management–are health and safety practitioners involved? *Safety Science, 91*, 84–92. DOI: 10.1016/j.ssci.2016.07.002

Lindøe, P. H., & Olsen, O. E. (2004). Implementing quality and health/safety systems in the hospitality industry. A comparison with the aluminium industry in Norway. *Scandinavian Journal of Hospitality and Tourism, 4*(2), 154–172. DOI: 10.1080/15022250410003843

Naji, G. M. A., Isha, A. S. N., Mohyaldinn, M. E., Leka, S., Saleem, M. S., Rahman, S. M. N. B. S. A., & Alzoraiki, M. (2021). Impact of safety culture on safety performance; mediating role of psychosocial hazard: An integrated modelling approach. *International Journal of Environmental Research and Public Health, 18*(16), 8568. DOI: 10.3390/ijerph18168568 PMID: 34444314

Nouaimeh, N., Pazhanthotta, R. T., Taylor, J. Z., & Bulatovic-Schumer, R. (2018). Measuring and improving food safety culture in a large catering company: A case study. *Worldwide Hospitality and Tourism Themes, 10*(3), 358–368. DOI: 10.1108/WHATT-02-2018-0011

Özay, M. E., Yılmaz, A., & Uçan, R. (2020). Occupational Safety Culture of Accommodation Sector Employees in Balıkesir Province: NOSACQ-50 Survey. *International Journal of Pure and Applied Sciences, 6*(2), 186–193. DOI: 10.29132/ijpas.807998

Singh, A. (2024). Quality of Work-Life Practices in the Indian Hospitality Sector: Future Challenges and Prospects. In Valeri, M., & Sousa, B. (Eds.), *Human Relations Management in Tourism* (pp. 208–224). IGI Global., DOI: 10.4018/979-8-3693-1322-0.ch010

Singh, A. (2025). Analysis of Perceptions on Employment of differently-abled People in the Indian Hotel Industry. *Pasos (El Sauzal)*, *23*(1), 167–183. DOI: 10.25145/j.pasos.2025.23.011

Singh, A., & Ansari, A. I. (2024). Role of Training and Development in Employee Motivation: Tourism and Hospitality Sector. In Mazurowski, T. (Ed.), *Enhancing Employee Motivation Through Training and Development* (pp. 248–261). IGI Global., DOI: 10.4018/979-8-3693-1674-0.ch011

Singh, A., & Hassan, S. C. (2024), "Identifying the Skill Gap in the Workplace and Their Challenges in Hospitality and Tourism Organisations", Thake, A.M., Sood, K., Özen, E. and Grima, S. (Ed.) Contemporary Challenges in Social Science Management: Skills Gaps and Shortages in the Labour Market (Contemporary Studies in Economic and Financial Analysis, Vol. 112B), Emerald Publishing Limited, Leeds, pp. 101-114. DOI: 10.1108/S1569-37592024000112B006

Singh, A., & Kumar, S. (2021). Identifying Innovations in Human Resources: Academia and Industry Perspectives. In Pathak, A., & Rana, S. (Eds.), *Transforming Human Resource Functions With Automation* (pp. 104–120). IGI Global. DOI: 10.4018/978-1-7998-4180-7.ch006

Singh, A., & Kumar, S. (2024). Effective Talent Management Practices Implemented in the Hospitality Sector. In Christiansen, B., Aziz, M., & O'Keeffe, E. (Eds.), *Global Practices on Effective Talent Acquisition and Retention* (pp. 126–144). IGI Global., DOI: 10.4018/979-8-3693-1938-3.ch008

Singh, A., & Sihmar, P. (2025). Higher Education Institution Sustainability: Role of Green Human Resource Management. In Adu-Gyamfi, M., Boateng, F., & Atiku, S. (Eds.), *Green HRM Awareness and Training in Higher Education Institutions* (pp. 1–14). IGI Global Scientific Publishing., DOI: 10.4018/979-8-3693-2956-6.ch001

Singh, A., & Singh, V. K. (2022). The impact of ergonomic practices on housekeeping employee retention and efficiency in hotels during COVID-19 in India. Turyzm/Tourism, 32(2), 29–50. DOI: 10.18778/0867-5856.32.2.02

Singh, A., & Supina, S. (2024). Talent Acquisition and Retention in Hospitality Industry: Current Skill Gaps and Challenges. In Christiansen, B., Aziz, M., & O'Keeffe, E. (Eds.), *Global Practices on Effective Talent Acquisition and Retention* (pp. 363–379). IGI Global., DOI: 10.4018/979-8-3693-1938-3.ch020

Stosic, K., Dahlstrom, N., & Boonchai, C. (2023). Applying lessons from aviation safety culture in the hospitality industry: A review and road map. *International Journal of Occupational Safety and Ergonomics, 29*(3), 1025–1036. DOI: 10.1080/10803548.2022.2108638 PMID: 35915910

Supina, S., & Singh, A. (2024). Exploring the Presence and Adoption of Quality of Work-Life Dimensions in the Hotel Industry? *African Journal of Hospitality, Tourism and Leisure, 13*(2), 421–429. DOI: 10.46222/ajhtl.19770720.524

Üngüren, E., Arslan, S., & Koç, T. S. (2017). The effect of fatalistic beliefs regarding occupational accidents on job satisfaction and organizational trust in hotel industry. [AHTR]. *Advances in Hospitality and Tourism Research, 5*(1), 23–56.

Zhou, Y., Mistry, T. G., Kim, W. G., & Cobanoglu, C. (2021). Workplace mistreatment in the hospitality and tourism industry: A systematic literature review and future research suggestions. *Journal of Hospitality and Tourism Management, 49*, 309–320. DOI: 10.1016/j.jhtm.2021.09.024

KEY TERMS AND DEFINITIONS

Occupational health and safety (OHS): It is a multi discipline area that takes care of the health, safety and well being of workers. In hospitality, OHS deals with the avoidance of injuries, diseases and accidents within the hotels, restaurants and tourism amenities by the use of policies, training and risk prevention (control).

Safety Culture The values: attitudes, and behaviors that are common and define how an organization handles and assigns safety levels of priority. A well-developed safety culture in hospitality will mean that all employees on different levels assume active responsibility to keep the work environment safe.

Hazard Detection and Risk Analysis (HIRA): A methodical procedure to identify possibleharmful (hazards) sources and assess the potential risk probability/severity. This in the case of hotels, involves an evaluation of kitchen burns, slip, chemical exposures or ergonomic hazards.

ISO 45001:2018 A global standard: which outlines the requirements of an Occupational Health and Safety Management System (OHSMS). It offers a guide to the hospitality organizations to systematically enhance the workplace safety and minimize risks.

Incident Reporting and Investigation: An established process of reporting investigating and reporting workplace accidents, near misses or hazardous conditions. Good reporting assists in detecting the root cause and elimination of repetition in the hospitality operations.

Behavior-Based Safety (BBS): A method that will involve analysing and shaping the behaviour of employees in order to foster safer work practice. BBS can be applied in the hospitality industry through watching the performance of the staff providing feedback, and supporting good safety practices.

Preparedness and Response to Emergencies: The training and the planning processes aimed at managing the emergencies like fires, food contamination, or natural disasters. Under this framework, hotels have evacuation drills, first aid stations and crisis communication plans.

Ergonomics in Hospitality: The art of creating the work environment to suit the physical ability of workers. Good ergonomics help in lessening the stress and risks of injury to housekeepers, cooks, and service staffs involved in routine or physically strenuous work.

Global Safety Benchmarks: The policies of hospitality OHS were adjusted to international norms and ethical standards through the international recognized standards and best practices (like international labour organization (ILO) standards or world health organization (WHO)) standards.

Ongoing Safety Management Enhancement (Kaizen): A philosophy of continuous improvement of safety systems, audits, employee feedbacks, and performance metrics are some of the ways used by hospitality organizations in maintaining and enhancing the results of OHS over time.

Chapter 14
Psychosocial Risk Management Among First Responders:
A Comparative Analysis of Burnout Prevention Strategies in Canada and the United States

Gayle Zenia Noronha
Adler University, Canada

Nola Adelasoye
http://orcid.org/0000-0002-6893-0648
Adler University, Canada

Jason Walker
http://orcid.org/0000-0003-1944-5432
Adler University, Canada

Hawwa Shiuna Musthafa
http://orcid.org/0000-0003-1048-1541
Adler University, Canada

ABSTRACT

This chapter focusses on burnout prevention strategies among first responders in Canada, and the United States, emphasizing the interconnected roles of transformational leadership, safety culture, and digital innovations in psychosocial risk management. Foundational theories such as the Job-Demands Resources (JD-R) Theory and Leader-Member Exchange Theory (LMX Theory) were examined. Case

DOI: 10.4018/979-8-3373-6370-7.ch014

studies compared Canada's Road to Mental Readiness program, and the United States' Departmental Wellness and Emerging Resilience programs were analyzed. Regulatory frameworks including Canada's Federal Framework on PTSD and decentralized, state-level statutes in the United States, including Connecticut's Senate Bill 1002/ House Bill 6595 and Senate Bill 42 of Vermont, among others, were strategically compared for structural divergences. CISM was found to be more effective when paired with other peer support programs, including digital ones. The chapter proposes a framework for a culturally responsive Occupational Health and Safety (OHS) model that emphasizes clarity in regulation, supportive leadership at organizational levels, built in peer support at the team level, and flexible technology developed in collaboration with real first responders for other first responders.

INTRODUCTION

The sound of sirens reminds us that somewhere, someone is facing an emergency. First responders, police, fire and EMS, are the professionals who step forward in those moments, working on the front lines to ensure that every situation is handled swiftly and that the safety and well-being of the broader community are protected. The very nature of their work is inherently high-risk and often carries a deeply human, often untold dimension. Given the intense demands of their respective roles, first responders frequently encounter distressing and traumatic situations, which can increase their vulnerability to complex trauma, stress, compassion fatigue, and burnout (Ceschi et al., 2022). A 2023 study among 1,101 first responders in Canada found that on average 40% experience burnout while working (Pike et al., 2023). First responders face multiple psychosocial risks, which worsen burnout and reduce job satisfaction (Walker & Stones, 2020). The emotional burden of burnout is not just heresy but has increasingly measurable negative impact.

Over 80% of first responders have experienced a traumatic event at work, with some studies citing figures as high as 85% (PTSD in First Responders, 2025) This frequent exposure to trauma places them at a higher risk for mental health conditions like Post-Traumatic Stress Disorder (PTSD), which affects roughly 1 in 3 first responders, compared to about 1 in 5 in the general population, according to SAMHSA (PTSD in First Responders, 2025). These events can vary considerably in nature and severity, ranging from immediate and violent incidents to those that unintentionally result in fatalities such as death. For instance, a study undertaken by Kaplan et al. (2024), designed to quantify the prevalence and predictors of burnout in emergency medical services personnel discovered that 60% of EMS personnel showed signs of burnout. The high prevalence rate indicates an alarming need to further study the organizational factors that add to burnout in EMS personnel determining that burnout within the emergency medicine field appeared to begin as early as their residency and training. The studies by Carleton et al. (2018) and Kaplan et al. (2024) raise important questions about the long-term implications of maladaptive psychological outcomes. To address the maladaptive psychological outcomes, the personal and organizational costs of burnout among first responders must be considered comprehensively.

Burnout among first responders is a serious concern with consequences at the individual, departmental, and societal levels. At the individual level, burnout has major implications for mental well-being, reflected in rising suicide rates among first responders, which now exceed on-duty deaths (Moslehi et al., 2025). For example, suicide rates among paramedics are 1.4 times higher than in the general population and first responders experiencing burnout have a 117% greater likelihood of contemplating suicide than peers not currently experiencing burnout. At the departmental level, burnout contributes to turnover, recruitment challenges, and retention issues (La Manna et al., 2025). National turnover among paramedics increased from 24% in 2021 to 36% in 2022 (Moore, 2022), with burnout among the top five reasons for leaving in a survey of more than 2,000 paramedics. Police turnover, though lower, remains perturbing at 10.8% annually. At the societal level, the financial toll of burnout is substantial, with related injuries costing over $121 million in direct and indirect expenses (La Manna et al., 2025). The findings have shown that there is an urgent need for continued research and effective crisis interventions to support first responders in both the US and Canada. One of the key strengths of this review is the proposal of a multi-level, culturally sensitive OHS model on the basis of the comparison of the regulatory frameworks between the United States and Canada.

CHAPTER OBJECTIVES

Given the high prevalence of burnout among first responders and the fragmented nature of existing interventions especially regarding cultural and technological responsiveness, there is an urgent need for comparative, policy-informed approaches to occupational health. This chapter therefore aims to:

- Analyze burnout in the first responder context in Canada and US
- Compare regulatory and organizational frameworks for first responder mental health
- Develop a comparative framework for a culturally responsive OHS model

FIRST RESPONDER CONTEXT: CANADA AND THE U.S.

First responders are a diverse group of highly specialized and well-compensated professionals and volunteers, including police, firefighters, and paramedics such as the emergency medical service-EMS who protect lives and property. These roles demand substantial emotional, physiological, and mental commitment, and first responders frequently encounter both tangible and psychological stressors (Carleton et al., 2018). Given the challenging nature of their work, teamwork and safety are essential components of the first responder role.

Exhaustion-related performance declines threaten not only individual well-being but also the safety of colleagues and the public (Cunningham et al., 2022). In the United States, an empirically validated fatigue-management manual has recently been developed (Patterson et al., 2018). In Canada, a national occupational guideline is being prepared to translate similar principles into workplace practices, either integrated into current management systems across paramedics, police, and firefighters or used as a standard across first responder groups (Yung et al., 2021). Crowe et al. (2018) in their study on burnout among EMS professional employed the Burnout Scale for Health Professionals, designed to measure emotional exhaustion, depersonalization, and reduced personal accomplishment and found burnout to be more prevalent among EMS providers compared to other emergency health professionals, specifically EMTs. Furthermore, two U.S. paramedic surveys using the Oldenburg Burnout Inventory and the Copenhagen Burnout Inventory reported work-related burnout rates of 30% among paramedics and 19% among EMTs, with burnout positively associated with turnover intentions and absenteeism (Fragoso et al., 2016). In the only available Maslach Burnout Inventory (MBI) report U.S. paramedics, mean scores were 19.2 for emotional exhaustion, 9.3 for depersonalization, and 28.1 for personal accomplishment emphasizing the urgent need for burnout-prevention strategies in first responder agencies in the U.S. and Canada (Fragoso et al., 2016).

FOUNDATIONAL THEORIES INFORMING FIRST RESPONDER'S WELL-BEING

First responders operate within environments defined by urgency, exposure to trauma, high public expectation, and limited recovery time. Understanding their well-being, therefore, requires a framework that accounts not only for individual stress responses but also for leadership behaviours, organizational culture, perceived safety, and the increasing role of digital systems in daily operations. For this reason, five interrelated concepts were selected as the core pillars of this chapter: burnout, transformational leadership, safety culture, safety climate, and digital innovations.

These concepts were chosen because together they capture the personal, relational, organizational, and technological dimensions that shape first responders' experiences at work.

The theoretical grounding for these concepts draws primarily on the Job Demands–Resources (JD-R) Model, Conservation of Resources (COR) Theory, Transformational Leadership Theory, Safety Culture and Safety Climate Theories, Psychological Safety, and Social Exchange Theory. Rather than functioning as isolated frameworks, these theories intersect to explain how stress accumulates, how resources are gained or depleted, and how leadership and organizational practices can either protect or erode well-being over time.

Psychosocial risk management among first responders requires more than identifying individual stress reactions. It requires understanding how work demands, leadership practices, organizational culture, and systems of support interact over time to either protect or erode psychological well-being. First responders in both Canada and the United States operate in environments characterized by chronic exposure to trauma, unpredictable demands, and high public accountability. The theories selected for this study were chosen because they collectively explain how burnout develops as an occupational risk and how it can be prevented through organizational design rather than individual endurance.

Job Demands–Resources (JD-R) Model and Psychosocial Risk Management

The Job Demands-Resources (JD-R) Model provides a practical foundation for psychosocial risk management because it frames burnout as a predictable outcome of work design rather than personal weakness. According to Bakker and Demerouti (2017), job demands such as trauma exposure, shift work, and sustained vigilance initiate a health impairment process when they exceed available resources. For first responders, these demands are inherent to the role and cannot be eliminated.

What makes the JD-R model especially relevant is its emphasis on organizational responsibility. Job resources such as supportive supervision, access to mental health services, training, and psychological safety buffer the impact of high demands and promote engagement aligning closely with psychosocial risk management frameworks that focus on prevention rather than crisis response. In a comparative context, the JD-R model allows examination of how differences in funding, occupational health policies, and support infrastructures between Canada and the United States shape the availability and effectiveness of these resources. The model was therefore selected to anchor the study's focus on structural prevention strategies rather than individual coping alone.

Conservation of Resources (COR) Theory and Cumulative Psychosocial Harm

While the JD-R model explains how demands and resources interact, the Conservation of Resources (COR) Theory explains why burnout often escalates once stress becomes chronic. COR theory posits that individuals strive to obtain and protect resources such as energy, time, social support, and emotional stability, and that stress occurs when these resources are threatened or lost (Hobfoll, 2011; Hobfoll et al., 2016).

For first responders, resource loss is rarely a single event. Instead, it accumulates through repeated exposure to trauma, disrupted sleep, moral strain, and limited recovery opportunities. COR theory is particularly useful for psychosocial risk management because it highlights the danger of loss spirals, where depleted resources reduce the capacity to cope with future stressors. In comparing Canada and the United States, this theory helps explain how system-level factors such as access to health care, staffing levels, and leave policies influence whether resource loss is interrupted or intensified. COR theory was chosen because it captures the long-term consequences of under-resourced systems and reinforces the importance of early organizational intervention.

Transformational Leadership Theory as a Protective Psychosocial Resource

Transformational Leadership Theory was selected because leadership consistently emerges as one of the most influential protective factors in high-risk occupations. Transformational leaders provide vision, emotional support, and individualized consideration, fostering trust and shared meaning in demanding environments (Bass, 1999; Bass & Riggio, 2006).

In first responder organizations, leaders often serve as the first point of contact following critical incidents. Their response shapes whether personnel feel validated or silenced. Transformational leadership mitigates psychosocial risk by normalizing emotional responses, encouraging help-seeking, and reinforcing purpose in difficult work. Kowalski's (2019) study of patrol supervisors in British Columbia illustrates how leadership presence and guidance following trauma influence recovery and team cohesion. Leadership training and supervisory practices vary across Canadian and U.S. agencies, making leadership an important comparative variable in burnout prevention strategies.

Leader-Member Exchange (LMX) Theory and Relational Safety

Leader-Member Exchange (LMX) Theory complements transformational leadership by focussing on the quality of everyday relationships between leaders and followers. High-quality exchanges characterized by trust, respect, and open communication function as relational resources that reduce stress and enhance well-being (Bass, 1999; Burns, 1978).

In first responder settings, where personnel may hesitate to disclose distress, strong leader–member relationships increase the likelihood that early signs of burnout are recognized and addressed. LMX theory helps explain why leadership effects are uneven across teams and why relational safety matters as much as formal policy. Psychosocial risk management depends not only on organizational intent but also on how support is experienced in daily interactions.

Safety Culture and Safety Climate as Organizational Signals

Safety culture and safety climate theories extend psychosocial risk management to organizational values and systems. Safety culture reflects the underlying beliefs about safety, while safety climate captures employees shared perceptions of how safety is prioritized in practice (Reason, 2016).

Traditionally, safety in first responder organizations has focused on physical harm. However, psychological harm is increasingly recognized as an occupational risk. A strong safety climate signals that mental health is taken seriously and that reporting stress or fatigue will not lead to punishment. Huang et al. (2016) demonstrated that safety climate influences job satisfaction, engagement, and retention through social exchange processes. Canada and the United States differ in how psychological health and safety are embedded in occupational standards, shaping organizational responses to psychosocial risk.

Psychological Safety and Social Exchange Theory as Mechanisms of Prevention

Psychological safety explains how organizational conditions influence help-seeking behaviour. Edmondson (1999) defined psychological safety as a shared belief that individuals can speak up without fear of negative consequences. In first responder cultures that often emphasize toughness, the absence of psychological safety can delay intervention and accelerate burnout.

Social Exchange Theory explains why supportive practices lead to engagement rather than withdrawal. When organizations demonstrate genuine care for employee well-being, employees reciprocate with trust and commitment (Huang et al., 2016).

These theories were selected because they clarify why policies alone are insufficient. Psychosocial risk management succeeds only when first responders feel safe enough to use available supports.

Rationale for Theoretical Integration

The integration of these theories reflects the complex, multi-level nature of psychosocial risk management among first responders. The JD-R model and COR theory provide the central framework for understanding burnout as a systemic outcome of sustained demands and resource loss. Transformational leadership, LMX, and safety culture theories identify organizational levers that mitigate risk. Psychological safety and social exchange explain how these levers influence behaviour. Socio-technical frameworks account for the expanding role of digital innovation.

Together, these theories support a comparative analysis of burnout prevention strategies in Canada and the United States by shifting the focus from individual resilience to organizational responsibility. This theoretical foundation positions psychosocial risk management as a core component of occupational health, public safety effectiveness, and long-term workforce sustainability.

To better understand first responder mental health and organizational outcomes requires examining five interrelated constructs and that is the burnout, transformational leadership, safety culture, safety climate, and digital innovations (Kelly & Hearld, 2020).

Burnout as a Central Outcome of Imbalanced Demands and Resources

Burnout was selected as a central construct because it represents one of the most documented and consequential outcomes of prolonged occupational stress in first responder populations. Burnout is not simply an individual failure to cope but a systemic response to sustained imbalance between job demands and available resources. The Job Demands–Resources (JD-R) Model provides a foundational lens for understanding this imbalance. According to Bakker and Demerouti (2017), job demands such as emotional strain, workload, and exposure to traumatic events increase the risk of exhaustion, while job resources such as supervisory support, autonomy, and safety systems buffer against this risk.

For first responders, demands are often non-negotiable. Emergencies cannot be postponed, and exposure to human suffering is inherent to the role. This makes the presence of organizational and relational resources especially critical. Conservation of Resources (COR) Theory further deepens this understanding by emphasizing that stress occurs when individuals perceive a threat to valued resources or experience actual resource loss (Hobfoll, 2011; Hobfoll et al., 2016). In high-risk professions, repeated exposure to trauma, insufficient recovery time, and lack of institutional support can trigger cycles of resource depletion that accelerate burnout. Within this chapter, burnout is therefore treated not as an endpoint but as a signal of systemic strain that reflects how well or poorly resources are being conserved within the organization.

Burnout is a psychological phenomenon characterized by emotional exhaustion, cynicism, or depersonalization and reduced personal accomplishment results from chronic workplace stress. It often co-occurs with secondary traumatic stress and moral injury, and it impairs the well-being and job performance of first responders (Maslach & Leiter, 2016). Occupational burnout, however, is characterized by exhaustion resulting from prolonged exposure to work-related problems (Canu et al., 2021). Transformational leadership plays a critical role in promoting positive organizational outcomes, emphasizing idealized influence, inspirational motivation, intellectual stimulation, and individualized consideration, with the special purpose of enabling leaders to communicative a vision, build trust, and develop their followers' capacities while safeguarding their interests and ensuring their safety (Bakker & Demerouti, 2017).

Paying attention to safety culture is critical, as it includes the shared values, beliefs, and norms that shape organizational behaviour during crises (Musthafa et al., 2023). Further, safety climate, which refers to employees' perceptions of those values at a specific point in time, shapes behaviours related to procedures, reporting, and learning in high-risk contexts. While culture reflects shared organizational values, climate reflects employees' perceptions of how those values are enacted. Building on this, digital innovations, such as mobile peer-support apps, tele-counselling, incident-tracking systems, wearable stress monitors, and analytics, serve as socio-technical interventions that enhance safety, communication, and mental health support by reshaping workflows and improving risk monitoring (Dodoo et al., 2024).

To understand how these factors influence worker wellbeing and performance, various theoretical perspectives offer useful explanations and guidance on this topic. However, our discussion will be restricted to the following: the Job Demands-Resources (JD-R) Model, the Conservation of Resources (COR) theory, the Transformational Leadership Theory, Safety culture and safety climate theories and Psychological Safety and Social Exchange theory. The JD-R model links workplace characteristics to burnout by showing how job demands, such as trauma exposure, shift work, and cognitive overload, create strain on the workers, while job resources, including support, feedback, and training, promote motivation and engagement. The COR theory complements JD-R by positing that individuals strive to obtain, retain, and protect valued resources, and that stress arises when these are threatened or lost (Hobfoll et al., 2016, 2018). The JD-R and COR frameworks explain resource augmentation and safety-buffering pathways, where transformational leaders and safety culture enhance resilience, while digital tools expand access to workers' support and recovery (Bakker & Demerouti, 2017; Hobfoll, 2011).

Transformational leadership theory emphasizes leaders to inspire, motivate, and empower followers to work toward shared goals, thereby enhancing psychological empowerment and organizational support (Kowalski, 2019). Within the first responder contexts, leaders who guide teams through trauma play a critical role in fostering resilience among their teams. Transformational leadership theory emphasizes how transformational leadership replenishes resources, the importance of a safety culture in preventing loss, and how digital tools can restore access and efficiency. However, poor implementation may further deplete resources. Transformational leadership and a strong safety culture improve these resources by providing support and nurturing a psychologically safe environment (Kowalski, 2019).

Safety culture and safety climate theories emphasize the importance of leadership and learning systems in promoting safe practice (Reason, 2016). These approaches emphasize on non-punitive reporting, incident learning, and recognition of safe behaviours reduces preventable stressors and promote psychological safety (Reiman & Rollenhagen, 2017).

Psychological Safety as a Bridge Between Leadership, Culture, and Well-Being

Psychological safety was selected because it explains how interpersonal dynamics shape learning, communication, and help-seeking behaviour in high-stakes environments. Edmondson (1999) defines psychological safety as a shared belief that the team is safe for interpersonal risk-taking. In first responder settings, psychological safety determines whether individuals feel able to speak up about errors, fatigue, mental health concerns, or unsafe practices without fear of punishment or ridicule.

Psychological safety is closely linked to transformational leadership and safety climate. Leaders who model openness and empathy create conditions where psychological safety can flourish, while organizations with strong safety climates reinforce these norms at a structural level. From a COR perspective, psychological safety protects emotional and relational resources by reducing fear-based depletion. Throughout this chapter, psychological safety is positioned as a critical mechanism through which leadership and culture translate into tangible well-being outcomes.

Psychological Safety and Social Exchange theory fosters an environment where individuals can seek help and discuss errors without fear of negative consequences. Supported by transformational leadership and a strong safety culture, psychological safety and social exchange theory is important to this topic because it enables early intervention, peer support, and reduced burnout. Social exchange theory and perceived organizational support further explain how reciprocal trust and care increase engagement and resilience (Edmondson, 2019).

Digital Innovations as Emerging Resources and Stressors

Within the JD-R framework, digital tools can function as job resources when they reduce workload or enhance safety, but they can also become job demands when poorly implemented or inadequately supported. For first responders already operating near capacity, technology that increases surveillance or administrative burden may exacerbate stress rather than relieve it. This chapter therefore treats digital innovation as a conditional resource whose impact depends on leadership decisions, training quality, and alignment with frontline realities.

Integrating the Theoretical Framework

Taken together, the JD-R Model and COR Theory explain how stress and burnout emerge from imbalances between demands and resources. Transformational Leadership Theory explains how leaders can actively generate resources and meaning. Safety Culture and Safety Climate Theories clarify how organizational values are communicated and perceived. Psychological Safety and Social Exchange Theory explain how trust, reciprocity, and voice develop within these systems. Finally, technology adoption theories illuminate how digital tools shape daily work experiences.

These theories collectively ground the chapter's examination of first responder well-being, offering a multidimensional lens that reflects the complexity of their work. Rather than viewing well-being as an individual responsibility, this framework emphasizes shared accountability across leadership, organizational systems, and technological design. In doing so, it provides a coherent foundation for understanding both the risks faced by first responders and the pathways through which sustainable well-being can be supported.

PATTERNS OF BURNOUT AND HEALTH OUTCOMES

Police, fire, and EMS personnel carry a higher risk of mortality by as much as 73% in comparison to other non-first responder roles (Lowery & Cassidy, 2022). During emergency events, first responders are at a greater risk of suffering from heart related issues, including instant death, as compared with the duration during which they undertake non-emergency obligations. First responders are also more likely to suffer early-onset cardiovascular disease and heart rate variability.

The frequent exposure to high-stress situations and cumulative exposure to trauma also increases first responders' risk for developing mental health consequences. The prevalence of depression (3.5%), PTSD (8%), anxiety (4.2%), sleep disturbances, and alcohol abuse (10.6%) in this profession is higher than among the general population (Lowery & Cassidy, 2022).

Cultural and Organizational Barriers to Help-Seeking

In order to fully explore burnout from a first responder's perspective, it is important to consider the culture of their workplace. The way EMS employees perceive workplace culture and approach their mental health, make decisions, and interact with resources fall under two subthemes: employee inability to talk about mental health and ask for help, and the provision and utilization of person-centered support personalized to the EMS context, including education and training (Johnston et al., 2025). First Responders, including law enforcement officers, firefighters, search and rescue officials, defense forces and paramedics are mostly male-dominated professions (Perrott, 2016). Female ratios of employees in these departments vary from about 4 to 32%. There remains a gap in the existing knowledge base regarding how females functionally cope in these male-dominated occupations. Of the little research that does exist pertaining to this, research reveals that females in male-centered occupations, pick up male behaviours, values and attitudes over time (Frazer et al., 2021).

The one exception to this involves first responder nurses, who include flight nurses, rural and or remote area nurses and emergency services nurses, who constitute a ratio of roughly 90% (Frazer et al., 2021). In a review of literature regarding occupational stress in these emergency department nurses, it was found that these staff might report higher levels of burnout and other mental health consequences stemming from higher occupational demands (Basu et al., 2017). Furthermore, this review of literature found that relationships with colleagues and managers might be important factors in coping with stress, including compassion fatigue. This, therefore hints that those supportive relationships maybe one of the protective factors in connection with retention among female first responders (Frazer et al., 2021).

One of the most commonly cited reasons first responders feel discouraged from revealing mental health concerns stems from the stigma surrounding mental health (Walker et al., 2021). In addition, first responder personnel are often apprehensive sometimes accurately that there will be unhelpful work-related consequences if these revelations are made (Bowers et al., 2022). While much of this reluctance is shaped by occupational culture and expectations of resilience, a first responder's ability to discuss their well-being is also influenced by personal histories they bring into their role. Life experiences preceding their service, including adverse or challenging childhood circumstances, may motivate some to join the field but can also increase vulnerability to certain emergencies or psychological distress and worsen obstacles to seeking assistance.

Incidents involving children and suicide, especially the suicide of a colleague, were seen as high-risk events for worsening employee mental health, particularly when poorly managed by the organization (Johnston et al., 2025). A lack of privacy when disclosing mental health concerns to peers was another challenge. Working in a policy-driven, client-focussed environment, many feel discouraged and lack the time or supportive policies needed to seek help (Johnston et al., 2025).

Providing and Applying Person-Centric Aid Customized to First Responders' Context

While conceptualizing personalized interventions for first responder mental health, researchers found a lack of quick and helpful sources of assistance and limited options suiting individual needs (Johnston et al., 2025). These disparities tend to demotivate employees from seeking help due to doubts about efficacy and fears that support might be unhelpful. Evidence-based interventions such as post-incident debriefing, were also less well known, and information about them was scarce. The approach allows EMS employees to share field experiences and emotions. Yet perspectives on its utility and safety were mixed some found it practical for dialogue, while others felt uneasy or unprepared to disclose experiences immediately after incidents (Johnston et al., 2025; National Survey on EMS Mental Health Services, 2016).

Accessibility of support was a critical concern, with procedures for obtaining help described as laborious and, in the U.S., often too costly without organizational financial backing (Loudoun, et al., 2020). Awareness of available resources and stigma also posed barriers, as some services required public acknowledgment of support needs (Johnston et al., 2025). Another important factor was the value placed on speaking with those who understood the first responder context. Many believed mental health professionals could not fully grasp their unique experiences or that standard approaches were ineffective. Timeliness of support was another key consideration. First responders accustomed to rapid emergency responses expected the same when seeking help. When assistance was delayed or unavailable, it fostered negative perceptions, dissatisfaction, and reluctance to use future services.

Knowledge Gaps in Effective Mental Health Support for First Responders

Despite the knowledge of potential stress disorders among first responders being invaluable in creating support systems for identifying and treating them, it is important to note that research in this field is relatively new (O'Connell et al., 2024). Fundamental differences such as perceptions of organizational stressors or greater resilience to burnout due to certain training procedures across departments may exist and enable researchers to develop extensive prevention and treatment methods (O'Connell et al., 2024).

Wellness programs designed for first responders may hold varying levels of efficacy across subgroups. For instance, Szeto et al. (2019) reviewed results from the Road to Mental Readiness resilience training program (Carleton et al., 2018), which builds psychological resilience, stress management skills, and mental health awareness among first responders. Their review showed the program was more effective for police officers than for other professions or first responder groups. These findings show that beyond program design and efficacy, regulatory and organizational frameworks also shape how mental health initiatives are implemented and supported.

The strength of this review rests in its foundational backdrop regarding the prevalence and consequences of burnout among first responders, which thereby adds to the existing knowledge base regarding the urgency of addressing this issue across varied first responder departments. The ground realities of changes in the operations of first responder departments such as increased time spent in the field, difficulty accessing in person support and a work culture that is burgeoning from the lens of burnout encultures the essence of why this review presents important notes for practitioners and policy makers. Furthermore, the review compares contemporary frameworks and legislations across the U.S and Canada, with the intent of proposing a sensitive and culturally responsive occupational health model that is rooted in the best practices across first responder departments for burnout prevention. One of the most significant standouts from this review is the review regarding the fact that Canada's regulatory frameworks apply across its provinces, while the U.S with its focus on state level statutes, also brings in an independent yet disjointed system of support for the first responder departments (Carleton et al., 2018; O'Connell et al., 2024). This knowledge is valuable for practitioners seeking to compare differences between the U.S and Canada while either developing new support mechanisms or introducing pertinent legislations.

COMPARING REGULATORY AND ORGANIZATIONAL FRAMEWORKS FOR FIRST RESPONDER MENTAL HEALTH

The Federal Framework on PTSD outlines the Government of Canada's vision, regulatory principles, and actions to address occupation-related PTSD, focusing on three key areas of significance (Public Health Agency of Canada, 2019). Each province also has workers' compensation legislation covering occupation-related PTSD for first responders, often including presumptive clauses that affect access to support and benefits (Union of British Columbia Municipalities, 2015). The framework aims to enrich knowledge creation, sharing, and collaboration across federal agencies and partners to inform evidence-based public health policies and programs, reduce stigma, and improve recognition of PTSD symptoms and impacts.

Vision of the Federal Framework on PTSD

The Federal Framework on PTSD was designed to give cognizance to and offer aid to those living with PTSD, those at risk of developing it and the ones close to them forward towards their path to recovery, hardiness, and flourishing. The underlying principles of Federal Framework on PTSD are to supplement present strategies and influence associations, enhance empathetic, unbiased and strengths-based techniques, originate action plans on the basis of evidence of what provides results or displays the potential of being effective, comprehend and address impartiality, heterogeneity and inclusiveness and to set in motion a health-focused initiative for the general masses.

The Priority Areas of the Federal Framework on PTSD

The Federal Framework on Post-Traumatic Stress Disorder Act (2018) identifies three priority areas, enhancing monitoring of PTSD prevalence and associated financial and social costs, establishing guidelines and best practices, and developing educational materials. The tracking and data priority area focuses on initiatives to strengthen national monitoring efforts and explore the use of health-related administrative data to record and report on PTSD (Public Health Agency of Canada, 2019). This includes ongoing improvements in PTSD data collection.

The establishment of guidelines and best practices emphasizes collaboration with partners and subject matter experts to develop a PTSD guideline and identify gaps in current approaches. The guideline aims to support the diagnosis, treatment, and management of PTSD and promote sharing of best practices. The development of educational materials focuses on creating standardized resources for use by Canadian public health and support providers to increase national awareness, diagnosis, treatment, and management of PTSD.

U.S. DECENTRALIZED, STATE-LEVEL STATUTES: DECENTRALIZED FRAMEWORK COMPARED TO CANADA

In the United States, several states have introduced regulations addressing the mental health impact of the COVID-19 pandemic on health care workers, first responders, and law enforcement officers. Key areas include workers' compensation, PTSD screening, and access to mental health programs. In Connecticut, Senate Bill 1002 and House Bill 6595 (2021), define PTSD health care eligibility, extending coverage to employees managing or responding to the COVID-19 emergency, including correctional officers, paramedics, and dispatchers. Nebraska's Legislative Bill 1133 extends the Workers' Compensation Act to cover mental injuries and illnesses among health care workers, recognizing PTSD as a legitimate work-related condition and natural human response to trauma requiring treatment without stigma. Nevada's Assembly Bill 315 (2021) requires employers of police, fire, and correctional officers to provide ongoing information on preventing and mitigating mental health challenges, and mandates counselling within three months of retirement. The Nevada Psychiatric Association also launched Curbside Nevada, a mental health emergency line for first responders.

In New York, Assembly Bill 6641 requires regular mental health screenings for first responders, with PTSD presumed work-related. Senate Bill 1301 (2025) directs the commissioner of mental health to form a workgroup on trauma-informed care for first responders. South Carolina's Senate Bill 94 seeks to restructure compensation claims by removing the need for first responders to prove stress or incapacity from atypical work conditions. Vermont's Senate Bill 42 (2021) created an emergency service provider wellness commission to assess worker needs and recommend targeted supports to improve well-being.

The key difference between Canada's Federal Framework on Post-Traumatic Stress Disorder Act and U.S. state legislation lies in Canada's unified, national focus on three core priorities, while U.S. statutes vary by state. Canada's framework also supports individuals close to those affected by PTSD, reflecting a community-based approach and informing national and provincial policies. Stakeholders include federal departments, NGOs, provincial and territorial governments, Indigenous organizations, and experts. In comparison, U.S. laws emphasize specific departmental initiatives for first responder mental health. Canada's framework promotes national PTSD data tracking, best-practice guidelines, and public education to reduce stigma, whereas U.S. bills outline localized mechanisms within departments. These differences reflect each country's governance structures and operational models. Despite differences, both nations recognize that first responder well-being requires systemic support, regulatory action, destigmatization, and improved access to care.

VIGNETTES HUMANIZING BURNOUT EXPERIENCES

In an interview with a public safety communicator by Lentz et al. (2022) on the mental health training needs of firefighters, paramedics, and public safety communicators in Canada, the communicator conveyed that the workload was never-ending. Although public safety communicators were thought to be able to take a short break, this was impractical as the phone never stopped ringing. Participants in the study highlighted the need for organization-wide training that sensitively and holistically considered the psychological weight of traumatic experiences and other stressors tied to first responder duties.

A firefighter in a focus group conducted by Lentz et al. (2022) observed that their department had undertaken some Critical Incident Stress Management training (CISM), a structured approach to help first responders cope with traumatic incidents. with the police and shared similar experiences. The firefighter found it more effective to receive CISM training from professionals with first responder experience and described this as a need. The courses were viewed as useful, with police stories being personal, realistic, and relatable. Participants also emphasized that current health education approaches underestimate the urgent need for practical, quick, and accessible training materials that first responders can apply directly in the field (Lentz et al., 2022).

CURRENT BURNOUT PREVENTION STRATEGIES

The literature on mitigating burnout prevention has evolved from reactive to proactive resilience-building for first responders and related high-stress professions, moving beyond the language of resilience, grit, and bravery. The focus has shifted to structures that normalize exhaustion cultures and reward emotional endurance, or what is commonly called emotional stoicism. There is also greater emphasis on multi-level engagement addressing not only the individual but also the team and organizational levels. Burnout is not merely an individual failure of coping but an institutional symptom of overexposure to trauma, moral injury, and organizational neglect (Roth et al., 2022). Models such as the Mitchell Model in the United States, a post-incident peer support and debriefing intervention, have been widely implemented, although evidence (Billings et al., 2023) indicates relatively low rates of success in preventing long-term burnout among first responders.

This reflects growing attention to occupational stressors and organizational culture (Corthésy-Blondin et al., 2022). In Canada, researchers explore strategies combining emotional resilience training, cognitive-behavioural techniques, and mindfulness (Khoury et al., 2025). Programs such as Before Operational Stress (BOS) (Stelnicki et al., 2021) supplement psychosocial prevention within organizations, reducing posttraumatic stress symptoms while improving coping skills and workplace culture through emotional regulation, psychoeducation, and group reflection. Johnson et al. (2020) highlight contextualization, noting that firefighters must understand fire culture to critically engage, as traditional skepticism toward therapy often undermines uptake. Literacy programs have also shown promise in reducing stigma and enhancing recovery post-crisis. Jones et al. (2024) found that workplace restoration training and peer-led initiatives meaningfully improve PSP attitudes toward mental health and reduce perceived isolation.

Peer Support Programs (Psps)

Digital transformations have reimagined these bonds in app-based environments. Apps have been advanced and relied upon to tackle the gap in in-person options for support. Therefore, governmental and public centric institutions in Canada have funded mobile mental health technology as a source of backing for employees. Some of these apps include PeerConnect (Foley & Ricciardelli, 2023) and PeerOnCall (Goraya et al., 2024; Moll et al., 2025), thereby extending unanimity into virtual spaces, offering immediate, confidential contact at any hour. The peer support programs are strongly established within the social support theory and is focussed on shared experience and trust among peers to lessen both the operational as well as secondary trauma (Goraya et al., 2024; Moll et al., 2025). These programs well align with PSP cultural norms of solidarity and collective resilience (Jones et al., 2024), uncovering that peer-led reintegration programs help regularize post-incident stress reactions and reduce stigma associated with help-seeking behaviours. For instance, peer support interventions among air medical crew have shown that it significantly decreased secondary traumatic stress and compassion fatigue (McCall, 2023). In yet another study, researchers who evaluated the PeerConnect app and found that participants valued the ability to access confidential support outside of formal structures (Foley & Ricciardelli, 2023).

This confirms that technology-enabled PSPs can complement as much the face-to-face models. This is in line with the research findings by (Moll et al., 2025) who found that individuals appreciated app-based models that resulted in not only increased adherence but also increased in accessibility for those situated in dispersed or rural locations. However, it is important to acknowledge the challenges as well. Goraya et al. (2024) found that culture around using such PSPs can prove to be a challenge, especially in workplaces where stigma or hierarchies persist. They also found that the same hierarchies that erode trust offline often shadow the digital realm, without psychological safety, no platform can foster genuine disclosure. In such cases, PSPs risk being perceived as tokenistic.

Moreover, peer supporters may also face emotional overload and in some cases role uncertainty without enough direction or debriefing mechanisms (Ricciardelli et al., 2022). What begins as support can transform into indirect shadowing of who is coping and who is cracking. And yet, where they are fostered within cultures of compassion, peer programs can become powerful acts of collective care, cultivating empathy as a professional competency rather than a private weakness. Thus, it is indomitable PSPs maybe most effective only when embedded in a multi-layered support system that acknowledges peer support and knowledge all the while maintaining professional supervision.

Critical Incident Stress Management (CISM)

CISM is noted to be one of the initial and organized approaches to burnout and trauma prevention in first responders and was developed by Jeffrey T.Mitchell, Ph.D in 1974 in the United States among first responder groups including paramedics, police officers and firefighters to address distress following exposure to traumatic events. CISM is grounded in crisis intervention and group literature and applications such as Eric Lindemann, Gerald Caplan and Irvin Yalom among others (Mitchell, 1974). CISM is also commonly referred to as the Mitchell model. Rooted in crisis-intervention models of the 1980s, CISM encompasses a continuum of care that includes pre-incident education, acute stress debriefing, and post-crisis follow-up and this approach CISM approaches continue to provide value when adapted to organizational context (Corthésy-Blondin et al., 2022; Mitchell & Everly., 2023). In the actual practice of CISM, often referred to as psychological first aid, it is a system of coordinated peer as well as professional support is offered after fatalities or even moral injuries (Jones et al., 2024). Khoury et al. (2025) also found that CISM can be even more effective when combined with resilience building training and continuous supervising and monitoring, rather than as a one-time mediation. CISM can also become compliance-oriented- simply implemented to meet policy and procedure obligations. The question is not merely whether debriefing works, but for whom, under what cultural conditions, and at what emotional cost (Corthésy-Blondin et al., 2022; Johnson et al., 2020).

Digital Mental-Health Platforms and Apps

Digital mental health platforms represent a quickly increasing frontier in burnout prevention. They offer, cost-effective interventions that can reach distributed personnel and address barriers to mental health related help-seeking. Internet-based cognitive behavioral therapy (ICBT) has showed strong evidence for reduction of symptoms related to depression, anxiety, and PTSI among populations (Hadjistavropoulos et al., 2024). Both transdiagnostic and disorder-specific digital CBT formats has led to significant improvements in PTSD symptoms (McCall, 2023). Similarly, digital mindfulness interventions have gained popularity with study findings indicating that it reduced burnout symptoms and users indicated satisfaction. Similar findings were reported by the studies undertaken by (Price et al., 2024) where clients reported that they found the case studies to be not only relatable but also that it helps to nurture understanding, compassion and self-reflection. Smith-MacDonald et al. (2021) likewise, identified digital trauma therapy as a suitable model for military and PSP populations, particularly when channelled through culturally competent clinicians. However, we must also acknowledge quality of digital interventions remains irregular at times with O'Toole & Brown (2021) observing that these apps remained widely varied across military users and hence they recommend a more robust evaluation frameworks and user co-design processes are essential to ensure both safety and relevance. Moreover, McCall et al. (2020) noted that during the COVID-19 pandemic, PSP increasingly relied on digital tools due to isolation and service disruptions, yet many platforms lacked trauma-sensitive adaptation. Successful implementation requires leadership buy-in, as well as consistent organizational alignment, and trust in data security.

Without these, even the most sophisticated digital programs risk underutilization or even attrition. Yet the potential is unquestionable. When culturally adapted, digital mental-health tools can democratize care. Cross-cultural adaptations of scalable measures tell us that future of intervention is not merely virtual but contextual and that technology must translate empathy across both language as well as labour. Building on these insights, the next section examines two recent case studies from Canada and the United States, outlining their frameworks, implementation strategies, and outcomes to identify effective practices for reducing psychosocial risk and strengthening mental health support in first responder professions.

COMPARATIVE CASE STUDIES

Canada – The Road to Mental Readiness and Peer Support Integration

The Road to Mental Readiness (R2MR) program was originally developed by the Canadian Forces Health Services Directorate of Mental Health (DND) for military personnel and adopted as a comprehensive, multi-tiered approach to psychosocial risk management for first responders. It was launched in 2008 and has since been adapted for police, fire, and emergency medical services. R2MR now provides performance, resilience, and mental health training embedded across the careers of Canadian Armed Forces (CAF) members.

R2MR aims to boost well-being, improve short-term performance, and reduce long-term psychological distress. Its objectives include increasing mental health literacy, reducing stigma, and equipping personnel with cognitive and behavioural tools to manage stress effectively. A meta-analysis by Szeto et al. (2019) found that R2MR improved participants' perceptions of resiliency and reduced stigmatizing attitudes, with effects maintained at a 3-month follow-up.

Central to R2MR is the Mental Health Continuum Model, which frames mental wellness along a spectrum from healthy to ill (Wood et al., 2025). It promotes early recognition of distress and proactive intervention. The curriculum integrates cognitive-behavioural techniques such as controlled breathing, cognitive reframing, and goal setting to strengthen resilience. Supervisors receive additional modules on identifying warning signs, supporting staff, and fostering psychologically safe workplaces (Carleton et al., 2018).

Complementing R2MR are peer-support initiatives in many first responder organizations. The Royal Canadian Mounted Police (RCMP) operate a national peer-to-peer program, and several provinces have mandated peer-support structures for paramedics and firefighters (Johnston et al., 2023). These programs emphasize confidentiality, empathy, and mutual understanding, offering trusted avenues for help. Peer supporters often receive training in psychological first aid and crisis communication to ensure timely referral to professional services (Ricciardelli et al., 2022). Currently, there is limited longitudinal research assessing the outcomes of such programs, highlighting the need for further study (Bowers et al., 2025).

Empirical evaluations indicate that R2MR and peer-support programs significantly improve mental health literacy and reduce stigma. Szeto et al. (2019) note that the program's broad adoption reflects the growing commitment to improving first responder mental health. Carleton et al. (2018) found that participants gained confidence in recognizing distress and initiating supportive conversations, while Szeto et al. (2019) reported lower perceived barriers to seeking psychological help.

Johnston et al. (2023), however, observed that long-term outcomes on burnout and PTSD remain mixed. Anderson et al. (2020) similarly concluded that while short-term coping improved, systemic issues such as excessive workload and cumulative trauma continued to drive burnout risk.

A key strength of R2MR lies in its national coordination and adaptability, with federal and provincial agencies integrating mental health training into standard protocols. Nonetheless, training alone cannot offset structural challenges. Some of the limitations of this approach include the fact that participants in a study by Johnston et al. (2023) on the effectiveness of the R2MR training among correctional officers felt that particular skills they had learned as part of the program and certain aspects of peer support were tricky to apply. These challenging elements included muscle relaxation or tactical breathing while working. They felt that some kinds of conflict resolution and the capacity reach out to coworkers who seemed to be going through stress was tricky too. Sustainable success depends on leadership commitment, continued evaluation, and access to clinical care. Canada's psychosocial risk management framework thus represents a balanced combination of preventive education and cultural change but requires ongoing investment to ensure lasting impact.

Limitations and Knowledge Gaps

Ricciardelli et al. (2022) in their research on testimonies following an investigation and the impact on molding the mental well-being of first responders noted that one of the major drawbacks in their analysis is the fact that they relied upon knowledge gained from the combination of experiences of both having testified as well as lived experiences. The researchers note the need for an empirically grounded study as they drew their sources from informal and formal conversation instead of an empirical study. Concerted efforts need to be put into protecting first responders who report stress following a psychologically traumatic experience, as the researchers noted multiple obstacles including mental health stigma and pressure to maintain a strong front for the organization.

In a meta-analytical study on program outcomes conducted by Szeto et al. (2019) on The Road to Mental Readiness for First Responders. The study involved a pre-post design with two follow ups- one at 3 months and the other at 5 months with varied first-responders at sixteen sites to analyze the effectiveness of the Road to Mental Readiness for First Responders program(R2MR), an anti-stigma and resiliency program. The researchers illuminated some noteworthy limitations. The study utilized a pre-test with an open trial as a follow-up. The researchers had to use this design as it was challenging to have organizations assent to a randomized control trial. The researchers went on to note that there is a need for a randomized control trial for these kinds of studies to guarantee that the results obtained are not caused by confounding factors or a sampling bias. Another drawback was a large dropout rate among participants at the time of the follow up, which was partly owing to a difficulty in aligning participants with a coding system. Future research studies need to ensure varied strategies to prevent dropouts among participants. A site coordinator to ensure appropriate matching of participants could be one such strategy. A follow-up assessment was only conducted at a 3-month period. It is uncertain what these results would like after that period of time. Therefore, longitudinal studies were highlighted as a need. Interventions on prevention might take longer to get sanctioned from organizations.

In an analytical study focused on assessing the Road to Mental Readiness (R2MR) training among correctional workers in Canada, Johnston et al. (2023) carried out a survey and obtained anonymous responses to open-ended questions presented by participants. Responses could not be sought after for added clarifications about their perceptions of and experiences with the R2MR program. Future qualitative research, they recommend, therefore, should entail space for further clarification following interviews or focus groups. Owing to a difficulty in distinguishing between organizational and union listservs, the exact response rate of the survey could not be illuminated for this study. Another listed drawback of the survey were smaller sample sizes in some provinces and territories. Furthermore, as the study was qualitative, generalizability was listed as a further limitation. The researchers indicate that an accurate response rate would greatly illuminate into the comfort level of correctional workers in providing information regarding mental health training and willingness.

CASE STUDY 2

United States – Departmental Wellness, Critical Incident Management, and Emerging Resilience Programs

In contrast to Canada's national model, psychosocial risk management in the United States is largely decentralized. Federal guidance from agencies such as NIOSH (2021) and SAMHSA provides direction, but implementation varies by jurisdiction. Individual departments design wellness and resilience initiatives tailored to their culture, leadership, and resources, including innovations such as the Health Disaster Response App (Seligman et al., 2015).

Historically, both the U.S. and Canadian programs focussed on Critical Incident Stress Management (CISM) and Critical Incident Stress Debriefing (CISD), developed in the 1980s (Everly & Mitchell, 1999). These aimed to provide immediate post-incident psychological support within 72 hours of a traumatic event (Devilly & Cotton, 2006). Modern U.S. programs now emphasize holistic wellness beyond critical incident care. Many departments integrate physical fitness, peer-support teams, mental health education, and Employee Assistance Programs (EAPs). The Fire Department of New York (FDNY) exemplifies this model through its Counselling Services Unit, offering confidential peer and clinical support. Similarly, the Los Angeles Police Department (LAPD) has implemented resilience training and psychological services to promote officer well-being.

During the COVID-19 pandemic, the Resilience and Prevention in Disaster (RAPID) study examined psychological outcomes among firefighters and EMTs (Pink, 2021), revealing heightened anxiety, exhaustion, and compassion fatigue. These findings show the need for ongoing psychosocial monitoring and adaptive interventions. Departments involved in RAPID used data-driven feedback to improve workload management, shift schedules, and telehealth access. Despite these advances, U.S. psychosocial management remains fragmented. Departments with strong funding and leadership support show positive outcomes, while smaller or rural agencies often lack resources for sustained implementation. Limited standardization also hampers cross-agency comparisons. The U.S. approach's strength lies in its flexibility and innovation, but its weakness is the absence of a unified national framework ensuring consistency and accountability. A good example of this is the American Academy of Experts in Traumatic Stress. This network sets the U.S. apart in terms of an innovative opportunity for professionals across varied fields to upskill to provide trauma informed support to survivors of traumatic events.

A comparison of Canadian and U.S. psychosocial risk management models reveals both shared values and structural differences. Canada's approach is nationally coordinated and uniform across provinces, primarily through the R2MR framework and federally supported peer-support programs. This coordination promotes consistency in training standards, evaluation, and data collection. In contrast, the U.S. model is highly decentralized, which has both advantages and disadvantages. The decentralized nature of the initiatives allows for innovation, local adaptation and innovative wellness and resilience initiatives that respond to dynamically evolving needs. However, the flipside of these decentralized initiatives is that they result in inconsistent access to evidence-based mental health resources. Both countries prioritize early intervention and destigmatization, with Canadian programs emphasizing cognitive-behavioural resilience training, while U.S. initiatives focus on immediate post-incident support and long-term wellness strategies. Each approach offers valuable lessons, with Canada demonstrating the effectiveness of centralized policy leadership. Both systems face similar challenges, including insufficient long-term evaluation and resource disparities. Training programs alone cannot offset organizational stressors, such as chronic understaffing, shift fatigue, and exposure to repeated trauma (Anderson et al., 2020). Therefore, effective psychosocial risk management requires systemic change, including leadership engagement, workload adjustment, and accessible mental health services.

Limitations and Knowledge Gaps

In a study conducted by Pink. (2021) on psychological distress and resilience in first responders and health care workers during the COVID-19 pandemic, the researcher took cognizance of some important limitations of the study. No pre-pandemic comparison data was readily at hand to afford comparisons with present standards of discomfort and resilience. Therefore, the study could only present a one-time glimpse of the psychological distress within first responders during the pandemic period. Further, the sampling was based on convenience and therefore may not have been fully representative of these service groups. Moreover, while resilience may account for some of the variations in the results obtained, the cross-sectional nature of the data may prove to be a confounder as some of the participants may have displayed greater resilience because they were in a good state of wellness at the time of the study. On the other hand, participants who were less resilient may either have not been able to complete the survey or not been willing to. The researcher opined that the study could be at an advantage from longitudinal surveys over an extended period following the COVID-19 pandemic and data post the pandemic. It would greatly add to the value of the study to analyze how predictive resilience is for reduced psychological distress over time. The researcher was also of the opinion that a qualitative element in upcoming longitudinal studies, might add depth by catching personal narratives of stress management and coping strategies. Adding this information may aid in elaborating on the variations in distress and resilience, which is present among the different first responder groups.

In their paper on Critical incident stress management -CISM-: A new era and standard of care in crisis intervention, Everly, G., & Mitchell, J. (1999) clarify some important misconceptions regarding critical incident stress management. Firstly, the efficacy of CISM programs has been empirically validated via multiple qualitative analyses, thoroughly conducted investigations, and meta-analyses. Sadly, this is often forgotten. Secondly, there exists a misconception that CISD/ CISM is evidenced to take away from its participants. The researchers clarify that there is no such evidence to argue that the "Mitchell model" CISD, or the CISM system has negatively impacted its participants. The research studies that suggest such negative impacts often basically did not use the CISD or CISM system as intended. This too is a fact that is often ignored.

COMPARATIVE ANALYTICAL FRAMEWORK

To synthesize the policy and practice differences identified in the case studies and literature, a comparative analytical framework will be presented to contrast Canada and the United States across key occupational health and safety (OHS) dimensions. This framework helps to illuminate variations in regulatory oversight, leadership models, use of digital tools, and underlying safety cultures that shape psychosocial risk management for first responders in each context.

In Canada, post-traumatic stress injury (PTSI) regulation operates through a dual-level structure involving both federal initiatives and provincial mandates, such as the National Action Plan on PTSI. In contrast, the United States relies on more decentralized, state-level statutes and agency-specific wellness policies. Leadership in Canadian contexts often involves centralized systems for Critical Incident Stress Management (CISM) and peer-support networks. In contrast, U.S. departments implement wellness interventions on a more localized, department-by-department basis.

Regarding digital interventions, Canadian provinces have begun funding telecounselling platforms to provide accessible mental health support. Meanwhile, U.S. agencies are experimenting with municipal-level pilot apps and early-stage artificial intelligence-driven risk screening tools. Also, safety culture in Canada emphasizes shared responsibility, routine debriefings, and institutional accountability. Equally, adherence to such practices in the U.S. varies widely, with some departments relying heavily on informal peer support networks and ad hoc responses to mental health challenges.

Table 1. Comparative Analytical Framework of Canada and the USA

Dimension	Canada	United States
Regulation	Federal/provincial PTSI mandates	State-level variation; agency-driven policies
Leadership	Centralized CISM & peer networks	Department-specific wellness initiatives
Digital Tools	Province-funded tele-counseling platform	Municipal pilot apps; AI-driven risk screening prototypes
Safety Culture	Emphasis on shared responsibility, routine debriefings	Variable adherence; reliance on informal peer networks

Note. CISM = Critical Incident Stress Management; PTSI = Post-Traumatic Stress Injury.

SIGNIFICANCE & SDG ALIGNMENT

The significance of this chapter lies in its dual contribution to academic scholarship and practical policymaking in the realm of occupational health and safety (OHS) for first responders. By situating North American best practices within a comparative framework, the chapter provides actionable insights for governments, institutions, and frontline services globally. These insights align closely with the United Nations Sustainable Development Goals (SDGs), particularly those that emphasize health, decent work, and equity in institutional response frameworks.

1. SDG 3 (Good Health & Well-being): Enhances access to mental health services, promotes early intervention, and reduces the treatment gap for high-risk workers (WHO, 2024).
2. SDG 8 (Decent Work & Economic Growth): Fosters safe, supportive work environments, improving retention and performance in critical public-service sectors (ILO, 2019).
3. Global Transferability: Lessons from North American first-responder OHS can inform culturally sensitive interventions in developing and transitional economies, bridging regulatory, organizational, and cultural dimensions.

Towards A Culturally Responsive Ohs Model

In Canada, workplace mental health is closely linked to public health and social safety nets, making it easier to connect mental-health programs with compensation and recovery services (Edgelow & Fecica, 2024). Meanwhile, in the U.S., responsibilities are split across cities, states, and federal agencies, which means support and funding can vary a lot quite a bit across these jurisdictions (Murphy et al., 2020). Canada can roll out large-scale online therapy and peer support programs much more smoothly while U.S. agencies are seen too often dependent on local leaders or grants to try new apps or peer programs (McCall et al., 2020).

Safety Culture and Stigma

A climate of safety involves not only adhering to procedures in place but also it is about whether people feel safe to be honest and show emotion without fear of being judged or punished. In Canada, peer-support and return-to-work programs try to break the stigma by making mental health conversations and support a normal part of workplace life (Edgelow & Fecica, 2024). It is evident that stigma is experienced in both the countries with first responders still admit to hiding their distress for fear of job consequences or worse job loss Stigma and barriers to seeking help are widely reported, with first responders often hiding their distress for fear of job consequences or worse, job loss (Lade et al., 2023). In the U.S., efforts to reduce stigma vary widely and often rely on strong local leaders or unions who take the initiative to support their teams (Murphy et al., 2020).

Technology Adoption

Digital interventions such as Peer Apps, ICBT and Telehealth, including the 988 Suicide Crisis Helpline seem to have been scaled in Canada with greater intentionality in Canada (Lade et al., 2023). In Canada, the PeerConnect projects show how peer support tools are built by listening to the people who actually use them (Foley & Ricciardelli, 2023b). In the U.S., new technology tends to roll out faster such as during the times of the pandemic when emergency workers tried new digital tools but how well these tools are tested or who gets access to them often varies (Murphy et al., 2020). Technology often reflects the same priorities and inequalities that already exists.

PROPOSED CULTURALLY RESPONSIVE OHS MODEL FOR BURNOUT PREVENTION

Based on the comparison between state level statutes in the U.S and the Federal Framework on PTSD in Canada, a good safety and health model for first responders should bring together four key parts: (1) clear rules and standards that is set at the system level, (2) strong and supportive leadership that is set at the organizational level, (3) peer support built into care and recovery at the team level, and (4) flexible technology that is designed with real people in mind at the individual level.

Clear System-Level Rules and Standards Set at the System Level

Every first responder, no matter where they work, should have access to trauma-informed care and clear pathways for returning to work after injury or stress. This means that linking workers' compensation, occupational therapy, and digital mental health tools so that care feels continuous and coordinated. Governments and related agencies should also share data on occupational risks, such as trauma exposure and heat stress, to help prevent harm before it happens.

Strong And Supportive Leadership Set at Organizational Level

Leaders and managers play a critical role in shaping workplace culture. Training in empathy, psychological safety, and inclusive leadership helps create environments where it is psychologically safe to ask for help. Leadership programs should also help address identity conflicts that come with high-pressure roles and actively challenge disciplinary or trivializing attitudes toward mental health concerns.

Peer Support Built into Care and Recovery at Team Level

Peers are often the first contact of trust at workplace. Establishing trained peer facilitators within teams can make support more immediate and less intimidating. These peers often trained in trauma-informed approaches (Choi et al., 2024), can connect colleagues to digital therapy, occupational rehab, or professional care, creating a no wrong door approach to getting help. Trauma-informed care helps ensure that all interactions acknowledge prior exposure to stress or trauma, reducing re-traumatization and fostering resilience. When peers, clinicians, and therapists work together, recovery becomes part of the culture and not an exception to it.

Flexible Technology Designed with Real People At Individual Level

Digital tools like ICBT and peer-support apps should be designed with first responders, not for them. When users are part of the design process, technology becomes more relevant, respectful, and effective. These tools should be trauma-sensitive, flexible for different work settings, and directly linked to follow-up care when more support is needed.

Figure 1. Proposed Culturally Responsive OHS Model for Burnout Prevention Among First Responders

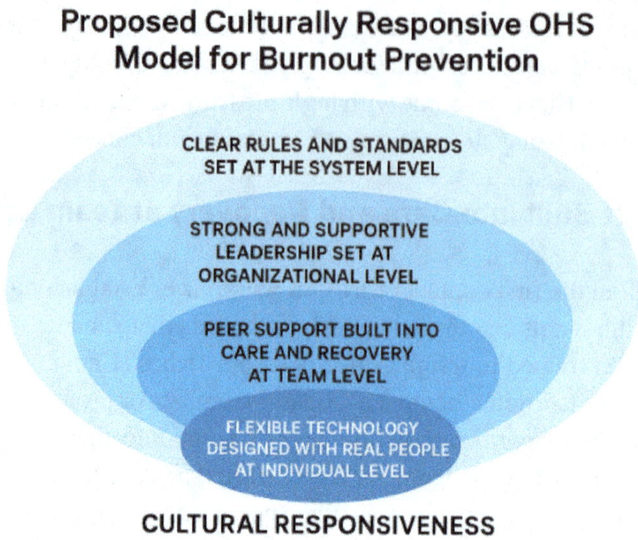

RECOMMENDATIONS FOR EMPLOYERS & LEADERS

First responder support systems need to be conceptualized and implemented in a manner that addresses the unique requirements of first responders on a person-centered basis (Johnston et al., 2025). It is important for departments to acknowledge that first responders, as officials of workplace positions and as humans, are susceptible to declining mental health (Lentz et al., 2022). These considerations should motivate leaders to nurture work cultures of psychological safety and continuous training. Leaders who proactively utilize evidence-based training in mental well-being can promote psychologically safe workplaces that are more resilient and capable of upholding first responder well-being (Lentz et al., 2022).

Leaders need to be trained to identify employees in crisis and how to respond when crises occur (Lentz et al., 2022). If such holistic changes are undertaken, they may reduce work-related stress and extend the well-being of personnel through psychologically safe workplace cultures supported by trauma-informed policies and practices (Knaak et al., 2022). Encouraging employee voice can help surface problems manageable through organization-wide initiatives. In a study among paramedics on peer-to-peer voice, a veteran employee recalled when shifts were divided between time in vehicles and at the department, allowing space for conversation and training. Over twenty years, newer employees have become accustomed only to busy phases, with little departmental time remaining (Lentz et al., 2022; Loudoun et al., 2020).

Workplace culture has not kept pace with maintaining practices such as peer dialogue and reflection as part of a proactive approach. Therefore, departments should prioritize establishing peer-to-peer support networks for first responders. The reduction in time spent with peers has created a need for personnel to self-initiate conversations rather than having structured reflection and dialogue (Lentz et al., 2022; Loudan et al., 2020). Furthermore, when employees do not receive ongoing training about available interventions and their benefits, they may be unaware of how to access or use them. Proactive systems should identify and support employees early, rather than intervening only when they are already distressed (Johnston et al., 2025). Fostering a climate of organizational safety can improve mental health, build trust, communication, training, performance, and overall commitment to the organization (Knaak et al., 2019).

Organizational stress remains a key element of mental health challenges in first responders. Staff shortages, leadership approaches, rotational work, sleep issues, stigma, and public scrutiny all significantly impact mental health outcomes following traumatic events. This highlights the need for improved workplace and daily stress management through essential support. Jessiman-Perreault et al. (2021) propose policy improvements that incorporate initiatives to enhance coping and resiliency.

Integrating resilience or field readiness training and development can profoundly uplift the mental well-being of first responders. Resilience building has long been at the core of researchers' and clinicians' efforts, and widening the understanding of factors that contribute to thriving could aid first responders individually and collectively (Papazoglou, 2023). Wellness programs can be financially demanding therefore, emergency service organizations could adopt cost-effective approaches offering prompt support through trained peers, such as anonymous peer networks, wellness-focussed materials, self-assessment and check-in tools, and access to department support resources. Candidates may also learn topics such as procedural justice, community relations, stress management, and psychological skills (Papazoglou, 2023). Embedding wellness into training enables first responders to proactively manage stress and discuss well-being, encouraging help-seeking and reducing stigma around mental health.

Another important step for policy makers is improving work-life balance through set schedules and enhanced schedule visibility (Bevan et al., 2022). Technology can also track incidents to help departments proactively mitigate burnout. Artificial intelligence tools can identify patterns and evaluate the links between interventions and outcomes to inform decisions during future emergencies (Amiot & Potier., 2025).

The role of AI and technology will become increasingly important in supporting first responder mental health.

While resilience and wellness training offer benefits, several limitations exist. Emphasizing individual resilience may shift responsibility from organizational systems to individual coping, overlooking systemic issues (Walker, 2025). Such training may not fully address structural workplace stressors like staffing shortages, leadership challenges, or shift schedules and may have limited impact without broader organizational reform. As the resilience paradox suggests, excessive focus on resilience can even fuel burnout if institutional causes remain unaddressed (Walker, 2025). Moreover, wellness programs and peer-support tools must be rigorously evaluated to ensure they essentially meet the diverse needs of all first responder groups.

HOLISTIC APPROACHES TO SUPPORTING FIRST RESPONDER MENTAL HEALTH AND RESILIENCE

Technology developers must consider key aspects when designing support applications. Online resources should be broad, accessible, and anonymous, as lack of anonymity can discourage use due to job-related stigma. Digital mental health training has shown meaningful post-intervention effects on well-being and work efficacy compared with control groups (Carolan et al., 2017). These approaches attract departments because they are economical and convenient, however, they may oversimplify complex mental health realities. Individualized and asynchronous designs risk high dropout rates and discontinuity (Di Nota et al., 2021), while insufficient work time for learning can further limit engagement (Carolan et al., 2017; Lentz et al., 2022).

Guided formats, interactive features like texting or self-monitoring, and customization to workgroup needs improve adherence (Carolan et al., 2017). Participants value anonymity and convenience but prefer engaging, in-person, job-specific sessions by professionals or peers familiar with their culture (Lentz et al., 2022). Therefore, developers should integrate first responder culture, field challenges, and engagement preferences into app design, collaborating with subject matter experts and those with firsthand trauma experience.

A key consideration is whether post-incident mental health support should be optional or mandatory. Johnston et al. (2025) found that mandatory sessions could reduce stigma and signal organizational commitment, though compulsion may lessen efficacy, with limited evidence confirming benefits. Mandating attendance after critical events, such as those involving children, might be valuable (Johnston et al., 2015). Exploring motivators and barriers to participation can clarify how to destigmatize support-seeking behaviors.

Pre-employment training often emphasizes technical skills (Smith-MacDonald et al., 2021), leaving gaps in communication and ethics. Respondents reported feeling underprepared to manage mental health issues (Johnston et al., 2025). Addressing these gaps through cognitive-behavioral and mindfulness training can equip employees to cope with stress (Bowers et al., 2025). Involving employees in decision-making and maintaining transparency increase engagement and commitment (Jiang & Shen, 2023).

CONCLUSION

This chapter significantly contributes to the existing knowledge base on burnout among first responders by analyzing burnout from the first responder context, comparing regulatory and organizational frameworks and foundational theories informing first responder mental health, evaluating existing psychosocial interventions and finally proposing a framework for a culturally responsive OHS model. Case studies from both the United States and Canada have highlighted just how demanding a toll first responder duty can exert on first responder mental health and the urgency for the enhancement of programs and initiatives that would uplift first responder resiliency as opposed to reactionary programs in the aftermath of traumatic incidents. Multiple studies have highlighted the significance of sensitivity and leadership training, a need for departmental wide regulations and consistency of availability of support for first responders. Future research needs to focus on how first responder departments and policy makers can take concrete steps to ensure that first responder mental health training would equip first responders with the skills they would need to tackle the challenges they are likely to face in the field. Motivational factors that would enhance the attractiveness of mental health training to first responders across varied levels of their professional journey and ensure the sustainability of initiatives across departments needs to be explored in depth. The tried and tested efficacy of specific first responder mental health initiatives and digital applications across particular first responder departments is still in its infancy and findings across studies have been inconsistent. This is an important gap that needs to be addressed through future research.

REFERENCES

Amiot, F., & Potier, B. (2025). Artificial Intelligence (AI) and Emergency Medicine: Balancing Opportunities and Challenges. *JMIR Medical Informatics*, *13*, e70903. DOI: 10.2196/70903 PMID: 40802997

Bakker, A. B., & Demerouti, E. (2017). Job demands–resources theory: Taking stock and looking forward. *Journal of Occupational Health Psychology*, *22*(3), 273–285. DOI: 10.1037/ocp0000056 PMID: 27732008

Bass, B. M. (1999). Two Decades of Research and Development in Transformational Leadership. *European Journal of Work and Organizational Psychology*, *8*(1), 9–32. DOI: 10.1080/135943299398410

Bass, B. M., & Riggio, R. E. (2006). *Transformational leadership* (2nd ed.). Lawrence Erlbaum Associates Publishers., DOI: 10.4324/9781410617095

Basu, S., Qayyum, H., & Mason, S. (2017). Occupational stress in the ED: A systematic literature review. *Emergency Medicine Journal : EMJ*, *34*(7), 441–447. DOI: 10.1136/emermed-2016-205827 PMID: 27729392

Bevan, M. P., Priest, S. J., Plume, R. C., & Wilson, E. E. (2022). Emergency First Responders and Professional Wellbeing: A Qualitative Systematic Review. *International Journal of Environmental Research and Public Health*, *19*(22), 14649. DOI: 10.3390/ijerph192214649 PMID: 36429361

Carleton, R. N., Afifi, T. O., Turner, S., Taillieu, T., Duranceau, S., LeBouthillier, D. M., Sareen, J., Ricciardelli, R., MacPhee, R. S., Groll, D., Hozempa, K., Brunet, A., Weekes, J. R., Griffiths, C. T., Abrams, K. J., Jones, N. A., Beshai, S., Cramm, H. A., Dobson, K. S., & Asmundson, G. J. G. (2018). Mental Disorder Symptoms among Public Safety Personnel in Canada. *Canadian Journal of Psychiatry*, *63*(1), 54–64. DOI: 10.1177/0706743717723825 PMID: 28845686

Carolan, S., Harris, P. R., & Cavanagh, K. (2017). Improving Employee Well-Being and Effectiveness: Systematic Review and Meta-Analysis of Web-Based Psychological Interventions Delivered in the Workplace. *Journal of Medical Internet Research*, *19*(7), e7583. DOI: 10.2196/jmir.7583 PMID: 28747293

Ceschi, G., Meylan, S., Rowe, C., & Boudoukha, A. H. (2022). Psychological Profile, Emotion Regulation, and Aggression in Police Applicants: A Swiss Cross-Sectional Study. *Journal of Police and Criminal Psychology*, *37*(4), 962–971. DOI: 10.1007/s11896-022-09548-0

Choi, J., Gabay, E. K., Rachel, A., Brown, A., Cruz, R., Yan, Q., & Cuccaro, P. M. (2024). Evaluation of a trauma-informed care (TIC) training program across community health centres in Texas: A qualitative study. *BMC Health Services Research*, *24*(1), 1360. DOI: 10.1186/s12913-024-11813-3 PMID: 39511566

Corthésy-Blondin, L., Genest, C., Dargis, L., Bardon, C., & Mishara, B. L. (2022). Reducing the impacts of exposure to potentially traumatic events on the mental health of public safety personnel: A rapid systematic scoping review. *Psychological Services*, *19*(Suppl 2), 80–94. DOI: 10.1037/ser0000572 PMID: 34338535

Crowe, R. P., Bower, J. K., Cash, R. E., Panchal, A. R., Rodriguez, S. A., & Olivo-Marston, S. E. (2018). Association of Burnout with Workforce-Reducing Factors among EMS Professionals. *Prehospital Emergency Care*, *22*(2), 229–236. DOI: 10.1080/10903127.2017.1356411 PMID: 28841102

Cunningham, T. R., Guerin, R. J., Ferguson, J., & Cavallari, J. (2022). Work-related fatigue: A hazard for workers experiencing disproportionate occupational risks. *American Journal of Industrial Medicine*, *65*(11), 913–925. DOI: 10.1002/ajim.23325 PMID: 35088430

Di Nota, P. M., Bahji, A., Groll, D., Carleton, R. N., & Anderson, G. S. (2021). Proactive psychological programs designed to mitigate posttraumatic stress injuries among at-risk workers: A systematic review and meta-analysis. *Systematic Reviews*, *10*(1), 126. DOI: 10.1186/s13643-021-01677-7 PMID: 33910641

Edgelow, M., & Fecica, A. (2024). Occupational therapy treatment of public safety personnel with work-related psychological injuries: Analyzing Ontario worker's compensation data from 2017–2021. *Frontiers in Psychiatry*, *15*, 1377157. Advance online publication. DOI: 10.3389/fpsyt.2024.1377157 PMID: 39834574

Edmondson, A. (1999). Psychological safety and learning behavior in work teams. *Administrative Science Quarterly*, *44*(2), 350–383. DOI: 10.2307/2666999

Everly, G., & Mitchell, J. (1999). *Critical incident stress management -CISM-: A new era and standard of care in crisis intervention*. https://www.semanticscholar.org/paper/Critical-incident-stress-management-CISM-%3A-a-new-of-Everly-Mitchell/743cbe594434e10734b7d91173 01a00d67b0da36

Federal Framework on Post-Traumatic Stress Disorder Act. S.C. 2018, c. 13. https://laws-lois.justice.gc.ca/eng/acts/f-7.38

Foley, G., & Ricciardelli, R. (2023). Views on the Functionality and Use of the PeerConnect App Among Public Safety Personnel: Qualitative Analysis. *JMIR Formative Research*, *7*(1), e46968. DOI: 10.2196/46968 PMID: 37930765

Fragoso, Z. L., Holcombe, K. J., McCluney, C. L., Fisher, G. G., McGonagle, A. K., & Friebe, S. J. (2016). Burnout and Engagement: Relative Importance of Predictors and Outcomes in Two Health Care Worker Samples. *Workplace Health & Safety, 64*(10), 479–487. DOI: 10.1177/2165079916653414 PMID: 27282979

Frazer, H., Van Hooff, M., Lawrence-Wood, E., & Searle, A. (2021). Factors influencing the recruitment and retention of female first responders: A mixed methods systematic review protocol. *JBI Evidence Synthesis, 19*(3), 721–726. DOI: 10.11124/JBISRIR-D-19-00427 PMID: 33141801

Goraya, N. K., Alvarez, E., Young, M., & Moll, S. (2024). *PeerOnCall*: Exploring how organizational culture shapes implementation of a peer support app for public safety personnel. *Comprehensive Psychiatry, 135*, 152524. DOI: 10.1016/j.comppsych.2024.152524 PMID: 39146608

Guseva Canu, I., Marca, S. C., Dell'Oro, F., Balázs, Á., Bergamaschi, E., Besse, C., Bianchi, R., Bislimovska, J., Koscec Bjelajac, A., Bugge, M., Busneag, C. I., Çağlayan, Ç., Cerni anu, M., Costa Pereira, C., Dernovšček Hafner, N., Droz, N., Eglite, M., Godderis, L., Gündel, H., & Wahlen, A. (2021). Harmonized definition of occupational burnout: A systematic review, semantic analysis, and Delphi consensus in 29 countries. *Scandinavian Journal of Work, Environment & Health, 47*(2), 95–107. DOI: 10.5271/sjweh.3935 PMID: 33258478

Hadjistavropoulos, H. D., McCall, H. C., Dear, B. F., Beahm, J. D., Carleton, R. N., & Titov, N. (2024). Outcomes of transdiagnostic internet-delivered cognitive behavioural therapy tailored to public safety personnel: A longitudinal observational study. *Journal of Anxiety Disorders, 104*, 102861. DOI: 10.1016/j.janxdis.2024.102861 PMID: 38640867

Hobfoll, S. E. (2011). Conservation of resources theory: Its implication for stress, health, and resilience. *The Oxford handbook of stress, health, and coping, 127*, 147.

Hobfoll, S. E., Tirone, V., Holmgreen, L., & Gerhart, J. (2016). Conservation of resources theory applied to major stress. In *Stress: Concepts, cognition, emotion, and behavior* (pp. 65–71). Academic Press., DOI: 10.1016/B978-0-12-800951-2.00007-8

Huang, Y. H., Lee, J., McFadden, A. C., Murphy, L. A., Robertson, M. M., Cheung, J. H., & Zohar, D. (2016). Beyond safety outcomes: An investigation of the impact of safety climate on job satisfaction, employee engagement and turnover using social exchange theory as the theoretical framework. *Applied Ergonomics, 55*(3), 248–257. DOI: 10.1016/j.apergo.2015.10.007 PMID: 26611987

Jessiman-Perreault, G., Smith, P. M., & Gignac, M. A. M. (2021). Why Are Workplace Social Support Programs Not Improving the Mental Health of Canadian Correctional Officers? An Examination of the Theoretical Concepts Underpinning Support. *International Journal of Environmental Research and Public Health, 18*(5), 2665. DOI: 10.3390/ijerph18052665 PMID: 33800869

Jiang, H., & Shen, H. (2023). Toward a relational theory of employee engagement: Understanding authenticity, transparency, and employee behaviors. *International Journal of Business Communication, 60*(3), 948–975. DOI: 10.1177/2329488420954236

Johnson, C. C., Vega, L., Kohalmi, A. L., Roth, J. C., Howell, B. R., & Van Hasselt, V. B. (2020). Enhancing mental health treatment for the firefighter population: Understanding fire culture, treatment barriers, practice implications, and research directions. *Professional Psychology, Research and Practice, 51*(3), 304–311. DOI: 10.1037/pro0000266

Johnston, M. S., Ricciardelli, R., Ghodrati, M., & Czarnuch, S. (2023). Assessing Road to Mental Readiness (R2MR) training among correctional workers in Canada. *Health & Justice, 11*(1), 2. DOI: 10.1186/s40352-023-00206-z PMID: 36683119

Johnston, S., Waite, P., Laing, J., Rashid, L., Wilkins, A., Hooper, C., Hindhaugh, E., & Wild, J. (2025). Why Do Emergency Medical Service Employees (Not) Seek Organizational Help for Mental Health Support? A Systematic Review. *International Journal of Environmental Research and Public Health, 22*(4), 629. DOI: 10.3390/ijerph22040629 PMID: 40283853

Jones, C., Spencer, S., O'Greysik, E., Smith-MacDonald, L., Bright, K. S., Beck, A. J., Carleton, R. N., Burback, L., Greenshaw, A., Zhang, Y., Sevigny, P. R., Hayward, J., Cao, B., & Brémault-Phillips, S. (2024). Perspectives and Experiences of Public Safety Personnel Engaged in a Peer-Led Workplace Reintegration Program Post Critical Incident or Operational Stress Injury: A Qualitative Thematic Analysis. *International Journal of Environmental Research and Public Health, 21*(7), 949. DOI: 10.3390/ijerph21070949 PMID: 39063525

Kaplan, G. R., Frith, T., & Hubble, M. W. (2024). Quantifying the prevalence and predictors of burnout in emergency medical services personnel. *Irish Journal of Medical Science (1971 -), 193*(3), 1545–1556. https://doi.org/DOI: 10.1007/s11845-023-03580-7

Kelly, R. J., & Hearld, L. R. (2020). Burnout and Leadership Style in Behavioral Health Care: A Literature Review. *The Journal of Behavioral Health Services & Research, 47*(4), 581–600. DOI: 10.1007/s11414-019-09679-z PMID: 32189184

Khoury, J. M. B., Teckchandani, T. A., Nisbet, J., Stewart, S. H., Asmundson, G. J. G., Afifi, T. O., McCarron, M. C. E., Kratzig, G. P., Sauer-Zavala, S., Neary, J. P., MacPhee, R. S., Brunet, A., Keane, T. M., & Carleton, R. N. (2025). Putative risk and resiliency factors after an augmented training program for preventing posttraumatic stress injuries among public safety personnel from diverse sectors. *Cognitive Behaviour Therapy, 54*(4), 457–476. DOI: 10.1080/16506073.2024.2420636 PMID: 39526850

Knaak, S., Luong, D., McLean, R., Szeto, A., & Dobson, K. S. (2019). Implementation, Uptake, and Culture Change: Results of a Key Informant Study of a Workplace Mental Health Training Program in Police Organizations in Canada. *Canadian Journal of Psychiatry, 64*(1, Suppl), 30S–38S. DOI: 10.1177/0706743719842565 PMID: 31056932

Kowalski, C. (2019). Leadership of first-responders following trauma. *Journal of Business Continuity & Emergency Planning, 13*(1), 81–90. https://pubmed.ncbi.nlm.nih.gov/31462365/. DOI: 10.69554/ZMLH9836 PMID: 31462365

Kowalski, C. (2019). Leadership of first-responders following trauma. *Journal of Business Continuity & Emergency Planning, 13*(1), 81–90. https://pubmed.ncbi.nlm.nih.gov/31462365/. DOI: 10.69554/ZMLH9836 PMID: 31462365

La Manna, A., Siddiqui, S., Gerber, G., Budesa, Z., Vance, K., Goulka, J., Beletsky, L., Marotta, P., & Winograd, R. (2025). Overdose and overwork: First responder burnout and mental health help-seeking in Missouri's overdose crisis. *Drug and Alcohol Dependence, 271*, 112590. DOI: 10.1016/j.drugalcdep.2025.112590 PMID: 40073804

Lade, S., Easterbrook, B., Brown, A., Millman, H., D'Alessandro-Lowe, A. M., O'Connor, C., & McKinnon, M. C. (2023). The mental health toll of service: An examination of self-reported impacts of public safety personnel careers in a treatment-seeking population. *European Journal of Psychotraumatology, 14*(2), 2269696. Advance online publication. DOI: 10.1080/20008066.2023.2269696 PMID: 37965795

Lentz, L., Smith-MacDonald, L., Malloy, D. C., Anderson, G. S., Beshai, S., Ricciardelli, R., Bremault-Phillips, S., & Carleton, R. N. (2022). A Qualitative Analysis of the Mental Health Training and Educational Needs of Firefighters, Paramedics, and Public Safety Communicators in Canada. *International Journal of Environmental Research and Public Health, 19*(12), 6972. DOI: 10.3390/ijerph19126972 PMID: 35742221

Loudoun, R., Townsend, K., Wilkinson, A., & Mowbray, P. K. (2020). The role of peer-to-peer voice in severe work environments: Organisational facilitators and barriers. *Industrial Relations Journal, 51*(6), 556–571. DOI: 10.1111/irj.12313

Lowery, A., & Cassidy, T. (2022). Health and well-being of first responders: The role of psychological capital, self-compassion, social support, relationship satisfaction, and physical activity. *Journal of Workplace Behavioral Health, 37*(2), 87–105. DOI: 10.1080/15555240.2021.1990776

Maslach, C., & Leiter, M. P. (2016). Burnout. In Fink, G. (Ed.), *Stress: Concepts, cognition, emotion, and behavior* (pp. 351–357). Elsevier Academic Press.

McCall, H., Beahm, J., Landry, C., Huang, Z., Carleton, R. N., & Hadjistavropoulos, H. (2020). How Have Public Safety Personnel Seeking Digital Mental Healthcare Been Affected by the COVID-19 Pandemic? An Exploratory Mixed Methods Study. *International Journal of Environmental Research and Public Health, 17*(24), 9319. DOI: 10.3390/ijerph17249319 PMID: 33322095

McCall, W. T. (2023). Piloting Peer Support to Decrease Secondary Traumatic Stress, Compassion Fatigue, and Burnout Among Air Medical Crewmembers. *Air Medical Journal, 42*(3), 157–162. DOI: 10.1016/j.amj.2023.01.004 PMID: 37150568

Mitchell, J. T. (1974). *Critical incident stress debriefing (CISD)* [PDF]. U.S. Army Corps of Engineers. https://corpslakes.erdc.dren.mil/employees/cism/pdfs/Debriefing.pdf

Moll, S. E., Ricciardelli, R., Carleton, R. N., MacDermid, J. C., Czarnuch, S., & MacPhee, R. S. (2025). PeerOnCall: Evaluating Implementation of App-Based Peer Support in Canadian Public Safety Organizations. *International Journal of Environmental Research and Public Health, 22*(8), 1269. DOI: 10.3390/ijerph22081269 PMID: 40869854

Moore, S. (2022, October 17). 4th Annual Study Shows Worsening EMS Turnover. *American Ambulance Association.* https://ambulance.org/2022/10/17/4th-annual-study-shows-worsening-ems-turnover/

Moslehi, S., Tavan, A., Khezeli, M., Soleimanpour, S., & Narimani, S. (2025). Silent crisis on the frontlines: A systematic review of suicidal behaviors among disaster responders - epidemiology, risk pathways, and evidence-based interventions. *Scandinavian Journal of Trauma, Resuscitation and Emergency Medicine, 33*(1), 161. DOI: 10.1186/s13049-025-01479-z PMID: 41044626

Murphy, D. L., Barnard, L. M., Drucker, C. J., Yang, B. Y., Emert, J. M., Schwarcz, L., Counts, C. R., Jacinto, T. Y., McCoy, A. M., Morgan, T. A., Whitney, J. E., Bodenman, J. V., Duchin, J. S., Sayre, M. R., & Rea, T. D. (2020). Occupational exposures and programmatic response to COVID-19 pandemic: An emergency medical services experience. *Emergency Medicine Journal : EMJ*, *37*(11), 707–713. DOI: 10.1136/emermed-2020-210095 PMID: 32958477

Musthafa, H. S., Moosa, S., Riyaz, A., & Abdul Raheem, R. (2023). *Stability of values and opinions during the COVID-19 crisis: Panel study data from the Maldives* (Vol. 11, No. 1). The Maldives National Journal of Research. https://doi.org/ DOI: 10.62338/vjwxxs02

NIOSH. (2021, July 8). *Emergency Response Safety and Health Database | NIOSH | CDC*. https://www.cdc.gov/niosh/ershdb/default.html

O'Connell, S., Troya, M. I., Arensman, E., & Griffin, E. (2024). That feeling of solidarity and not being alone is incredibly, incredibly healing: A qualitative study of participating in suicide bereavement peer support groups. *Death Studies*, *48*(2), 176–186. DOI: 10.1080/07481187.2023.2201922 PMID: 37092540

O'Toole, K., & Brown, C. A. (2021). Evaluating the quality of resilience apps for military members and public safety personnel. *Journal of Military, Veteran and Family Health*, *7*(1), 87–101. DOI: 10.3138/jmvfh-2020-0002

Papazoglou, K. (2023). Stress, Prevention, and Resilience among First Responders. *International Journal of Environmental Research and Public Health*, *20*(24), 7174. DOI: 10.3390/ijerph20247174 PMID: 38131724

Perrott, T. (2016). Beyond 'token' firefighters: Exploring women's experiences of gender and identity at work. *Sociological Research Online*, *21*(1), 51–64. DOI: 10.5153/sro.3832

Pink. (2021). *Psychological distress and resilience in first responders and health care workers during the COVID-19 pandemic—Pink—2021—Journal of Occupational and Organizational Psychology—Wiley Online Library*. https://bpspsychub.onlinelibrary.wiley.com/doi/10.1111/joop.12364

Price, J. A., Gregory, J., McCall, H. C., Landry, C. A., Beahm, J. D., & Hadjistavropoulos, H. D. (2024). Client Perspectives of Case Stories in Internet-Delivered Cognitive Behavioral Therapy for Public Safety Personnel: Mixed Methods Study. *JMIR Formative Research*, *8*(1), e64454. DOI: 10.2196/64454 PMID: 39453705

PTSD in First Responders. Institutes of Health. (2025, March 7). https://institutesofhealth.org/ptsd-in-first-responders/#references

Public Health Agency of Canada. (2019, January 22). *Federal Framework on Post-traumatic Stress Disorder: Recognition, collaboration and support* (94 pp.). Government of Canada. https://www.canada.ca/en/public-health/services/publications/healthy-living/federal-framework-post-traumatic-stress-disorder.html

Reason, J. (2016). *Managing the risks of organizational accidents*. Routledge., DOI: 10.4324/9781315543543

Ricciardelli, R., Carleton, R. N., Anschuetz, B., Gravel, S., & McKay, B. (2022). Testifying after an Investigation: Shaping the Mental Health of Public Safety Personnel. *International Journal of Environmental Research and Public Health, 19*(20), 13643. DOI: 10.3390/ijerph192013643 PMID: 36294223

Roth, S. L., Andrews, K., Protopopescu, A., Lloyd, C., O'Connor, C., Losier, B. J., Lanius, R. A., & McKinnon, M. C. (2022). Mental health symptoms in Public Safety Personnel: Examining the effects of adverse childhood experiences and moral injury. *Child Abuse & Neglect, 123*, 105394. DOI: 10.1016/j.chiabu.2021.105394 PMID: 34808481

Seligman, J., Felder, S. S., & Robinson, M. E. (2015). Substance Abuse and Mental Health Services Administration (SAMHSA) Behavioral Health Disaster Response App. *Disaster Medicine and Public Health Preparedness, 9*(5), 516–518. DOI: 10.1017/dmp.2015.84 PMID: 26165522

Smith-MacDonald, L., Lentz, L., Malloy, D., Brémault-Phillips, S., & Carleton, R. N. (2021). Meat in a Seat: A Grounded Theory Study Exploring Moral Injury in Canadian Public Safety Communicators, Firefighters, and Paramedics. *International Journal of Environmental Research and Public Health, 18*(22), 12145. DOI: 10.3390/ijerph182212145 PMID: 34831899

Stelnicki, A. M., Jamshidi, L., Fletcher, A. J., & Carleton, R. N. (2021). Evaluation of Before Operational Stress: A Program to Support Mental Health and Proactive Psychological Protection in Public Safety Personnel. *Frontiers in Psychology, 12*, 511755. Advance online publication. DOI: 10.3389/fpsyg.2021.511755 PMID: 34484013

Szeto, A., Dobson, K. S., & Knaak, S. (2019). The Road to Mental Readiness for First Responders: A Meta-Analysis of Program Outcomes. *Canadian Journal of Psychiatry, 64*(1, suppl), 18S–29S. DOI: 10.1177/0706743719842562 PMID: 31010293

Union of British Columbia Municipalities. (2015). *Workers Compensation Act – Presumptive clause for first responders* [Resolution C33]. https://www.ubcm.ca/convention-resolutions/resolutions/resolutions-database/workers-compensation-act-presumptive-clause

Walker, J., & Stones, A. (2020). Impact of workplace bullying amongst first responders. A systematic review. *International Journal of Psychological Research and Reviews*, *8*, 3–24. DOI: 10.28933/ijprr-2020-01-1205

Walker, J., Stones, A., & Stephenson, L. (2021). Real trouble: The mental health impact of workplace bullying and sexual harassment amongst first responders. *International Journal of Case Studies in Clinical Research*, *5*(3), 27–44.

World Health Organization. (2024, April 25). Protecting health and care workers' mental health and well-being: Technical consultation meeting. https://www.who.int/news/item/25-04-2024-202404_protecthw_mentalhealth

World Health Organization. (2024, August 21). *Why are so many health and care workers suffering poor mental health and what can be done about it? – perspectives from Finland.* WHO Regional Office for Europe. https://www.who.int/europe/news-room/feature-stories/item/why-are-so-many-health-and-care-workers-suffering-poor-mental-health-and-what-can-be-done-about-it----perspectives-from-finland

Yung, M., Du, B., Gruber, J., & Yazdani, A. (2021). Developing a Canadian fatigue risk management standard for first responders: Defining the scope. *Safety Science*, *134*, 105044. DOI: 10.1016/j.ssci.2020.105044

Chapter 15
Nonverbal Incivility in the Workplace:
Rethinking Organizational Politics

Renu Vij
http://orcid.org/0000-0001-9202-8390
Chandigarh University, India

Aashna Narula
http://orcid.org/0009-0002-7761-8573
Pyschopedia Journal, India

ABSTRACT

Workplace incivility, often minimized as trivial, exerts significant adverse effects on organizational dynamics especially when enacted through nonverbal channels such as gestures, posture, spatial exclusion, and facial expressions. This chapter reconceptualizes nonverbal incivility as a strategic dimension of organizational politics, grounded in theories of social power and communication. By integrating robust organizational behaviour and occupational health research, along with empirical findings from India, this study highlights how nonverbal cues contribute to psychological distress, diminished psychological safety, turnover intention, and weakened productivity. Through a proposed conceptual framework, it identifies non-verbal cues as early indicators of toxic cultural undercurrents and suggests proactive interventions. Ultimately, the chapter aims to widen the discourse on occupational well-being by illuminating the nuanced, often invisible role of nonverbal incivility in shaping workplace climates and power relations.

DOI: 10.4018/979-8-3373-6370-7.ch015

INTRODUCTION

Workplace incivility, often dismissed as minor or ambiguous behavior, has far-reaching consequences for organizational health, particularly when expressed nonverbally. This chapter explores nonverbal incivility, including gestures, facial expressions, and spatial behaviour, as a subtle but potent dimension of workplace dynamics that frequently intersects with organizational politics. Drawing upon contemporary organizational behaviour theories and empirical evidence, the discussion repositions nonverbal incivility as not merely an interpersonal concern but a strategic expression of power, control, and resistance within hierarchical settings.

Workplace incivility, characterized "by low-intensity deviant behaviours with ambiguous intent to harm, such as ignoring colleagues, delivering dismissive glances, or excluding others, undermines organizational wellbeing through subtle yet pervasive means" (Andersson & Pearson, 1999). Huang, Wei and Ang (2022) defined workplace incivility as "the seemingly insignificant behaviours that are rude, disrespectful, discourteous or insensitive where the intent to harm is ambiguous or unclear". Uncivil behaviour may include behaviours such as talking down to others, making demeaning remarks and not listening to somebody (Pearson, 2010).

Though research frequently emphasizes verbal incivility, nonverbal incivility (e.g., gestures, body posture, spatial distancing, and facial expressions) remains underexplored, despite exerting critical influence over power dynamics and psychological safety in professional environments.

The cost of incivility is profound. Porath & Pearson (2013) states that 98% of employees experience incivility, 50% experiences such conduct weekly. Additionally, it has been found that monetary cost of experiencing incivility is approximately 14000$ per employee annually, the reason being cognitive distractions and project delays at work. Workers may further, try to avoid coming to workplace, withdraw from work, avoid the instigator and even be angry with their customers (Porath & Pearson, 2013). The researchers have revealed that targets of negative workplace behaviour engage less in organisational citizenship behaviour, have high turnover intentions and experience more stress (Dalal, 2005; Chiaburu & Harrison, 2008; Bowling & Beehr, 2006).

Globally, it diminishes job performance, erodes engagement, increases turnover, and damages customer relationships (McKinsey, 2016). Longitudinal evidence further reveals that incivility is not harmless; it serves as a precursor to more severe harassment and burnout-related outcomes (Vartia & Hyyti, 2022).

In conclusion, nonverbal incivility is an underappreciated yet significant driver influencing relationships at work. Subtle body language, facial expressions and spatial exclusion frequently function below the surface to affect psychological health and team relationships. By considering non-verbal cues as a component of organisational politics, one can gain deeper understanding of how routine interactions lead to more significant patterns of stress, disengagement and conflict. The goal of this chapter is to provide a more sophisticated framework for identifying, analysing and dealing with these kinds of behaviours in work environments.

Socio-cultural Significance

In the Asian and Indian context, workplace incivility is shaped by socio-cultural forces. Researchers note its prevalence in Asian corporate settings, including India, where uncivil triggers often derive from blurred boundaries between professional decorum and hierarchical norms (Ghosh, 2017). In frontline service industries in India, incivility correlates negatively with job satisfaction and positively with turnover intentions (Sharma & Singh, 2016). Broader Indian service sectors also reveal that demographics like age, gender, and educational background affect both the perpetration and experience of workplace incivility (Singh & Verma, 2021).

Moreover, the undermining effect of incivility on psychological safety, a foundational element of effective teamwork and innovation is well documented. When individuals feel uncertain about protection from ridicule or exclusion, they withhold voice, resist collaboration, and disengage from the organizational mission (Edmondson, 1999; 2019).

PURPOSE AND SCOPE

This chapter reorients nonverbal incivility from a minor interpersonal annoyance to a strategic form of organizational politics, where nonverbal signals serve as instruments of dominance, resistance, and exclusion, particularly potent in hierarchical or collectivist workplaces.

Conceptual Clarification: Distinguishing Nonverbal Incivility from Related Constructs

Nonverbal incivility is a subtle but significant phenomenon within workplace dynamics. It can be defined as rude or disrespectful behaviours communicated through nonverbal channels such as facial expressions, gestures, tone of voice, posture, or spatial positioning that violate interpersonal norms of respect without overtly hostile verbal content (Andersson & Pearson, 1999; Giumetti et al., 2013). Unlike verbal incivility, which is conveyed explicitly through language, nonverbal incivility is often ambiguous, making it easier for perpetrators to deny intent and harder for recipients to articulate or report (Kern & Grandey, 2009).

While related to micro aggressions, nonverbal incivility differs in scope and intent. Micro aggressions often stem from unconscious biases related to social identity categories such as race, gender, or age, and may manifest both verbally and nonverbally (Sue et al., 2007). Nonverbal incivility, however, may be entirely unrelated to identity-based prejudice, arising instead from power struggles, workplace politics, or interpersonal tensions (Schilpzand, De Pater & Erez, 2016). Similarly, it must be distinguished from bullying, which is typically repeated, targeted, and involves a clear power imbalance with an intent to harm (Einarsen et al., 2020). Nonverbal incivility, in contrast, may be situational, less intense, and lacking a sustained campaign against an individual—yet still harmful to workplace climate and well-being.

Theoretical Foundations

The study of nonverbal incivility can be enriched through multiple theoretical lenses:

1. ***Communication Theory***: Drawing from interpersonal communication research, nonverbal behaviours are understood as powerful conveyors of relational meaning, often superseding verbal content in emotional and attitudinal impact (Burgoon et al., 2016). Nonverbal incivility exploits these channels to convey disregard or contempt without verbal confrontation. Communication theory explains how small indicators such as avoiding eye contact during a meeting can indicate exclusion or disapproval without displaying overt animosity. The relational impact of such cues is heightened when recipients view them as deliberate acts of contempt. For instance, an employee who is regularly denied eye contact by co-workers may interpret this as intentional social rejection, even if others are ignorant of their actions.
2. ***Expectancy Violation Theory (EVT)*** EVT posits that individuals hold expectations about appropriate communication behaviour, and violations particularly negative ones evoke arousal and cognitive appraisal processes (Burgoon, 1993). Nonverbal incivility represents a form of negative expectancy violation, where ambiguous cues breach professional norms, generating discomfort and reduced trust. Training programs can assist employees in identifying their implicit expectations and interpreting nonverbal clues within context, rather than reacting hastily. For example, fostering conscious monitoring of communication patterns can help to avoid misinterpretation.
3. ***Power Theory***: From a political perspective, nonverbal incivility can function as a micro-expression of dominance or resistance within organizational hierarchies

(Keltner et al., 2003). Controlling gaze, dismissive gestures, or physical exclusion from space are nonverbal acts that reinforce or challenge existing power relations. Encouraging leaders to engage in transparent, supportive non-verbal behaviours such as nodding, leaning in and maintaining persistent eye contact can mitigate the negative consequences of power driven incivility.

4. ***Occupational Health Framework***: Occupational health literature recognizes incivility as a psychosocial hazard, with potential effects on stress, burnout, and job satisfaction (Cortina et al., 2013). Integrating this perspective positions nonverbal incivility as not merely a behavioural concern but a legitimate workplace health and safety issue requiring intervention. Integrating non verbal behaviours into occupational health assessments, whether through observation methods, or anonymous feedback mechanisms, ensures that psychosocial hazards are addressed before they become official complaints or health crises.

Conceptual Clarification: Distinguishing Nonverbal Incivility from Verbal Incivility, Microaggressions, and Bullying

Nonverbal incivility refers to subtle, indirect behaviours communicated without spoken words that convey disrespect, disregard, or hostility. These may include eye-rolling, dismissive gestures, deliberate avoidance of eye contact, exaggerated sighs, or the purposeful invasion of personal space. Unlike verbal incivility, which is conveyed through spoken or written language, nonverbal cues often operate below the threshold of explicit recognition, making them harder to confront or address (Andersson & Pearson, 1999; Giumetti et al., 2013). While both verbal and nonverbal incivility share the qualities of low-intensity, ambiguous intent, and norm violation, the latter's covert nature allows it to evade organizational detection and policy intervention.

Micro aggressions differ from nonverbal incivility in that they are typically linked to bias, prejudice, or discrimination against marginalized social groups. They can be verbal, nonverbal, or environmental, but their distinguishing feature is the reinforcement of systemic inequities through subtle interpersonal interactions (Sue et al., 2007). For example, consistently failing to acknowledge a colleague's presence in meetings may function both as nonverbal incivility and as a racial or gender-based micro aggression, depending on context and power relations.

Bullying, in contrast, is characterized by repeated, targeted acts—verbal, nonverbal, or both—that are intended to harm, intimidate, or isolate the target (Einarsen et al., 2020). Unlike incivility, bullying is typically high in intensity, sustained over time, and involves an overt intent to cause harm. The critical conceptual boundary lies in the intensity, frequency, and purpose of the act, with nonverbal incivility existing at the lower end of the aggression spectrum but still capable of significant psychological and organizational harm.

DIMENSIONS AND MANIFESTATIONS OF NONVERBAL INCIVILITY

Incivility as a concept is defined as "an action that is ill-mannered, impolite, irritated and demonstrating a gesture of lack of respect" (Kane & Montgomery, 1998). Whereas workplace civility is defined as "respectful and courteous behaviour exhibited by individuals towards their colleagues in the workplace" (Clark & Walsh, 2016). Workplace civility requires one to showcase behaviours such as active listening, being empathic, and constructive communication. This boosts job satisfaction, improving organisational outcomes and fostering positive relationships.

Nonverbal incivility in workplace contexts manifests through a variety of subtle yet impactful behaviors that communicate disrespect, disregard, or exclusion without the use of explicit language. Unlike verbal incivility—which can often be documented and directly addressed, nonverbal acts are frequently ambiguous, context-dependent, and open to interpretation. This ambiguity not only complicates detection but also facilitates their normalization in organizational cultures where such behaviours become embedded in everyday interactions (Andersson & Pearson, 1999).

1. Gestural Expressions

Gestures such as eye-rolling, dismissive hand waves, or abrupt interruptions during another's speaking turn can communicate condescension or impatience. These actions, although fleeting, can erode interpersonal trust and create an environment of psychological threat (Burgoon, Guerrero, & Floyd, 2016).

2. Facial Cues

Facial expressions, smirking, glaring, or pointedly avoiding eye contact—serve as potent indicators of status hierarchies and emotional disapproval. Research shows that facial displays are often more memorable than verbal insults, thereby having a prolonged psychological impact on targets (Ekman & Friesen, 2003).

3. Spatial Behaviour

The use of physical space in organizational settings can communicate inclusion or exclusion. For example, deliberately positioning oneself at a distance in meetings, physically turning away, or blocking access to shared resources are spatial strategies that subtly reinforce power differentials (Hall, 1966).

4. Paralinguistic Signals

Elements such as tone, pitch, and speech rate—although technically linked to verbal communication—are paralinguistic cues that can carry nonverbal incivility. A sarcastic tone, exaggerated sighs, or prolonged silences can communicate disdain or resistance without breaching formal language protocols (Knapp, Hall, & Horgan, 2014).

5. Organizational Impact

These behaviours, when persistent, influence organizational dynamics by undermining psychological safety, reducing team cohesion, and escalating political manoeuvring (Cortina et al., 2017). Their subtle nature allows perpetrators to maintain plausible deniability while the cumulative effect on targets mirrors, and at times exceeds, the harm caused by overt aggression.

The different aspects of non-verbal incivility such as facial cues, spatial behaviour, paralinguistic signals, gestural expressions, show that workplace incivility can take many different forms. These cues present particular difficulties for monitioring and action shince they are frequently unclear and subject to interpretation. Organisations can better promote **psychological safety** and discourage little acts of incvil behaviour from turning into more significant disturbances by identifying these actions as manifestations of exclusion, dominance or resistance.

MEASUREMENT AND ASSESSMENT OF NONVERBAL INCIVILITY

Measuring nonverbal incivility presents unique challenges because of its subjective interpretation, situational ambiguity, and dependence on contextual norms. Unlike overt forms of aggression, which can be reliably identified through explicit acts or verbal content, nonverbal incivility often operates beneath the threshold of formal organizational policy frameworks (Cortina et al., 2017). Therefore, methodological rigor in its assessment is critical to advancing research and informing organizational interventions.

1. Self-Report Surveys and Perception Scales

Self-report instruments remain the most widely used approach for measuring workplace incivility. Adaptations of established scales, such as the Workplace Incivility Scale (WIS), have incorporated items that specifically capture nonverbal behaviors, for example, "coworkers glaring at you" or "colleagues avoiding eye contact when you speak" (Andersson & Pearson, 1999). While these tools offer valuable insight into subjective experiences, they are vulnerable to recall bias, social desirability bias, and individual differences in sensitivity to nonverbal cues (Matthiesen & Einarsen, 2010). WIS has recently gone substantial changes and now comprises of 12 items which include studying of experiences such as "being interrupted, being targeted with angry outbursts, or receiving hostile looks from co-workers or supervisors over the past year" (Cortina et al., 2013). In The year 2005, Martin and Hine (2005) developed and validated the "Uncivil Workplace Behaviour Questionnaire" which is a 20 item questionnaire. The scale inquired participants to report how often they experience behaviours such as raised voices, eye-rolling, being interrupted, being excluded and being gossiped about work, keeping all these in mind for past 1 year. Penney and Spector (2005)and constructed three different instruments, the WIS, the Leymann Inventory of Psychological Terror and the Workplace Research Questionnaire (Cortina et al. 2001; Griffin 2010; Neuman & Keashly, 2002).

2. Behavioural Observation Methods

Direct observation, either in real-time or via recorded workplace interactions, enables researchers to code nonverbal acts with higher objectivity. Coding frameworks often draw upon established nonverbal communication taxonomies (Ekman & Friesen, 2003; Knapp, Hall, & Horgan, 2014). While labour-intensive, this approach captures the nuanced forms of nonverbal incivility that participants may not consciously register or recall.

3. Experience Sampling Method (ESM)

The Experience Sampling Method, in which participants report incidents close to the time of occurrence, reduces memory decay and contextual distortion (Bolger & Laurenceau, 2013). Using mobile-based reporting applications, employees can log perceived nonverbal slights, such as exclusionary body language or avoidance behaviours, in real time. This allows researchers to capture situational triggers and emotional responses with greater precision.

4. Physiological and Biometric Indicators

Emerging research in occupational health integrates physiological measures such as heart rate variability, galvanic skin response, and facial expression recognition software to detect stress responses triggered by nonverbal incivility (Ganster & Rosen, 2013). These methods provide an objective complement to self-reports, particularly in understanding the psychophysiological toll of subtle aggression.

5. Multimethod Triangulation

Given the limitations of any single approach, triangulating self-report, observational, and physiological data enhances validity and reliability. This mixed-methods approach allows for the cross-verification of findings and offers a holistic view of the phenomenon, capturing both perceived and actual occurrences of nonverbal incivility.

IMPLICATIONS FOR RESEARCH AND PRACTICE

Accurate measurement is not merely an academic exercise it shapes organizational awareness, policy development, and prevention strategies. Without precise assessment, nonverbal incivility risks being dismissed as misinterpretation or interpersonal sensitivity, thereby allowing harmful patterns to persist under the radar of formal intervention mechanisms.

Antecedents and Triggers of Nonverbal Incivility in Organizational Settings

Nonverbal incivility rarely occurs in a vacuum; it is embedded within a network of interpersonal, organizational, and societal dynamics. Understanding its antecedents is essential for both theoretical development and effective intervention strategies. These antecedents can be conceptualized across individual, interpersonal, and structural-organizational levels, each interacting to shape the likelihood and intensity of such behaviours.

1. *Individual-Level Antecedents*

Certain personality traits and affective dispositions predispose individuals to either engage in or perceive nonverbal incivility. For instance, individuals high in trait hostility or low in agreeableness may display negative nonverbal signals such as eye-rolling, dismissive gestures, or physical withdrawal (Douglas & Martinko,

2001). Conversely, employees with heightened sensitivity to social cues, such as those with high neuroticism or low emotional stability, may interpret ambiguous behaviours as intentionally uncivil. Emotional states—especially stress, burnout, or frustration—can also lower thresholds for both perpetrating and perceiving nonverbal slights.

2. *Interpersonal and Relational Dynamics*

Power asymmetries play a central role in the expression of nonverbal incivility. Supervisors, for example, may use subtle body language delayed responses, controlled eye contact, or dismissive posture—to signal dominance or disapproval without verbalizing conflict. Peer-to-peer competition for resources, recognition, or advancement may similarly fuel micro-behaviours of exclusion, such as turning away during discussions or deliberately omitting someone from informal social exchanges. Existing interpersonal tensions, prior conflicts, and low trust levels intensify the likelihood that such behaviours will be interpreted as hostile.

3. *Structural and Organizational Context*

Workplace environments characterized by high workload, role ambiguity, or competitive cultures often normalize subtle forms of disrespect. In such contexts, nonverbal incivility can emerge as an adaptive, albeit toxic, communication strategy to manage time, preserve status, or avoid direct confrontation. Weak enforcement of civility norms, inadequate training on respectful communication, and tolerance of ambiguous behaviours create organizational climates where subtle disrespect thrives under plausible deniability. Additionally, open office layouts and digital communication channels (e.g., video calls) have created new contexts in which nonverbal cues such as inattentive gaze or muted camera participation—can signal disengagement or disregard.

4. *Cultural and Societal Influences*

Cultural differences in nonverbal communication norms complicate the identification of incivility. Gestures or expressions considered neutral in one culture may be perceived as offensive in another. In multicultural workforces, misinterpretations can arise from differing expectations about interpersonal space, eye contact, or facial expressiveness. Broader societal narratives such as gender stereotypes or hierarchical traditions—can further shape who becomes the target of subtle disrespect and how such acts are interpreted.

5. *Triggering Events*

Nonverbal incivility often surfaces during high-stress situations, such as organizational change, performance evaluations, or conflict escalation. Even seemingly minor events, like a missed deadline or miscommunication, can serve as flashpoints. In these moments, individuals may resort to indirect expressions of displeasure folded arms, avoidance behaviours, or sarcastic facial expressions rather than direct confrontation.

Integrative Perspective

From an integrative standpoint, antecedents and triggers of nonverbal incivility cannot be understood in isolation. Individual predispositions interact with relational histories and organizational climates, all within a cultural frame. This multi-layered interplay suggests that prevention efforts must address not only personal awareness and skill-building but also systemic organizational reforms and culturally responsive training.

Table 1. Instigated Workplace Incivility: Antecedents and Outcomes (Adapted from Schilpzand, De Pater & Erez, 2016).

Study	Sample	Targets	Antecedents/Outcomes	Key Findings
Blau & Anderson (2005)	Working adults	Co-workers, supervisors	Distributive justice, job satisfaction, exhaustion	Low distributive justice and job satisfaction at Time 1 predicted higher instigated incivility at Time 2, exhaustion had positive effect
Blau (2007)	Manufacturing plant	Co=workers, supervisors	Contract violation, distributive & procedural justice, depression, strain	Higher contract violation, strain, depression linked with more instigated incivility, lower justice and legitimacy related to increased incivility
Cortin et al (2001)	Federal court employees	Co-workers, supervisors	Power	Greater power associated with higher instigated incivility.
Kirk et al. (2011)	Working Adults	Co-workers	Expressive writing intervention	Participants completing a 2 week writing intervention exhibited reduced instigated incivility

continued on following page

Table 1. Continued

Study	Sample	Targets	Antecedents/Outcomes	Key Findings
Meier & Semmer (2013)	Various organisations	Co-workers, Supervisors	Lack of reciprocity, trait and state anger	Lack of reciprocity and anger predicted instigated incivility, anxiety served as a mediator
Scott et al (2013)	Call centre employees	Co-workers	Distrust, exchange partner quality, exclusion	Instigators experiences exclusion mediated by distrust, especially when seen as low quality exchange partners
Trudel & Reio (2011)	Manufacturing & Healthcare	Co-workers, supervisors	Experienced incivility, conflict style	Being targeted, having dominating conflict style predicted incivility, integrated style was negatively associated
Van Jaarsveld et al. (2010)	Call Centre employees	Customers	Job demands, emotional exhaustion, experienced incivility	Customer incivility, paired with high job demands and exhaustion, predicted instigated incivility towards customers.

The table summarizes previous empirical research that examined the causes and consequences of instigated workplace incivility, as complied in this table. It provides a summary of the circumstances in which workers act rudely towards co-workers by highlighting the study's background, objectives, important factors and main conclusion.

Supervisor incivility includes behaviour such as supervisor losing his/her temper, showcasing cold behaviour and intimidating which can have a negative impact on employee's health (Potipiroon & Ford, 2019). Additionally co-worker mistreatment includes uncivil behaviours such as social undermining and emotional abuse (Leiter, 2013 & Viotti et al., 2018). Customer incivility includes rude or disrespectful behaviour from customers to employees and can cause emotional stress (Han et al., 2016).

CONSEQUENCES OF NONVERBAL INCIVILITY: PSYCHOLOGICAL, BEHAVIOURAL, AND ORGANIZATIONAL OUTCOMES

Nonverbal incivility, although subtle in form, can have profound consequences that extend beyond the immediate interaction. Unlike overt verbal hostility, its indirect nature often creates ambiguity, which can intensify psychological strain and complicate resolution efforts. The repercussions manifest across individual, interpersonal, and organizational domains, cumulatively eroding workplace climate and productivity.

1. Psychological Consequences

At the individual level, repeated exposure to nonverbal incivility can trigger stress responses similar to those caused by overt aggression. Ambiguity in interpreting these behaviours such as whether a dismissive gesture was intentional can generate rumination, emotional exhaustion, and anxiety (Schilpzand, De Pater & Erez, 2016). Employees may experience diminished self-worth, feelings of social exclusion, and lowered confidence in their professional competence. Over time, these micro-experiences can accumulate into chronic workplace stress, which is a well-established risk factor for burnout and other occupational health concerns. Employees attitude which is determined by how well they get along with their co-workers and conduct themselves in a productive manner inside their workspace, is regarded as psychological well-being (Holm et al., 2015). Heavy workloads and hostile behaviour at work are related, which have a negative impact on employee's psychological well-being (Illies, Schwind, Wagner, 2007). Liu et al. (2020) have also emphasised that rude behaviour that occurs in the workplace will result in a decline in psychological well-being and an increase in the likelihood that an employee will resign. Additionally, work alienation as defined by Kongar (1979) refers to "decrease in the harmony of the individuals with social, cultural and natural environment and especially decrease in his control over his immediate environment causing loneliness and hopelessness". Erjem (2005) states that the signs of alienation include indifference, withdrawal, isolation, feeling disconnected at work and an inability to be attached to the work. Target's perception of the relationship between work and themselves is reflected in work alienation which can lead to a feeling of incomprehensibility regarding their employment and its significance (Nair & Vohra, 2009).

2. Behavioural Consequences

Victims of nonverbal incivility often adjust their behaviour in ways that undermine collaboration and organizational trust. Some may withdraw socially, avoid meetings, or limit interactions with perceived perpetrators. Others may retaliate with reciprocal incivility, perpetuating a cycle of subtle disrespect. Reduced engagement, increased absenteeism, and declines in discretionary effort are common outcomes, especially when employees feel that such behaviour is condoned by management. In certain cases, nonverbal incivility leads to turnover intentions, particularly among high-performing employees who expect mutual respect as a baseline professional norm. Porath and Pearson (2012) in their study titled "*Emotional and behavioural responses to workplace incivility and the Impact of Hierarchical Status*" used appraisal theory to examine emotional reactions to rudeness at work affect employees behavioural reactions. Anger, anxiety and dejection were higher among employees. The workplace can be an environment that makes individuals angry and rude behaviour from co-workers is an indicator for same (Domagalski, 1998).

3. Interpersonal and Team Dynamics

On a relational level, nonverbal incivility corrodes trust and hinders the development of cohesive team cultures. Small acts of exclusion—avoiding eye contact in discussions, physically turning away from colleagues, can signal relational devaluation, making team members feel less valued or heard. In collaborative environments, where shared decision-making and open communication are critical, such micro-behaviours can fragment teams, fostering silos and reducing knowledge sharing. Over time, this weakens group cohesion and impairs collective problem-solving capabilities. Vickers (2006) found that disrespect in the workplace deters teamwork and that a lack of team spirit is followed by a feeling of isolation.

4. Organizational-Level Outcomes

When pervasive, nonverbal incivility becomes embedded in the organizational culture, creating a climate where subtle disrespect is normalized. This erosion of civility can undermine strategic goals by reducing employee engagement, impairing innovation, and damaging the employer brand. From a risk management perspective, unresolved incivility can escalate into formal grievances, legal disputes, or reputational damage. Additionally, the organizational political climate may intensify, as ambiguous acts of disrespect can be leveraged as tools for power manipulation and informal influence.

5. Link to Organizational Politics

The connection between nonverbal incivility and organizational politics is particularly significant. In politically charged environments, nonverbal cues can serve as covert mechanisms for exercising influence—signalling inclusion, exclusion, or disapproval without verbal confrontation. This indirect mode of communication allows individuals to shape perceptions, build alliances, or marginalize rivals while maintaining plausible deniability. As a result, nonverbal incivility not only reflects interpersonal strain but also becomes a strategic resource in organizational power dynamics.

6. Occupational Health and Safety Implications

From an occupational health perspective, sustained exposure to nonverbal incivility can contribute to a psychosocial hazard profile within the workplace. The cumulative emotional toll can increase the risk of depression, anxiety disorders, and stress-related physical ailments, thereby affecting both employee well-being and organizational performance. Recognizing and addressing nonverbal incivility is thus not only a matter of interpersonal courtesy but also an essential component of occupational health and safety strategy.

LIMITATION OF STUDY

This chapter is grounded in a conceptual and theoretical synthesis rather than original empirical data collection. While this approach enables a broad integration of frameworks from communication, power dynamics, and occupational health, it is constrained by the availability and cultural scope of existing studies. Much of the available literature originates from Western contexts, and although Indian and Asian perspectives have been incorporated, cultural variations in the interpretation of nonverbal behaviour may not be fully captured. The inherently ambiguous and context-dependent nature of nonverbal cues also presents challenges for objective measurement and operationalization. Furthermore, the analysis focuses predominantly on professional, service-sector workplaces, limiting its applicability to informal, industrial, or gig-economy settings. The reliance on largely cross-sectional studies further restricts the ability to assess the evolving and cumulative effects of nonverbal incivility over time.

CONCLUSION

This chapter reframes nonverbal incivility as more than an interpersonal annoyance—it is a subtle yet potent mechanism of organizational politics with significant consequences for occupational health and workplace culture. By synthesizing insights from communication theory, expectancy violation theory, power theory, and occupational health perspectives, the discussion reveals how gestures, spatial positioning, facial expressions, and paralinguistic signals can convey exclusion, dominance, or resistance.

Three key contributions emerge from this analysis. First, nonverbal incivility warrants recognition as a legitimate occupational health concern capable of eroding psychological safety, trust, and engagement. Second, its covert nature enables it to persist beneath formal reporting thresholds, normalizing patterns of subtle disrespect. Third, it can be intentionally leveraged as a strategic tool to shape power relations within organizations.

Addressing this phenomenon demands a dual approach: cultivating individual awareness and interpersonal competence, alongside implementing systemic measures such as explicit behavioural standards, leadership role modelling, culturally responsive communication training, and confidential feedback mechanisms. Reframing nonverbal incivility as both a psychosocial hazard and a political instrument broadens the scope of occupational health discourse and provides a foundation for more comprehensive interventions.

The conceptual insights presented here have direct implications for organizational leaders, human resource practitioners, and policymakers. Organizations seeking to address nonverbal incivility should:

a) Establish clear guidelines addressing both verbal and nonverbal interactions.
b) Integrate cultural sensitivity and nonverbal communication training into employee development programs.
c) Encourage leaders to exemplify respectful nonverbal behaviour in daily interactions.
d) Create confidential reporting mechanisms for subtle acts of incivility.
e) Incorporate observation of nonverbal conduct into workplace climate assessments and leadership evaluations.

Embedding such measures into policy and practice can help create inclusive, respectful, and psychologically safe workplace environments.

FUTURE RESEARCH DIRECTIONS

Despite growing awareness, the empirical study of nonverbal incivility remains limited, offering several opportunities for future research:

1. **Longitudinal Studies**: Exploring the long-term effects of nonverbal incivility on individuals and teams.
2. **Cross-Cultural Analysis**: Examining how cultural norms influence perceptions, interpretations, and impacts of nonverbal behaviour.
3. **Sectoral Variations**: Investigating nonverbal incivility across informal, manufacturing, and gig-economy contexts.
4. **Mixed-Method Approaches**: Employing observational research, experimental designs, and qualitative interviews to capture both patterns and lived experiences.
5. **Intervention Assessments**: Evaluating the effectiveness of targeted training and leadership development initiatives in reducing nonverbal incivility.

Pursuing these lines of inquiry will deepen theoretical clarity, expand cross-cultural understanding, and generate actionable, evidence-based strategies to mitigate the silent yet pervasive effects of nonverbal incivility in organizational settings.

REFERENCES

Andersson, L. M., & Pearson, C. M. (1999). Tit for tat? The spiralling effect of incivility in the workplace. *Academy of Management Review, 24*(3), 452–471. DOI: 10.2307/259136

Bolger, N., & Laurenceau, J. P. (2013). *Intensive longitudinal methods: An introduction to diary and experience sampling research*. Guilford Press.

Bowling, N. A., & Beehr, T. A. (2006). Workplace harassment from the victim's perspective: A theoretical model and meta-analysis. *The Journal of Applied Psychology, 91*(5), 998–1012. DOI: 10.1037/0021-9010.91.5.998 PMID: 16953764

Burgoon, J. K. (1993). Interpersonal expectations, expectancy violations, and emotional communication. *Journal of Language and Social Psychology, 12*(1–2), 30–48. DOI: 10.1177/0261927X93121003

Burgoon, J. K., Guerrero, L. K., & Floyd, K. (2016). *Nonverbal communication*. Routledge., DOI: 10.4324/9781315663425

Chiaburu, D. S., & Harrison, D. A. (2008). Do peers make the place? Conceptual synthesis and meta-analysis of coworker effects on perceptions, attitudes, OCBs, and performance. *The Journal of Applied Psychology, 93*(5), 1082–1103. DOI: 10.1037/0021-9010.93.5.1082 PMID: 18808227

Clark, O. L., & Walsh, B. M. (2016). Civility climate mitigates deviant reactions to organizational constraints. *Journal of Managerial Psychology, 31*(1), 186–201. DOI: 10.1108/JMP-01-2014-0021

Cortina, L. M., Kabat-Farr, D., Leskinen, E. A., Huerta, M., & Magley, V. J. (2013). Selective incivility as modern discrimination in organizations: Evidence and impact. *Journal of Management, 39*(6), 1579–1605. DOI: 10.1177/0149206311418835

Cortina, L. M., Kabat-Farr, D., Magley, V. J., & Nelson, K. (2017). Researching rudeness: The past, present, and future of the science of incivility. *Journal of Occupational Health Psychology, 22*(3), 299–313. DOI: 10.1037/ocp0000089

Cortina, L. M., Magley, V. J., Williams, J. H., & Langhout, R. D. (2001). Incivility in the workplace: Incidence and impact. *Journal of Occupational Health Psychology, 6*(1), 64–80. DOI: 10.1037/1076-8998.6.1.64 PMID: 11199258

Cortina, L. M., Magley, V. J., Williams, J. H., & Langhout, R. D. (2001). Incivility in the workplace: Incidence and impact. *Journal of Occupational Health Psychology, 6*(1), 64–80. DOI: 10.1037/1076-8998.6.1.64 PMID: 11199258

Cortina, L. M., Magley, V. J., Williams, J. H., & Langhout, R. D. (2013). Incivility in the workplace: Incidence and impact. *Journal of Occupational Health Psychology, 8*(1), 64–80. DOI: 10.1037/1076-8998.6.1.64 PMID: 11199258

Dalal, R. S. (2005). A meta-analysis of the relationship between organizational citizenship behavior and counterproductive work behavior. *The Journal of Applied Psychology, 90*(6), 1241–1255. DOI: 10.1037/0021-9010.90.6.1241 PMID: 16316277

Domagalski, T. A. (1998). *Experienced and expressed anger in the workplace.* University of South Florida.

Douglas, S. C., & Martinko, M. J. (2001). Exploring the role of individual differences in the prediction of workplace aggression. *The Journal of Applied Psychology, 86*(4), 547–559. DOI: 10.1037/0021-9010.86.4.547 PMID: 11519640

Edmondson, A. C. (1999). Psychological safety and learning behavior in work teams. *Administrative Science Quarterly, 44*(2), 350–383. DOI: 10.2307/2666999

Edmondson, A. C. (2019). *The fearless organization: Creating psychological safety in the workplace for learning, innovation, and growth.* Wiley.

Einarsen, S. V., Hoel, H., Zapf, D., & Cooper, C. L. (2020). *Bullying and harassment in the workplace: Developments in theory, research, and practice* (3rd ed.). CRC Press., DOI: 10.1201/9780429462528

Ekman, P., & Friesen, W. V. (2003). *Unmasking the face: A guide to recognizing emotions from facial clues* (Reprint ed.). Cambridge, MA: Malor Books.

Erjem, Y. (2005). Eğitimde Yabancilaşma Olgusu ve Öğretmen: Lise Öğretmenleri Üzerine Sosyolojik Bir Araştirma. *Türk Eğitim Bilimleri Dergisi, 3*(4), 395–417.

Ganster, D. C., & Rosen, C. C. (2013). Work stress and employee health: A multidisciplinary review. *Journal of Management, 39*(5), 1085–1122. DOI: 10.1177/0149206313475815

Ghosh, R. (2017). Workplace incivility in Asia: How do we take a socio-cultural perspective? *Human Resource Development International, 20*(4), 263–267. DOI: 10.1080/13678868.2017.1336692

Giumetti, G. W., Hatfield, A. L., Scisco, J. L., Schroeder, A. N., Muth, E. R., & Kowalski, R. M. (2013). What a rude e-mail! Examining the differential effects of incivility versus support on mood, energy, engagement, and performance in an online context. *Journal of Occupational Health Psychology, 18*(3), 297–309. DOI: 10.1037/a0032851 PMID: 23834445

Griffin, B. (2010). Multilevel relationships between organizational-level incivility, justice and intention to stay. *Work and Stress*, *24*(4), 309–323. DOI: 10.1080/02678373.2010.531186

Hall, E. T. (1966). *The hidden dimension*. Doubleday.

Han, S. J., Bonn, M. A., & Cho, M. (2016). The relationship between customer incivility, restaurant frontline service employee burnout and turnover intention. *International Journal of Hospitality Management*, *52*, 97–106. DOI: 10.1016/j.ijhm.2015.10.002

Holm, K., Torkelson, E., & Bäckström, M. (2015). Models of workplace incivility: The relationships to instigated incivility and negative outcomes. *BioMed Research International*, *2015*(1), 920239. DOI: 10.1155/2015/920239 PMID: 26557714

Huang, Y. S., Wei, S., & Ang, T. (2022). The role of customer perceived ethicality in explaining the impact of incivility among employees on customer unethical behavior and customer citizenship behavior. *Journal of Business Ethics*, *178*(2), 519–535. DOI: 10.1007/s10551-020-04698-9

Kane, K., & Montgomery, K. (1998). A framework for understanding dysempowerment in organizations. *Human Resource Management*, *37*(3-4), 263–275. DOI: 10.1002/(SICI)1099-050X(199823/24)37:3/4<263::AID-HRM8>3.0.CO;2-U

Keltner, D., Gruenfeld, D. H., & Anderson, C. (2003). Power, approach, and inhibition. *Psychological Review*, *110*(2), 265–284. DOI: 10.1037/0033-295X.110.2.265 PMID: 12747524

Kern, J. H., & Grandey, A. A. (2009). Customer incivility as a social stressor: The role of race and racial identity for service employees. *Journal of Occupational Health Psychology*, *14*(1), 46–57. DOI: 10.1037/a0012684 PMID: 19210046

Knapp, M. L., Hall, J. A., & Horgan, T. G. (2014). *Nonverbal communication in human interaction* (7th ed.). Wadsworth Cengage Learning.

Kongar, E. (1979). *Toplumsal değişme kuramları ve Türkiye gerçeği. (No Title). Leiter, M. (2013). Analysing & theorizing the dynamics of the workplace incivility crisis*. Springer.

Martin, R. J., & Hine, D. W. (2005). Development and validation of the uncivil workplace behavior questionnaire. *Journal of Occupational Health Psychology*, *10*(4), 477–490. DOI: 10.1037/1076-8998.10.4.477 PMID: 16248694

Matthiesen, S. B., & Einarsen, S. (2010). Bullying in the workplace: Definition, prevalence, antecedents, and consequences. *International Journal of Organization Theory and Behavior, 13*(2), 202–248. DOI: 10.1108/IJOTB-13-02-2010-B004

Nair, N., & Vohra, N. (2009). Developing a new measure of work alienation. *Journal of Workplace Rights, 14*(3), 293–309. DOI: 10.2190/WR.14.3.c

Neuman, J. H., & Keashly, L. (2002). Workplace aggression research questionnaire (WAR-Q). *Unpublished manuscript.*

Pearson, C. (2010). The cost of bad behavior: How incivility is damaging your business and what to do about it. *Human Resource Management International Digest, 18*(6). Advance online publication. DOI: 10.1108/hrmid.2010.04418fae.002

Penney, L. M., & Spector, P. E. (2005). Job stress, incivility, and counterproductive work behavior (CWB): The moderating role of negative affectivity. *Journal of Organizational Behavior: The International Journal of Industrial. Journal of Organizational Behavior, 26*(7), 777–796. DOI: 10.1002/job.336

Porath, C., & Pearson, C. (2013). The price of incivility. *Harvard Business Review, 91*(1-2), 114–121. PMID: 23390745

Porath, C. L., & Pearson, C. M. (2012). Emotional and behavioral responses to workplace incivility and the impact of hierarchical status. *Journal of Applied Social Psychology, 42*(S1), E326–E357. DOI: 10.1111/j.1559-1816.2012.01020.x

Potipiroon, W., & Ford, M. T. (2019). Relational costs of status: Can the relationship between supervisor incivility, perceived support, and follower outcomes be exacerbated? *Journal of Occupational and Organizational Psychology, 92*(4), 873–896. DOI: 10.1111/joop.12263

Schilpzand, P., De Pater, I. E., & Erez, A. (2016). Workplace incivility: A review of the literature and agenda for future research. *Journal of Organizational Behavior, 37*(S1), S57–S88. DOI: 10.1002/job.1976

Sharma, N., & Singh, V. K. (2016). Effect of workplace incivility on job satisfaction and turnover intentions in India. *South Asian Journal of Global Business Research, 5*(2), 234–249. DOI: 10.1108/SAJGBR-02-2015-0020

Singh, L. B., Srivastava, S., & Bhardwaj, B. (2023). I am done now! Linking workplace incivility to job search behaviour and employee silence. *International Journal of Conflict Management, 34*(4), 717–745. DOI: 10.1108/IJCMA-11-2022-0201

Sue, D. W., Capodilupo, C. M., Torino, G. C., Bucceri, J. M., Holder, A. M. B., Nadal, K. L., & Esquilin, M. (2007). Racial microaggressions in everyday life: Implications for clinical practice. *The American Psychologist*, *62*(4), 271–286. DOI: 10.1037/0003-066X.62.4.271 PMID: 17516773

Vartia, M., & Hyyti, J. (2022). Workplace incivility as a risk factor for workplace bullying and psychological well-being: A longitudinal study of targets and bystanders in a sample of Swedish engineers. *BMC Psychology*, *10*(1), 182. Advance online publication. DOI: 10.1186/s40359-022-00996-1 PMID: 36503593

Vickers, M. H. (2006). Writing what's relevant: Workplace incivility in public administration-A wolf in sheep's clothing. *Administrative Theory & Praxis*, *28*(1), 69–88. DOI: 10.1080/10841806.2006.11029525

Viotti, S., Converso, D., Hamblin, L. E., Guidetti, G., & Arnetz, J. E. (2018). Organisational efficiency and co-worker incivility: A cross-national study of nurses in the USA and Italy. *Journal of Nursing Management*, *26*(5), 597–604. DOI: 10.1111/jonm.12587 PMID: 29318688

Vishwanath, R., Singh, A., & Verma, P. (2021). Workplace incivility and its socio-demographic determinants in India. *International Journal of Conflict Management*, *32*(3), 369–386. DOI: 10.1108/IJCMA-02-2021-0023

About the Contributors

Kaan Koçali serves as a faculty member and department head in the Occupational Health and Safety Department of the Istanbul Gelişim Vocational School at Istanbul Gelişim University. Koçali completed his undergraduate education at Istanbul University-Cerrahpaşa Mining Engineering Department and earned his master's and doctoral degrees in Occupational Health and Safety from Istanbul Aydın University. He also has a multidisciplinary academic background in entrepreneurship, business administration, and engineering. In his academic career, Dr. Koçali has focused primarily on occupational health and safety, ergonomics, social security policies, and risk management. He has published numerous articles in national and international journals, presented papers at numerous scientific congresses, and contributed to various academic books. He has also served as a project manager on TUBITAK-supported projects and conducted studies on occupational safety practices in critical areas such as the mining industry, disaster management, and logistics. Koçali holds international qualifications such as ISO standards, NEBOSH, and IOSH. He plays an active role in education and quality committees at various universities and participates in international academic mobility programs as part of the Erasmus program. He has been deemed worthy of many national and international research and achievement awards for his academic achievements.

* * *

Nola Adelasoye is a PhD student in Industrial and Organizational Psychology at Adler University and a registered clinical counsellor with the British Columbia Association of Clinical Counsellors. With a deep passion for mental health and well-being, she is dedicated to advancing psychological health in both organizational and individual contexts. Nola currently serves as a Mental Health Counsellor, where she supports clients in navigating emotional challenges, fostering resilience, and promoting personal growth. Her research and professional interests center on workplace mental health, burnout prevention, and the intersection of leadership and employee well-being. Driven by empathy and evidence-based practice, Nola seeks to bridge clinical insight with organizational strategy to create environments where people can thrive both personally and professionally

Gaurav Bathla is having over fourteen years of experience in Academia & Industry and presently working as Professor in School of Hotel Management, Airlines and Tourism at CT University Punjab. He has received his Doctorate (PhD) in Hotel Management and Tourism from GNA University, Punjab. Dr. Bathla has authored more than 30 research papers in Scopus Indexed, UGC Approved and Peer Review Journals. He has been associated with many universities and hotel management institutions for their curriculum development and examination system. Dr. Bathla is editorial / reviewer board member of more than 8 National & International Journals. He has also organized various conferences and Faculty Development Programs in the field of Hospitality and Tourism.

Rajashree K. C is a Biomedical Engineer with a Master's in Business Administration from Saveetha University, specializing in Hospital Management and Marketing, and holds a Ph.D. in Healthcare Management. An energetic and enthusiastic professor, she excels in academics and as a multitasker, organizing seminars, workshops, cultural programs, and other extra-curricular activities. She serves as the Coordinator of NSS Unit of Saveetha School of Management, leading them to achieve multiple accolades including the Best Club award. A prominent researcher with publications in Web of Science, Scopus, and UGC-listed journals, she actively enhances her expertise through conferences, workshops, MOOCs, and certifications, including Quality Coordination in Hospitals. Awarded the "Young Academician Award" and "Young Teacher Award," she is passionate about innovation, student skill development, and inspiring young minds through exemplary teaching and mentorship.

Rajiv Gurugopinath serves as an Assistant Professor of Law at Ramaiah College of Law, Bengaluru, and is also enrolled as an advocate with the Karnataka State Bar Council. He began his academic journey with a B.A. LL.B. degree from B.M.S. College of Law, Bengaluru, and went on to complete his Master of Laws (LL.M.) in Constitutional and Administrative Law from University Law College, Bangalore University, where he authored a thesis on "Constitutional Implications of the Appointment of Judges of the Higher Judiciary in India – A Critical Evaluation." He further pursued a Master of Business Laws (M.B.L.) at the National Law School of India University, Bengaluru. Dr. Gurugopinath earned his doctorate from the School of Law, Vel Tech Rangarajan Dr. Sagunthala R&D Institute of Science and

Technology, Chennai, with his research focusing on "Right to Health – A Study on the Legal Framework and Its Implementation with Special Reference to Bangalore Urban District." An active academic and practitioner, he has presented and published widely in national and international conferences, delivered invited lectures, judged moot courts and legal events, and engaged as a resource person in diverse forums. His teaching and research interests span Constitutional Law, Administrative Law, Jurisprudence, Criminal Law, Business and Intellectual Property Law, as well as Health and Medical Laws. Drawing from both advocacy and academia, he emphasizes a balanced, practice-oriented approach to legal education.

Junaid Hassan is an Assistant Professor in the School of Management and Commerce at Sanskriti University. He earned his PhD from the Central University of Kashmir and has since dedicated himself to teaching, research and academic mentorship. Dr. Hassan's research spans consumer behavior, tourism economics, entrepreneurship, and regional development, reflecting his interest in understanding how economic and social dynamics influence business and society. He has presented his work at numerous national and international conferences and has published in reputed journals, contributing significantly to the academic community. Beyond research, Dr. Hassan is committed to nurturing critical thinking and professional growth in his students, fostering an environment of learning, innovation, and scholarly excellence.

Subramaniam Seshan Iyer is a distinguished academic and Professor at the Datta Meghe Institute of Management Studies, Nagpur, with extensive experience in teaching, research, and academic leadership. Known for his commitment to excellence in management education, he has played a significant role in mentoring students and contributing to curriculum development and institutional growth. Dr. Iyer's scholarly interests span key areas of management studies, and he has actively participated in research, publications, seminars, and conferences. His dedication to academic rigor, ethical values, and professional development has earned him respect among students and peers alike, making him a valued member of the academic community.

Ramya Krishnappa is an accomplished legal academic and Assistant Professor of Law at the School of Law, Christ University, Bengaluru. A member of the Karnataka State Bar Council, she graduated with a B.A. LL.B. from B.M.S. College of Law, Bengaluru. She went on to pursue her Master of Laws (LL.M.) in Constitutional and Administrative Law at KLE's Law College, Bengaluru, where she was awarded a Gold Medal for academic excellence. Her doctoral research was undertaken at Christ University, Bengaluru, where she earned a Ph.D. in Law on the theme "Study on the Legal Framework and Working of Panchayat Institutions with Special Reference to Mandya District." Her scholarship reflects a deep commitment to grassroots governance and constitutional development. With over six years of combined teaching, research, and practice experience, Dr. Krishnappa has established herself as a dedicated educator who prioritizes interactive learning, critical analysis, and student engagement. She is passionate about mentoring young law students, advancing legal research, and contributing meaningfully to both academia and practice.

T.C. Manjunath was born on 6th February 1967. He did his B.E. from RV College of Engg. in 1989 (B'lore Univ), M.E. in Robotics & Automation from Govt.'s LD College of Engg., (Gujarat Univ., Ahmedabad) in 1992 with FCD-First Rank & PhD in Systems & Control Engg. from IIT Bombay in the year 2007, got best research scholar award for the entire IITB. He is currently working as the Dean Research (R & D) and as Professor in Dept. of Computer Science & Engg. with specialization in IoT, Cyber Security, Blockchain Technology in Rajarajeswari College of Engineering, Bengaluru, Karnataka. He has got a teaching (academic), research & administrative experience of more than 35+ years (10 years as Principal & 8 years as HoD, 1 year as Dean Research & rest as faculty) and has published more than 1500+ research papers in various national & international journals & conferences, with 175+ patents, 100+ book-chapters, 25 awards, 100s of Keynote talks and also published 25 text books in various field of science, engineering & technology, completed projects worth Rs. 9 Crores till date & guided 10 PhD scholars in VTU. He is a member of 30 Professional Society Chapters and is a fellow of IE, IETE, AMIEE, Sr. Member of IEEE and a chartered engineer.

Hawwa Shiuna Musthafa, Ph.D. is an accomplished researcher, academic, and educator with over 15 years of experience in the higher education sector. She previously served as an Assistant Professor in the Department of Leadership and People Management at University Canada West (UCW), where she also held the position of Vice Chair of the Research Ethics Board (REB). She currently serves as an Adjunct Professor at Adler University and as a Board Member of the Adler Research Ethics Board, teaching Capstone courses in the Master's programs in Industrial-Organizational Psychology and Health and Wellness Psychology, as well as Research Methods courses. She also serves as Associate Faculty at City University of Seattle, where she teaches Research Methods and Statistics. With extensive expertise in social sciences, research methods, educational leadership, research ethics, management, educational effectiveness, curriculum development, higher education policy, and matters concerning to children, youth, women, her passion is for research that makes impact and brings meaning and empowerment to ordinary people and informs effective practices improve livelihoods. She serves as a reviewer for leading Q1 journals, including Studies in Educational Evaluation and Cogent Education, and also serves as reviewer to international academic conferences such as European Association for Research on Learning and Instruction (EARLI). An experienced educator, Dr. Musthafa has led course development, curriculum design, and professional workshops across multiple programs, demonstrating expertise in teaching, learning strategies, and student-focussed pedagogies. Dr. Musthafa's contributions to academia have been recognized through awards, including a Presidential Award by the President of the Maldives.

Gayle Zenia Noronha recently completed her Master's in Industrial & Organizational Psychology at Adler University. She currently works as a Senior Outreach Coordinator with Vancouver Second Mile Society. She is passionate about wellbeing among seniors and empowering first responders to thrive. Gayle's love of research blossomed through her childhood love of reading and continued to bloom through school and high school level essay competitions. Research and writing appeal to Gayle as she explores various mental health issues among her populations of interest. Gayle's background in counselling and her interest in Psychology are a key motivator that propel her research interests.

Ayşenur Öktem Özgür graduated in 2010 from the Department of Labor Economics and Industrial Relations, Faculty of Economics and Administrative Sciences, Muğla Sıtkı Koçman University. In the same year, she was appointed as a research assistant at the Department of Labor Economics and Industrial Relations, Human Resources and Management Program, Karamanoğlu Mehmetbey University. Between 2011 and 2018, she was assigned to Dokuz Eylül University to complete her graduate studies. In 2014, she completed the Master's Program with Thesis in Human Resources, Department of Labor Economics and Industrial Relations, Institute of Social Sciences, Dokuz Eylül University. She received her Ph.D. in Labor Economics and Industrial Relations from the same university in 2021. Since 2023, she has been serving as an Assistant Professor at the Department of Political Science and Public Administration, Faculty of Economics and Administrative Sciences, Karamanoğlu Mehmetbey University. Her research interests include career anchors, employability, elderly employment, the silver economy, decent work, and gender issues.

P. Praveen Kumar holds an MCom, MBA, and PhD in Finance from Pondicherry University. He secured the University 1st Rank in his Under Graduation. His teaching interests include Financial Accounting, Financial Management, Mergers & Acquisitions, Security Analysis & Portfolio Management, and other finance-related courses. His research interests extend to interdisciplinary areas.

P. Selvakumar is currently an Associate Professor in the Department of Science and Humanities at Nehru Institute of Technology, Coimbatore, Tamil Nadu, India. Prior to joining Nehru Institute of Technology, he held positions as an Assistant Professor in the Department of Science and Humanities various engineering colleges in Coimbatore, Tamil Nadu, India, spanning from 2011 to 2023. Dr.P. Selvakumar's academic and research interests lie primarily in the field of phytochemistry. He has made significant contributions to his field, with an impressive record that includes 30 journal publications, 12 patents, and 20 conference papers. Additionally, he has organized 5 seminars and conferences, further demonstrating his commitment to academic and scientific exchange. Throughout his career, Dr. P.Selvakumar has been recognized for his dedication and excellence in teaching, having been honored with several awards, including the Best Teacher Awards. He holds a prominent position in the academic community, contributing actively to research.

Rishabh Patel is a student at the Datta Meghe Institute of Management Studies, Nagpur, pursuing his studies in management with a strong interest in academic learning and professional development. He is dedicated to building a solid foundation in management concepts and actively engages in coursework and institutional activities. Rishabh is committed to developing analytical, leadership, and practical skills that will support his future career, reflecting enthusiasm for continuous learning and personal growth.

Kalai Selvi R Assistant Professor in the Department of Computer Science and Engineering at Kuppam Engineering College. She received the B.Tech degree in Information Technology from Anna University and the M.Tech degree in Information Technology from Anna University; 2015.Her research interests include (list 3–6 areas such as Machine Learning, Artificial Intelligence, Data Science, Advanced Data Structure, Compiler Design, etc.). She has published research papers in reputed journals and international conferences, including IEEE, Springer and has guided several undergraduate projects in emerging technology domains.

Shobarani Ranganathan is currently working as a Professor and Addl.HoD in the Department of Artificial Intelligence at Dr. M.G.R. Educational and Research Institute, Maduravoyal, Chennai, TamilNadu. She has got 24 years of experience in teaching and ten years of experience in Research. She has completed her Ph.D. degree from Mother Teresa Women's University, Tamil Nadu, India in Feb 2014 and her areas of interest are Image Processing, Machine Learning and Cyber Security. Moreover, she has published research papers in various UGC, Scopus indexed, SCI journals, and patents published at the international level. Currently she is guiding 8 Ph.D scholars and 8 scholars received their PhD degree under her supervision..

Also acting as Doctoral Committee members for internal and external scholars. She can be contacted at email: shobarani.cse@drmgrdu.ac.in.

Undabatla Rambabu is an Associate Professor in the Department of MBA at Lakireddy Bali Reddy College of Engineering, Mylavaram, with over 16 years of teaching and research experience. He specializes in Managerial Economics, Financial Management, Security Analysis, and Financial Accounting. A Ph.D. holder in Commerce & Business Administration from Acharya Nagarjuna University, he has published extensively in reputed journals, presented at national and international conferences, and received multiple awards for academic excellence, including the NPTEL Discipline Star Award and Best Faculty Award for 2021–2023.

Varinder Singh Rana Passionate, resourceful, and innovative professional, with 22 years Academic and Industry career demonstrating consistent success as an Administrator and Educator at higher education levels. Outstanding track record in assuring student success. Recognized for teaching excellence based on technical, research, and design fundamentals. Strong follow-through, administrative and time management capabilities. Highly articulate, demonstrating refined interpersonal skills across all levels and backgrounds. Effective communicator with excellent planning, organizational, and negotiation strengths as well as the ability to lead, reach consensus, establish goals, and attain results. Received Doctorate (PhD) in Hospitality from Amity University, Noida and postgraduate in Hotel Management and Catering Technology from St Peters University, Chennai and obtained his Bachelors in Hotel Management from Bangalore University in 2001. 21 years of experience in Academia and Industry. Provided services to Lovely Professional University, GNA University, Punjab Tourism Development Corporation and various reputed hotels. Have 11 Limca Records in account. General Secretary of Chefs Association of Five Rivers which is the nodal body of Indian Federation of Culinary Associations (IFCA). Member of Board of Management and Academic Council of GNA University and Board of Studies member of various Universities. Authored 12 books and 30 Research papers. Served as head chef to honourable President of India and President of Republic of Afghanistan during their visit to Punjab. Certified Master trainer of Food Safety and Standards Authority of India. Awarded by Phagwara Administration for outstanding professional work in 2017. Star Achiever Award for regional culinary development from Indian Federation of Culinary Associations. Honoured by Tourism Minister of Punjab for initiating a Bakery Unit at Central Jail, Ludhiana. Presence as Jury of Master Chef Punjabi-A Cookery Reality Show of DD National Punjabi. Cooking shows on National Television. Represented State of Punjab in Republic day Celebration at Red Fort, New Delhi. Jury in ICF and IFCA events at National and International Level. He has been awarded with Best Culinary Educator award by Indian Federation of Culinary Association (A Nodal Body of Ministry of Tourism, Government of India). Won Two silver Medals and one bronze medal for Republic of India (Our Nation) at International Culinary Challenge, Uzbekistan where 22 countries participated. Associated with many projects of Punjab Heritage and Tourism Promotion Board.

Ninad Satish Rithe is a dedicated management student currently pursuing his Master of Business Administration (MBA) in the Department of MBA at Datta Meghe Institute of Management Studies, Nagpur, Maharashtra, India. He demonstrates a strong interest in Human Resource, leadership development and collaborative teamwork, along with a keen focus on effective communication, analytical thinking, and problem-solving. With a commitment to continuous learning and professional excellence, Ninad aims to apply academic knowledge to real-world business challenges while contributing meaningfully to organizational growth and team success.

Shubhashree Sahoo is a distinguished faculty member in the Department of Computer Science and Engineering (Artificial Intelligence & Machine Learning) at Sphoorthy Engineering College, Hyderabad, Telangana. She earned her Ph.D. from Biju Patnaik University of Technology, with her research carried out in collaboration with the Defence Research and Development Laboratory (DRDL), focusing on advancements in Artificial Intelligence. Dr. Sahoo brings a wealth of knowledge and expertise to the field of AI & ML, contributing significantly to both academic and applied research.

Amrik Singh is working as Professor in the School of Hotel Management and Tourism at Lovely Professional University, Punjab, India. He obtained his Ph.D. degree in Hotel Management from Kurukshetra University, Kurukshetra. He started his academic career at Lovely Professional University, Punjab, India in the year 2007. He has published more than 65 research papers in UGC and peer-reviewed and Scopus/Web of Science) journals. He has published 12 patents and 01 patent has been granted in the inter-disciplinary domain. Dr. Amrik Singh participated and acted as a resource person in various national and international conferences, seminars, research workshops, and industry talks. His area of research interest is accommodation management, ergonomics, green practices, human resource management in hospitality, waste management, AR VR in hospitality, etc. He is currently guiding 8Ph.D. scholars and 5 Ph.D. scholars have been awarded Ph.D.

Vijayasree Srinivasan is a Manager – Operations at Geri Care, Chennai, with a strong background in hospital operations and healthcare administration. She is a final-year PhD scholar, focused on advancing her expertise in healthcare systems and operational excellence. Passionate about improving care delivery and organizational efficiency, she is committed to driving meaningful and sustainable impact within the healthcare industry.

Vijay Uprikar is a passionate Faculty of Marketing Management at Datta Meghe Institute of Management Studies, Nagpur. With over 20 years of blended experience in academics and pharmaceutical industry, he demonstrates a unique blend of corporate insights and academic expertise to the world of marketing education. A firm believer in innovative pedagogy, have developed study materials that hold six copyrights, designed to enhance conceptual clarity in marketing management. His interdisciplinary research explores Mahatma Jotirao Phule's contributions to Entrepreneurial Social Responsibility (ESR), integrating historical and social perspectives into contemporary management frameworks. With a strong foundation in management education, Dr. Uprikar has contributed significantly to the academic and professional development of students through his expertise and dedication. His work focuses on fostering innovation, leadership, and excellence in the field of management. A respected figure in the academic community, Dr. Uprikar continues to play a vital role in shaping future business leaders.

Hiremath V.M.R. is a committed academician and emerging researcher currently serving as an Assistant Professor in the Department of MBA at Rao Bahadur Y. Mahabaleswarappa Engineering College (RYMEC), Ballari, Karnataka. With a strong academic foundation and a growing research profile, he is dedicated to advancing management education through effective teaching, student mentorship, and scholarly engagement. He holds a Master of Business Administration (MBA) with specializations in Human Resource Management and Marketing, equipping him with interdisciplinary knowledge across people management and market strategy. Further strengthening his academic journey, he is presently a Research Scholar at Srinivas University, where he is pursuing his Ph.D., focusing on advanced management and research domains. His doctoral pursuit reflects his commitment to contributing meaningful insights to business and management studies. As an educator, HIREMATH V.M.R. is known for his practical and student-centered teaching approach. He integrates conceptual frameworks with real-world case discussions, encouraging students to develop analytical thinking, leadership ability, and problem-solving skills. He actively supports experiential learning, presentations, and collaborative academic activities that prepare students for professional environments. In addition to his academic role, he possesses strong technical and administrative skills, including proficiency in MS Office tools, which support his work in teaching, research documentation, and academic coordination. He is also multilingual, enabling him to communicate effectively with diverse student groups and contribute to inclusive learning environments. Earlier in his academic and social engagement journey, he was honored with the "Best Volunteer" award in the National Service Scheme (NSS), highlighting his spirit of community service, leadership, and social responsibility. This recognition reflects his balanced commitment to both academic excellence and societal contribution.

Renu Vij is a seasoned educator with a robust 16-year career in teaching at the master's level Strategic HR courses . She has made significant contributions to academic research, authoring over 15 research papers published in Scopus-indexed journals and contributing chapters to Scopus-indexed books. Her expertise extends beyond academia, as she is a certified trainer in HR analytics and an accomplished workplace psychologist. Additionally, she holds certification from AON, further solidifying her credentials in the field. Her work reflects a deep commitment to advancing knowledge in her areas of specialization, making her a respected figure in both academic and professional circles.

Jason Walker is an internationally recognized, award-winning scholar, practitioner, and author, widely known for his influential voice in trauma-informed leadership and workplace mental health. A contributing writer for Forbes, he leads the Graduate Programs in Health and Wellness Psychology and Industrial-Organizational Psychology at Adler University, and serves as a clinical adjunct professor in the School of Health and Social Sciences at City University of Seattle Master of Counselling Program. His work has been featured in Newsweek, MSN, Yahoo!, and other major outlets. A sought-after speaker and trusted advisor, he is known for translating complex psychological challenges into practical, evidence-based strategies that create meaningful change.

Ozlem Yurtsever graduated from the Department of Chemical Engineering at the Faculty of Engineering, Marmara University, in 2006. She received her Ph.D. in Industrial Engineering from the same university in 2019. Between 2011 and 2022, she served as a lecturer at the Vocational School of Technical Sciences of Marmara University and has been working as an Assistant Professor since 2022. Her research interests include sustainability, carbon footprint, energy, and occupational safety.

Index

A

B

C

D

E

F

G

H

I

J

L

M

N

O

www.ingramcontent.com/pod-product-compliance
Lightning Source LLC
LaVergne TN
LVHW061529290926
849162LV00004B/726